U.S. Marine Corps Wilderness Medicine Survival Course

Plus

Desert Operations

Department of Defense

Pentagon
Publishing

UNITED STATES MARINE CORPS
Mountain Warfare Training Center
Bridgeport, California 93517-5001

WILDERNESS MEDICINE COURSE
STUDENT HANDBOOK

TABLE OF CONTENTS

TOC

TABLE OF CONTENTS (CONT.)

UNITED STATES MARINE CORPS
Mountain Warfare Training Center
Bridgeport, California 93517-5001

FMST.07.18
04/02/02

STUDENT HANDOUT

MOUNTAIN SAFETY (SUMMER)

TERMINAL LEARNING OBJECTIVE: Given a unit in a wilderness environment and necessary equipment and supplies, apply the principles of mountain safety to prevent death or injury per the reference. (FMST.07.18)

ENABLING LEARNING OBJECTIVE: Without the aid of references and given the acronym "BE SAFE MARINE", list in writing the 12 principles of mountain safety, in accordance with the references. (FMST.07.18a)

OUTLINE

1. **PLANNING AND PREPARATION:** (FMST.07.18a) As in any military operation, planning and preparation constitute the keys to success. The following principles will help the leader conduct a safe and efficient operation in any type of mountainous environment. We find this principle in the acronym "BE SAFE MARINE". Remember the key, think about what each letter means and apply this in any type of environment.

B - Be aware of the group's ability.

E - Evaluate terrain and weather constantly.

S - Stay as a group.

A - Appreciate time requirements.

F - Find shelter before storms if required.

E - Eat plenty and drink lots of liquids.

M - Maintain proper clothing and equipment.

1

A - Ask locals about conditions.

R - Remember to keep calm and think.

I - Insist on emergency rations and kits.

N - Never forget accident procedures.

E - Energy is saved when warm and dry.

a. Be Aware of the Group's Ability. It is essential that the leader evaluates the individual abilities of his men and uses this as the basis for his planning. In his evaluation, the leader must include the group's overall physical conditioning, and the consideration of change in climate and how long the unit has had to acclimatize.

(1) Mental attitude of your group. Is morale high? How much tactical training has the group had in a particular type of terrain?

(2) Technical aspect of your group. Have they been on skis, snowshoes, etc.?

(3) Individual skills. At this point, you must choose who is most proficient at the individual skills that will be required for your mission, navigation techniques, security, call for fire, rope installations, track plans, bivouac site selection, skijoring, etc.

b. Evaluate Terrain and Weather Constantly.

(1) Terrain. During the planning stages of your mission, the leader must absorb as much information as possible on the surrounding terrain and key terrain features involved in your area of operation. Considerations to any obstacles must be clearly planned for. Will you need such things as fixed ropes, rope bridges, climbing gear, etc?

(a) Stress careful movement in particularly dangerous areas, such as loose rock and steep terrain.

(b) Always know your position. Knowing where you are on your planned route is important.

(2) Weather. Mountain weather can be severe and variable. Drastic weather changes can occur in the space of a few hours with the onset of violent storms, reduced visibility, and extreme changes. In addition to obtaining current weather data, the leader must plan for the unexpected "worst case". During an operation he must diagnose weather signs continually to be able to foresee possible weather changes.

(a) Constantly evaluate the conditions. Under certain conditions it may be advisable to reevaluate your capabilities. Pushing ahead with a closed mind could spell disaster for the mission and the unit.

c. Stay as a Group. Individuals acting on their own are at a great disadvantage in this environment.

(1) Give the unit adequate rest halts based upon the terrain and elevation, physical abilities of the unit, combat load and mission requirements.

(2) Remember to use the buddy system in your group.

(3) Maintain a steady pace so that it will allow accomplishment of the mission when all members of the unit reach the objective area.

d. Appreciate Time Requirements. Efficient use of available time is vital. The leader must make an accurate estimate of the time required for his operation based on terrain, weather, unit size, abilities, and on the enemy situation. This estimate must take into account the possibility of unexpected emergencies and allow sufficient leeway to make unplanned bivouacs in severe conditions.

(1) Time-Distance Formula (TDF). This formula is designed to be a guideline and should not be considered as the exact amount of time required for your movement. Furthermore, this formula is for use in ideal conditions:

3 km/mph + 1 hour for every 300 meters ascent; and/or + 1 hour for every 800 meters descent.

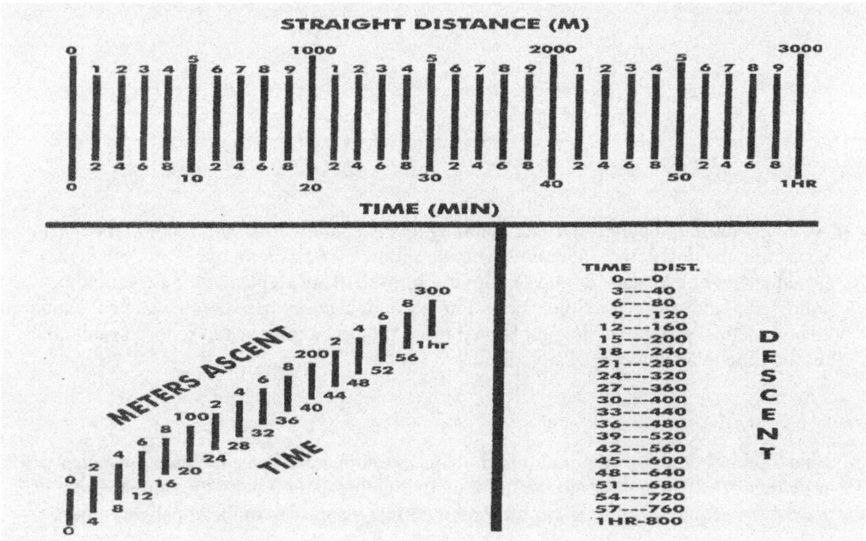

(2) Route Planning. Route cards are not to be used in place of an overlay, but as a tool to be used in route planning. Overlays/Route cards should contain the following information at the minimum:

Unit Designation:
Unit Commander.
Number of personnel.
Inclusive dates and times of movement.
Grid coordinates of each checkpoint and bivouac.
Map references.
Azimuth and distances for each leg.
Elevation gain/loss per leg.
Description of the ground.
ETA and ETD.

ROUTE CARD

UNIT I.D.		UNIT COMMANDER		NUMBER OF PERSONNEL		DATE AND TIME	MAP REFERENCE
LEG	AZM	DIST	GRID	ETA	ETD	ELEVATION	DESCRIPTION

(3) As in any military operation, route planning and execution are of vital importance. Prior to departure, the unit commander must submit a route card or patrol overlay to his higher headquarters and keeps a duplicate copy for himself. This preplanned route should be followed as closely as possible, taking into account changes based on the tactical situation. In non-tactical situations, the preplanned route should be followed to reduce search and rescue time in an emergency situation.

e. Find Shelter before Storms if required. Under certain conditions, inclement weather can provide tactical advantages to the thinking unit commander, but by the same token it can reduce the efficiency of a unit to nil if an incorrect evaluation of the situation is made.

(1) Bivouac. If the group decision is to bivouac, then it's vital that we know the principles for an unplanned bivouac.

(a) Unplanned bivouac. The principles and techniques discussed here apply both to unplanned and tolerated bivouacs. In any survival situation, especially in a mountainous environment, the most immediate danger is from exposure to the elements. Being lost will not directly kill an individual. Starvation takes time, but hypothermia can manifest itself in a matter of hours resulting in death. Adhering to the following principles will give an individual the best chance to spend a relatively safe bivouac with the prospect of continued effort toward mission accomplishment.

1. Make shelter. The requirements for expedient shelters and the building procedures will be covered in another section. The basic requirement for protection from the elements is essential.

2. Keep warm. The retention of body heat is of vital importance; any action in which body heat is lost should be avoided. The following points should be considered:

a. Adequate shelter.

b. Insulation from the ground using branches, a rucksack, etc.

c. Wear extra clothing.

d. Use extra equipment for insulation.

e. Produce external heat while trying to conserve fuel for future use.

3. Keep dry. Being wet causes the loss of body heat 24 times faster than when dry. Adequate protection from the elements is of prime importance to prevent the onset of hypothermia.

f. Eat Properly and Drink Plenty of Fluids

(1) Food. The human body can be compared to a furnace, which runs on food to produce energy (warmth). By planning the consumption of food to suit the specific situation, adequate nutrition and extra warmth can be supplied.

(2) Water. The intake of adequate amounts of water will maintain the body in proper working order. Danger from dehydration is as high in mountain regions as in hot dry areas. Loss of liquids is easily seen and felt in hot climates; whereas in the

mountains, the loss of body fluids is much less noticeable. High water intake, at least 6 quarts per day when in bivouac, 8 quarts per day when active, will help to prevent dehydration.

g. Maintain Proper Clothing and Equipment

(1) Clothing

(a) Our clothing has to perform an important function in our mission; therefore, when choosing our clothing we have to take into account some essential requirements.

1. Protection against wind and rain.

2. Layered and easily adjustable.

3. Lightweight and durable.

(b) To help us remember how to maintain and wear our clothing, we use the acronym, "COLD".

C - Keep clothing Clean.

O - Avoid Overheating.

L - Wear clothing loose and in Layers.

D - Keep clothing Dry.

NOTE: In the mountains a man should never be separated from his gear. Here are some basic and essential items that should be considered during your planning stage.

(2) Required equipment:

(a) A daypack should always be carried per squad and one Marine should carry a combat load.

(b) Map and compass. Every individual in a leadership position and his assistant should carry a map and compass. The maps should be weatherproofed and extra maps should be distributed throughout the unit.

(c) Repair kit. This kit should include those items necessary to do emergency repairs on your equipment.

(d) Survival Kit. Always carried on your person. The contents of a survival kit will be covered in another period of instruction dealing specifically with survival kits.

h. Ask Locals About Conditions. An often-overlooked source of information is the indigenous population of an area. Local weather patterns, rockslide areas, watering points, and normal routes can all be obtained by careful questioning. The leader must obtain current information of the actual conditions along his intended route. Of particular importance are recent precipitation and enemy sightings.

i. Remember to Keep Calm and Think.

(1) Emergency situation. Having recognized that you are lost and that an emergency situation exists, the following principles should be followed:

(a) Keep calm and do not panic. At this point you must make every effort to conserve body heat and energy.

(b) Think. When an individual is cold, tired, hungry or frightened he must force himself to organize his thoughts into a logical sequence.

(c) The group must try to help itself by either finding the way back to safety or by preparing shelters and procuring food.

(d) Above all else, the group must act as a tight-knit unit. In emergency situations, individual dissension can cause a total loss of control and unit strength.

(2) If the decision is reached that the group should seek its way back to safety, several possibilities exist. In most situations, the safest approach will be to retrace the route to the last known point and continue from there. The other course of action is to get a group consensus on the present location and send out a small search party to locate a known point. This party must ensure that they mark their trail adequately to return to the group. If all attempts at finding a way back to known terrain fails, a definite survival situation exists and actions discussed later in this section must be instituted.

j. Insist on Emergency Rations and Kits. Just like what was covered in the SUMMER MOUNTAIN WARFIGHTING LOAD REQUIREMENTS class, survival rations and a survival kit should always be carried.

k. Never Forget Accident/Emergency Procedures.

(1) Causes of accidents. The general procedures used to handle accidents differ little in this environment, but several distinct points should be kept in mind. The most frequent causes of accidents are as follows:

(a) Overestimation of physical and technical abilities.

(b) Carelessness.

(c) General lack of observation of one's surroundings.

(d) Lack of knowledge and experience by leaders.

(e) The failure to act as a group.

(f) Underestimation of time requirements to move through mountainous terrain and underestimation of the terrain itself.

(2) Preventive measures. The only truly effective preventive measures for the above lie in the education and experience of leaders at all levels. Too often, leaders sit by watching during training and as a result have no concept of the requirements involved in the mountainous environment. Only by active involvement can a leader gain the knowledge and experience needed to effectively lead in this environment.

(3) General procedures for handling an accident. These require only a good dose of common sense as outlined below.

(a) Perform basic first aid.

(b) Protect the patient from the elements to include insulation on top and bottom.

(c) Evacuate if necessary.

(d) Send for help if required. If possible, never send a man for help alone.

(e) Send the following information regarding the accident:

1. Time of accident.
2. Nature and location of accident.
3. Number injured.
4. Best approach route to accident scene.

(4) If one man of a two-man team is injured, the injured man must be given all available aid prior to going for help. If the injured man is unconscious, he should be placed in all available clothing and sleeping gear and anchored if on steep terrain. A note explaining the circumstances, and reassuring him, should be left in a conspicuous spot. This note must also contain the following information:

(a) When you expect to return.

(b) Where you went.

(c) What you did before you left (medication, etc.).

(5) International distress signal (whistle):

(a) Six short blasts in 1 minute from person requesting help.

(b) The return signal is three blasts in 1 minute from the respondent.

(6) Other methods if help is required:

(a) Red pyrotechnics.

(b) SOS, (... --- . . .).

(c) "Mayday" by voice communications.

l. Energy is saved when warm and Dry. With the previous 11 principles in mind this one should fall right into place. Save your heat and energy by following these steps:

(1) Dress properly.

(2) Eat properly.

(3) Drink properly.

(4) Ensure shelter meets criteria.

(5) Produce external heat (fires, stove, extra clothing, etc.) to save body heat and energy for future use.

(6) Don't lose body heat by getting wet.

UNITED STATES MARINE CORPS
Mountain Warfare Training Center
Bridgeport, California 93517-5001

FMST.07.01
03/08/02

STUDENT HANDOUT

NUTRITION

TERMINAL LEARNING OBJECTIVE: Given a unit in a mountainous environment, and the necessary equipment and supplies, apply the principles of nutrition in a mountainous environment to prevent death or injury per the reference. (FMST.07.01)

ENABLING LEARNING OBJECTIVES:

(1) Without the aid of references and from a given list, choose the correct caloric intake needed per day in a mountainous environment, in accordance with the references. (FMST.07.01a)

(2) Without the aid of references and from a given list, choose which type of diet increases your tolerance to cold, in accordance with the references. (FMST.07.01b)

(3) Without the aid of references and from a given list, choose one of the preventive measures of excessive water loss in a mountainous terrain, in accordance with the references. (FMST.07.01c)

OUTLINE.

1. **STARVATION FACTORS.** Factors that combine to result in "Near Starvation" during military operations in any mountainous environment:

   a. Difficulty in carrying sufficient food.

   b. Lack of time for preparing meals and beverages.

   c. Loss of appetite and low food palatability resulting from hypoxia/hypothermia.

   d. Fat intolerance resulting from altered metabolism.

2

2. **PHYSIOLOGICAL ASPECT TO COLD STRESS.**

a. Man's adaptation to cold is essentially behavioral. Shelters and clothing are utilized to provide a micro-environment, which will allow survival. Body composition adjustments such as increases in body weight and fat are also examples of the body's adaptation to the cold. The winter months lend themselves to greater periods of inactivity, leading to these changes.

b. In the Falkland War, constipation was a major concern. Dehydration was found to be the culprit in the majority of cases, but diet had a significant role as well. MRE's are low in fiber, containing less than 10 grams. Man needs at least 31 grams of fiber per day to ease passage of stool. Constipation can be a debilitating condition, affecting overall unit readiness.

(1) Low fiber diets can lead to diverticulosis, diverticulitis and Irritable Bowel Syndrome (IBS). IBS flare-ups tend to occur during upsetting situations and emotional stress. Educate your troops on the importance of fiber, especially in a cold weather environment.

(2) Fiber is supplied by foods derived from plants. The best sources are:

(a) Whole grain cereals.

(b) Whole wheat breads.

(c) Fresh and dried fruits.

NOTE: Vegetables also help prevent constipation, but the fiber from whole grain cereals and whole wheat bread is more effective than vegetables.

3. **CALORIC REQUIREMENTS.** (FMST.07.01a) Recommended caloric intake (Naval Medical Command Instruction 10110.1) for troops operating in a cold weather environment is 4,500 calories per day.

a. Energy Expenditure. Military operations in a cold weather environment usually involve strenuous activity. Walking on snow expends roughly twice the energy that walking on hard ground does. If troops are unfamiliar with personal over-snow mobility equipment, i.e. snowshoes and skis, awkward movement and thus wasted energy will result.

(1) The heavy clothing necessary for functioning in the cold increases energy expenditures by 5-15%. You will have the opportunity to experience this effect in the upcoming field evolution.

(2) Properly clothed, experienced individuals should have negligible heat loss from radiation, conduction, and convection. However, losses associated with respiration and evaporation can be significant.

(3) In intense cold, particularly with high winds, heat loss to the environment may be inevitable.

4. **NUTRITION IN A HIGH ALTITUDE ENVIRONMENT.** A diet high in carbohydrates and high in fat is the most beneficial, while a high protein diet is generally the least desirable.

a. The marked weight loss seen in individuals at high altitude is largely due to increased metabolic demands and anorexia secondary to hypoxia. Total caloric intake commonly does not meet energy demands.

b. Altered metabolic processes that have been reported at high altitude include delayed gastric emptying, intestinal malabsorption and decreased gastric enzyme secretion.

5. **BEST DIET FOR COLD WEATHER**. (FMST.07.01b)

a. High carbohydrate and high fat foods should be included rather than high protein foods.

b. Small meals (snacks) spaced at time intervals of two hours or less should be consumed rather than the usual three meals-a-day.

c. During a typical eight-hour day of cold exposure, three high-fat (unsaturated fat) meals providing 60% of the day's calories, coupled with two normally scheduled meals (breakfast and dinner) should provide good cold tolerance.

6. **BASIC CONSTITUENTS OF FOOD**. The three basic constituents of food are carbohydrates, fats and proteins.

a. Carbohydrates. Also known as the quick energy foods that provide energy to produce heat with the byproducts of carbon dioxide and water when oxidized by the body.

(1) Carbohydrates are present in food mainly as sugars and starches and are broken down during digestion into simple sugars that are converted into glucose.

(2) Carbohydrates are stored as glycogen in the liver and muscles and can be broken down quickly into glucose to provide a rapid source of energy. However, these stores are not large and are markedly depleted by fasting for as little as 24 hours.

(3) Eating a high carbohydrate diet for several days will double one's glycogen stores and can increase endurance by up to three times compared to the ordinary balanced diet.

(4) The main carbohydrate food sources are: CARBOHYDRATE SOURCES

(a) Fruits.

(b) Vegetables.

(c) Cereals.

(d) Sugar.

b. Fats: In the body, fat serves as the main storage form of energy. One gram of fat has the equivalent of 9.3 Cal/gm as opposed to 4.1 Cal/gm for carbohydrates and proteins. They also produce energy, heat, CO_2 and H_2O.

(1) In cases of starvation, body fat tends to be broken down into acidic compounds. If these compounds accumulate in the blood faster than they can be burned, they cause the body tissues and blood to become excessively acidic (acidosis).

(2) During brief physical activity, energy is derived equally from fat and carbohydrates, but as the duration of the activity lengthens, the percent of energy supplied by fat increases as carbohydrate energy is quickly depleted.

(a) Fat sources:

1. Butter.
2. Lard.
3. Cooking oil.
4. Mayonnaise.
5. Ice cream.

FAT SOURCES

(4) Fat is also found in lesser amounts in:

(a) Dairy products. (d) Nuts.

(b) Meat. (e) Vegetables.

(c) Eggs. (f) Cereals.

c. Proteins: Protein is a reparative food of complicated molecules composed of chains of amino acids.

(1) The protein structure is made up of 25 different amino acids. Eleven of these are important for growth while eight are essential for normal tissue maintenance.

(2) Common protein food sources are:

PROTEIN SOURCES

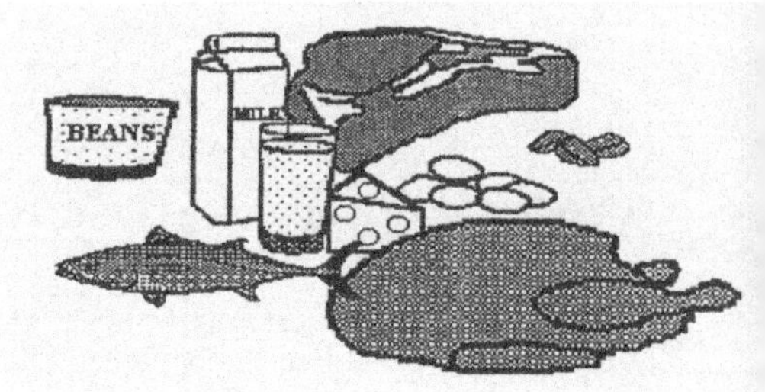

(a) Eggs.

(b) Dairy products.

(c) Meat.

(d) Poultry.

(e) Fish.

(f) Legumes (peas and beans, nuts and cereals with animal products, especially eggs, being better sources of complete protein).

(3) A pure protein diet may cause fatalities in 3-8 weeks from "rabbit starvation". This refers to those who have tried to live on relatively fat-free rabbit meat only. Fat must be a part of a normal diet.

7. **PERCENTAGE OF FATS, PROTEINS AND CARBOHYDRATES IN A SUMMER MOUNTAINOUS ENVIRONMENT**.

a. Requirements:

(1) 20% Protein.

(2) 35% Fats.

(3) 45% Carbohydrates.

b. When troops move to higher altitudes and/or are doing strenuous activity, carbohydrates should definitely be increased to at least 70 - 80% of energy intake. (Experienced mountaineers find carbohydrates more palatable at altitude).

8. **VITAMINS.** Vitamins are essential to the metabolic functioning of the body and are even more important in a cold weather environment because of an increase in metabolism and stress. Because the body cannot make vitamins, they are supplied in the food we consume. A daily multivitamin tablet supplement is recommended in a cold weather environment, but mega doses (more than 10 times the RDA) may be harmful.

a. Water-soluble vitamins are not stored in the body and need to be consumed daily. They are not stable to cooking and oxidation.

(1) B complex, Niacin, Pantothenic acid, Biotin, Choline, Folic Acid and C. (Note: Vitamin C facilitates absorption of iron from food.)

b. Fat-soluble vitamins can be stored in the body and are more stable to cooking and oxidation.

(1) A, D, E and K.

9. **FOODS THAT MAKE YOU FEEL WARM**.

a. Hot Beverages. Increases peripheral vasodilatation.

b. Adequate Dietary Iron. Mild iron deficiency increases chilling, thus decreasing cold tolerance.

c. Fat in the Diet. Fat slows gastric emptying and anecdotal evidence suggests less chilling and more restful sleep at night.

d. Carbohydrates. Keeps energy levels up. "Tired" is associated with "Cold".

e. Spices. Red pepper (Tabasco sauce) and monosodium glutamate increase peripheral vasodilatation, while capsicum and capsaicin, which are found in Cayenne, produces the same result when applied externally to the feet.

f. Water. Dehydration leads to lethargy, which tends to reduce food intake resulting in less heat production.

10. **INCREASING COLD TOLERANCE**.

a. Frequent Feedings. A high carbohydrate diet is superior to a high protein diet in increasing tolerance to cold.

(1) Diets deficient in both calories and protein reduce physical work capacity in the cold. The provision of both components is required to maintain work capacity.

11. **HIGH FAT DIETS AND SUSCEPTIBILITY TO FROSTBITE**.

a. High fat diets consumed at frequent (2 hour) intervals, maintain body temperatures better than high carbohydrate diets, but not if normal (4 hour) intervals are observed between meals.

b. Fat enriched diets lead to a propensity toward increased arterial thrombosis and increased platelet aggregation, which translates into increased potential for thrombosis.

c. Frostbite injuries are associated with disruption of the microcirculation, progressive micro vascular thrombosis, alteration of platelet function, and fibrin deposition in affected areas.

d. Therefore, diets high in (saturated) fat may predispose to frostbite.

CAUTION! A high-fat diet is important in a high-altitude, cold-weather environment. The benefits associated with the high-energy source called Fat far outweigh the risks of arteriosclerosis from the diet itself. A high-fat diet is appropriate in this specific environment during the time frame of cold, high-altitude exposure. In garrison, at sea level, a high fat diet should be avoided in order to maintain a healthy lifestyle.

12. **IRON DEFICIENCY AND THERMOREGULATION**.

a. Iron deficiency results in 9% less heat production.

b. Consuming only 1/3 of the RDA for iron results in a 29% greater heat loss during cold exposure. "The furnace is going, but the windows are open" is a phrase that helps describe this effect.

c. Iron absorption is increased by consuming iron rich foods:

(1) Lean meat.

(2) Poultry.

(3) Beets.

(4) Beans.

(5) Green leafy vegetables.

(6) Fish.

(7) Vitamin C.

d. Bottom line, IRON improves the body's response to cold.

13. **WATER REQUIREMENTS**.

a. On average, water makes up about 60% of the one's entire body weight. The average sedentary person excretes about 2.7 quarts of water each day. Of this total, 1.3 quarts are lost in the urine, 1.1 quarts through the skin and lungs and 10 ounces in the stool.

b. At high altitude, and during strenuous exercise, the amount of water lost through the skin and lungs due to the cold, dry air increases greatly.

c. Cold weather decreases the sense of thirst, which may lead to a state of chronic, mild dehydration. At temperatures below freezing and at elevations above the snow line, the lack of liquid water and the time and effort required to melt snow compound the problem.

d. Prevention of excessive water loss: (FMST.07.01c)

(1) Force fluids. Drink 6-8 quarts of water daily.

(2) Observe for dark urine and other symptoms of dehydration. (Headache, nausea/vomiting, dizziness, constipation, etc.)

(3) Limit caffeine to 1-2 cups per day.

(4) Do not rely on thirst as an indicator of hydration status.

NOTE: Eating snow or ice will cause extreme inflammation of the mucous membranes of the mouth after two days. Warm canteens of water placed in your sleeping bag will give you water to drink in the morning. They will initially act as hot water bottles. (Place them upright, as they tend to leak ...keep bag dry.)

e. Health Problems Associated with Excessive Water Loss:

(1) Dehydration.

(2) Constipation.

(3) Hypothermia.

(4) Frostbite.

(5) Heat Exhaustion.

f. The importance of hydration is essential while in a cold weather environment. If these guidelines are followed then complications arising from poor hydration should be prevented.

14. **SPECIAL PURPOSE MILITARY RATIONS**.

RATION	APPLICATION
Ration, Cold Weather, RCW.	Cold weather operations where weight and freezing are a problem.
MRE Supplement Packet.	Cold weather calorie supplements for the MRE.
Ration, Light Weight, RLW.	Long range patrols, reconnaissance, and assault.
Nutrition Sustainment Module, NSM.	Calorie dense food, bar ration for mission lasting 3-5 days where soldier must carry his own food.

Family of Operational Rations, FOR.	Modular rations that allow an optimum mix of meals for tactical operations under all climatic conditions.

15. **FIELD STRIPPING OF MRE'S.**

a. Tips on how to conserve space and weight:

(1) Remove the entire contents from the plastic wrapper.

(2) Remove the cardboard, and all materials that you will not use.

(3) Be sure to place the toilet paper and other materials from the accessory pack either back in your pocket or into the brown packaging.

(4) Carefully repack the stripped meal back into the wrapper and seal with tape.

(5) Your meal will be half its original size and two-thirds its original weight.

UNITED STATES MARINE CORPS
Mountain Warfare Training Center
Bridgeport, California 93517-5001

FMST.07.09
03/31/02

WILDERNESS PATIENT ASSESSMENT

TERMINAL LEARNING OBJECTIVE: Given a casualty in a cold weather environment and the necessary equipment and supplies, perform a patient assessment in a mountainous environment to determine the nature and extent of injury per the reference. (FMST.07.09)

ENABLING LEARNING OBJECTIVES:

(1) Without the aid of references, from a given list, choose when CPR may be legally discontinued, in accordance with the references. (FMST.07.09a)

(2) Without the aid of references and from a given list, choose which situations CPR may not be started in a remote wilderness environment, in accordance with the references. (FMST.07.09b)

(3) Without the aid of references, in the proper sequence, correctly perform a patient assessment in the proper sequence in accordance with the references. (FMST.07.09c)

OUTLINE

1. **INTRODUCTION.**

   a. An organized prioritized assessment of a wilderness trauma victim is essential to proper care. A modified Advanced Trauma Life Support (ATLS) assessment system will be introduced in this chapter.

2. **TRIAGE.**

   a. Two situations of triage may exist in a wilderness environment.

(1) The number of patients and the severity of their injuries do not exceed the ability of the care-giver and his/her equipment to render care. In this situation, treat life-threatening problems first.

3

(2) The number of patients and the severity of their injuries exceed the capacity of the care-giver and his/her equipment to render care. In this situation, those with the greatest chance of survival are treated first

3. **CPR: URBAN vs. WILDERNESS SPECIFICS.**

a. In a remote wilderness situation where advanced life support assistance may be hours or days away, CPR probably has no chance of success. In addition, administering CPR under wilderness conditions may put group members in serious danger, due to physical hazards and the effects of exhaustion.

b. Based on the above, several suggestions regarding CPR in remote wilderness areas can be made. Even though the following suggestions seem reasonable, they should be viewed as recommendations, without legal force, and not doctrine. At the present time, a rescuer who begins CPR is legally obligated to continue unless or until one of the following conditions is fulfilled:

(1) CPR may be discontinued legally when: (FMST.07.09a)

(a) The patient revives.

(b) ACLS is started by higher authority.

(c) Relieved by another rescuer.

(d) Too exhausted to continue.

(e) Pronounced dead by proper authority.

(2) In a remote wilderness environment, CPR may not be started under one of the following situations: (FMST.07.09b)

(a) The patient is in cardiac arrest caused by trauma.

(b) The patient is a drowning victim who has been immersed for over an hour.

(c) The patient is in cardiac arrest and advanced life support is more than an hour away, especially if the patient must be carried out.

(d) The patient's cardiac arrest was unwitnessed and the time of onset is unknown.

(e) The patient appears to be dead, based on rigor mortis, lethal injuries, or a body temperature below 60°F.

(f) Giving CPR would be hazardous to rescuers.

(3) After 30 minutes of CPR with no signs of life, further CPR is probably useless and may reasonably be discontinued. Administering CPR to a victim who is being evacuated by litter or sled is very difficult, if not impossible. Unless an ambulance or helicopter can be brought in rapidly, chances of survival are slim.

(4) Exceptions include patients in cardiac arrest caused by hypothermia; patients with another illness or injury complicated by hypothermia (avalanche burial and near drowning in cold water), and patients in cardiac or respiratory arrest caused by lightning injury. In these cases, the outlook is probably more favorable and CPR should be aggressively administered.

4. **PATIENT ASSESSMENT** (FMST.07.09c). There are different considerations to think about in a wilderness setting. The environment plays a major role in treatment, care and even the extraction of the patient. Not only do we have to be mindful of the enemy situation, by keeping tactical. We also have to take into consideration the elements. Snow, rain ice, and the terrain can and will pose a great danger to the patient and the rescuers. We must protect the patient from the elements. We can do this by placing an insulating barrier between the patient and the ground. Also exposing one body part at a time and recovering as you move on will help to fight hypothermia.

A) SCENE SIZE-UP.

1) Determine scene safety:

a. Rescuers: Your safety is priority. You are no good to the team as a casualty.

b. Patient: Protect the patient from further injuries, enemy, and the elements.

c. Bystanders: They can cause many accidents. Limit the personnel on scene.

d. Tactical: Have the Marines maintain security and find cover for the rescuer and the patient.

e. Fire: Forest fires can pose a formidable threat.

f. Electrical: Make sure to shut off electricity first. Rushing in injures many rescuers.

g. Special Rescues: Swift water and high angle rescues are very dangerous and only trained professionals should attempt the rescue.

2) Determine the mechanism of injury / nature of illness:

a. Obtain SAMPLE and Mechanism of Injury (MOI) history from patient, witnesses, or other caregivers.

S = Signs and/or Symptoms.
A = Allergies.
M = Medications.
P = Past Illnesses.
L = Last Meal.
E = Events of Injury.

MOI = What, when, where, and how injury occurred.

b. Environment: Survey the area to get clues. (i.e. falling rocks, next to a cliff, broken branches, etc…)

c. Bystanders: Ask them what they saw.

3) Determine number of patients:

a. Survey area for number of patients.

b. Look for the hidden patients. Patients could be buried in rock, sand or snow. They also could be spread out in bushes and trees. One of the most over looked places is in the bystanders. Be sure to have people search and identify all patients.

4) Request additional help / Organize help if needed:

a. Medical assistance: Due to the lack of medical personnel and equipment, you might need to request additional help.

b. Technical assistance: When the scope of care is over your ability or higher than the casevac corpsman can provide, you should request higher trained personnel on the casevac bird.

c. Organize help at the scene: Use the people around you to help out.

5) Consider stabilization of the spine:

a. If indicated from: Falls of 10 feet or greater, high speed collisions, unwitnessed loss of consciousness, etc…

b. Designate an assistant: Use the bystanders / marines. Tell them exactly what you want them to do and show them how to do it.

c. Improvise control: If you are alone you can use the patient's boots, sand bags, rocks, etc as forms of temporary control.

B) INITIAL ASSESSMENT.

1) Make general impressions of the patient: What can you obtain from the patient's condition as you walk up? Is the patient talking, or unconscious?

2) Determine responsiveness / level of consciousness.

a) Use AVPU.

b) Assess pupils for size, equality, and reactivity.

3) Determine chief complaint / Apparent life threats.

4) Assess Airway and breathing.

a) Assessment of airway.

- Perform chin lift or jaw thrust maneuver, with C-Spine precautions.
- Clear the airway of foreign bodies.
- Insert an oropharyngeal or nasopharyngeal airway.
- Use of other airways as needed, and as equipment and training allow.

b) Assessment of breathing.

- Expose the neck and chest, assure immobilization of the head and neck.
- Determine the rate and depth of respirations.
- Auscultate the chest bilaterally. (Ear to chest wall, if necessary)
- Check for symmetry. (Rise and fall of the chest)

c) Management.

- Ventilate. (Face-mask, bag-valve-mask, if available)
- Alleviate tension pneumothorax.
- Seal open pneumothorax.
- Administer high concentration oxygen, if available.

5) Assess Circulation.

a) Assess and control major bleeding.

- Identify and control major bleeding.
- Pulse: Quality, rate, and regularity.
- Pulse Pressure: Check radial, then carotid.
- Skin color: cyanosis, pallor and temperature.

b) Management.

- Insert two large bore IV's with saline locks. IV fluid replacement, if indicated.
- Place in shock position.
- Prevent hypothermia.

6) Identify priority patients and make transportation decisions.

a) Triage patients at this point.

b) Make transportation decisions. In this type of environment it is possible that a casevac vehicle might not gain access to your position. At this point you will have to organize how the patient will be transported out. Also, you might have to set up relay points and rewarming stations.

C) DETAILED PHYSICAL EXAM: This is the head to toe assessment. We will find and treat all secondary injuries at this time. When performing this exam in the wilderness, you will need to expose one body part at a time and cover it back up as you go. When assessing the patient, you are looking for deformities, contusions, abrasions, penetrations/punctures, bruising, tenderness, lacerations, and swelling (DCAP-BTLS).

1) Head.

a. Assessment.

- Inspect and palpate entire head and face for lacerations, contusions, and fractures.
- Re-evaluate pupils.
- Re-evaluate level of consciousness.
- Assess eyes for retinal hemorrhage, optic disc bulging, visual acuity disturbances, and contact lenses.
- Evaluate cranial nerve function.
- Inspect ears and nose for CSF leakage, hemorrhage
- Inspect mouth, tongue, and teeth.

b. Management.

- Maintain airway, continue ventilation, and oxygenate as indicated.
- Control hemorrhage.
- Prevent secondary brain injury.
- Remove Contact Lenses.

2) Assess the neck.

a. Assessment.

- Inspect for signs of blunt trauma, penetrating injuries, tracheal deviation, and use of accessory muscles during respiration, and jugular vein distention.
- Palpate for tenderness, deformity, swelling, or subcutaneous emphysema.

b. Management.

- Maintain open airway.
- Maintain immobilization and protection of cervical spine.
- Mark trachea.

3) Assess the chest.

3

a) Assessment.

- Inspect chest wall for signs of blunt trauma, penetrating injuries, and use of accessory muscles during respiration.

- Auscultate the chest wall and posterior bases for breath sounds.
- Auscultate for heart sounds.
- Palpate the entire chest wall for evidence of blunt trauma, penetrating injuries, flail chest, tenderness, crepitus, or subcutaneous emphysema.
- Percuss for hyperresonance or dullness.

c. Management.

- Needle thoracenthesis, if tension pneumothorax is suspected and proper training and equipment are present.
- Three-sided dressing for open chest wounds.
- Pericardiocentesis, if cardiac tamponade is suspected and proper training and equipment are present.

4) Abdomen.

a. Assessment.

- Inspect abdomen for signs of blunt trauma, penetrating injuries, and internal bleeding.
- Auscultate for the presence / absence of bowel sounds.
- Palpate the abdomen to elicit subtle tenderness, rebound, or guarding.

- Check pelvis for stability.

b b. Management.

c

- Correctly dress wounds.
- Keep patient NPO until definitive care is delivered.

5) Perineum / Rectum / Penis / Vagina.

d a) Assessment.

e

(1) Perineum.

- Contusions / hematomas.
- Lacerations.

(2) Rectum.

- Blood. Sphincter tone.
- High-riding prostate.

(3) Penis.

- Urethral bleeding.

(4) Vagina.

- Bleeding.
- Lacerations.

6) Musculoskeletal.

a) Assessment.

- Inspect extremities for evidence of blunt / penetrating injury. (deformities, altered range of motion).
- Palpate extremities for tenderness, crepitus, and abnormal movements.
- Palpate all pulses.
- Assess pelvis for stability. (Fractures) - Palpate spine for tenderness, crepitus, step-off's, depressions, and abnormal movements.

b) Management.

- Splint all suspected fractures in the position of function or comfort.

- Reduce dislocations as soon as possible, depending on training.
- Maintain immobilization and alignment of cervical, thoracic, and lumbar spine.

7) Reassess the vitals signs.

a) Assessment.

- Assess the vitals, pulse rate, breathing rate, pulse pressure.
- Re-evaluate the pupils and level of consciousness (GCS).
- Evaluate extremities for motor and sensory deficits.

b) Management.

- Maintain adequate immobilization of entire patient, as indicated.

5. **PATIENT PACKAGING AND EVACUATION.**

a. See the Mountain Casualty Evacuation chapter.

UNITED STATES MARINE CORPS
Mountain Warfare Training Center
Bridgeport, California 93517-5001

FMST.07.13
03/30/02

STUDENT HANDOUT

HIGH-ALTITUDE ILLNESSES

TERMINAL LEARNING OBJECTIVE: Given a simulated casualty, in a high altitude wilderness environment, treat high altitude health problems, in accordance with the references. (FMST.07.13)

ENABLING LEARNING OBJECTIVES:

(1) Without the aid of references, select from a given list the definition of acute mountain sickness, in accordance with the references. (FMST.07.13a)

(2) Without the aid of references, select from a given list the four main mechanisms of acute mountain sickness, in accordance with the references. (FMST.07.13b)

(3) Without the aid of references, select from a given list the most common and prominent symptom of acute mountain sickness, in accordance with the references. (FMST.07.13c)

(4) Without the aid of references, select from a given list the field management of acute mountain sickness, in accordance with the references. (FMST.07.13d)

(5) Without the aid of references, select from a given list the seven high altitude health preventive measures, in accordance with the references. (FMST.07.13e)

(6) Without the aid of references, select from a given list the correct definition of high altitude cerebral edema, in accordance with the references. (FMST.07.13f)

(7) Without the aid of references, select from a given list the pathophysiology of high altitude cerebral edema, in accordance with the references. (WMC.6.4g)

4

(8) Without the aid of references, select from a given list five of the symptoms of high altitude cerebral edema, in accordance with the references. (FMST.07.13h)

(9) Without the aid of references, select from a given list the field management of high altitude cerebral edema, in accordance with the references. (FMST.07.13i)

(10) Without the aid of references, select from a given list the correct definition of high altitude pulmonary edema, in accordance with the references. (FMST.07.13j)

(11) Without the aid of references, select from a given list three of the nine symptoms of high altitude pulmonary edema, in accordance with the references. (FMST.07.13k)

(12) Without the aid of references, select from a given list the field management of high altitude pulmonary edema, in accordance with the references. (FMST.07.13l)

(13) Without the aid of references, select from a given list choose the most important treatment for both high altitude cerebral edema and high altitude pulmonary edema, in accordance with the references. (FMST.07.13m)

OUTLINE

1. **ACCLIMATIZATION**.

a. Successful acclimatization depends on three factors:

(1) Degree of hypoxic stress (i.e. altitude)

(2) Rate of onset of hypoxic stress (i.e. ascent rate)

(3) Individual Physiology (i.e. genetic differences between individuals)

b. Physiologic Changes:

(1) Respiratory - Hypoxic Ventilatory Response (HVR)

- Hypobaric hypoxia
- Triggers carotid body
- Central respiratory center stimulated
- Ventilatory rate increases

HVR can be measured:

- "High" leads to successful acclimatization and superior performance.
- "Low" tends to be found in victims of altitude illnesses.
- HVR can be influenced by depressants (ETOH) and stimulants. (caffeine), and progesterone.
- **May** be negatively influenced by physical conditioning.

(2) Respiratory – Fast response: minutes to hours

- HVR leads to increased Ventilatory rate.
- Increased ventilations "blows off' C02."
- Low C02 levels yield a respiratory alkalosis and rising pH.
- Elevation of pH feeds back to depress the central respiratory center.
- No further immediate rise in ventilatory rate (a brake is placed on any further increase in respiration.)

(3) Renal – Slower response: kidneys respond over the next 24 - 48 hours:

- Kidneys respond to alkalosis by inducing a bicarbonate diuresis.
- PH slowly declines - brake on central respiratory center eases.
- Respiratory rate gradually rises over 4 - 7 days to new baseline.
- Final respiratory rate is altitude dependent.

(4) Cardiovascular - changes are a result of increased sympathetic tone (due to hypoxia).

- Increased: Blood Pressure, venous tone and Cardiac output (increased heart rate and contractility.)
- Stroke Volume declines in first 1 - 3 days due to bicarbonate diuresis.
- Left ventricular pressures are normal, usually; therefore, wedge pressures are usually normal.

(5) Pulmonary Circulation - critical in understanding the pathophysiology of HAPE.

- Ascent results in global Hypoxic Pulmonary Vasoconstriction.
- Elevated Pulmonary Artery Pressure (PAP) occurs.
- Exertion at altitude worsens this Pulmonary Hypertension.
- Damage to pulmonary vascular endothelium leading to leaky capillaries.

(6) Cerebral Circulation - critical in understanding pathophysiology of HACE

- Hypoxia -----------Vasodilatation
- Hypocapnia ----------Vasoconstriction
- Hypoxia is the dominant influence and vasodilatation results.
- On average, Cerebral Blood Flow (CBF) increases 20-40% on ascent to 12.000ft.

(7) Hemopoietic Response

- Erythropoietin levels increase within hours of ascent.

- New immature Red Blood Cells (RBC) seen within days.
- New mature RBC's present within 4-10 days
- RBC mass increase seen over weeks to months (altitude dependent).
- Early changes due to hemoconcentration.
- Polycythemia is a potential danger – this can lead to a condition known as CMS, or Chronic Mountain sickness.

(8) Muscle Tissue

- Intramuscular capillary and mitochondrial density increases after three weeks exposure to altitude.

(9) Sleep Pattern

- Stage 1 increase, Stage 2 unaffected
- Stage 3 & 4 decreases
- Rapid Eye Movement (REM) decreases
- Arousals increase
- Periodic Breathing (Cheyne-Stokes).
    Increased respiratory rate leads to respiratory alkalosis
    Elevated pH depresses central respiratory center
    Apnea
    02 saturation declines, C02 levels rise
    Carotid body stimulated
    Ventilatory rate rises

(10)Summary: Physiologic changes seen at altitude result from the body's responses to hypo baric hypoxia. It has been speculated that man would be limited to an 8000-foot elevation limit without supplemental oxygen if not for the acclimatization process. As Messner and Haebler demonstrated on Mt. Everest, the process of acclimatization enables man to explore the entire planet.

2. **ACUTE MOUNTAIN SICKNESS (AMS).**

a. Definition - AMS is an acute, self-limiting illness, which results when an unacclimatized individual ascends rapidly to high altitude. (FMST.07.13a)

(1) It is rare below 8,000 feet but will occur in approximately 42% of persons rapidly exposed to altitudes greater than 10,000 feet in elevation.

(2) Aerobic fitness is no predictor for risk of AMS with exposure to altitude. There is some evidence that prior aerobic fitness leads to increased AMS incidence, most likely due to excessively rapid ascent or over-exertion at altitude. Elite athletes may also be at increased risk due to depressed cardiovascular and Ventilatory responses to exertion.

(3) Prior ascents to altitude without symptoms of AMS are no guarantee against having symptoms of AMS with future ascents. However, those with past AMS are at an increased risk of AMS with repeat exposure to high altitude. Generally, there is no correlation between the severity of the illness and increasing altitudes. The incidence is slightly increased in younger persons.

(4) There is still much we do not know about altitude-induced physiologic changes. Current explanations in the literature center around genetic differences in hypoxic ventilatory response and vascular endothelium release of inflammatory mediators and Nitric Oxide.

b. Pathophysiology of AMS: The physiologic changes associated with the symptom complex known as AMS are linked to: (FMST.07.13b).

(1) Poor ventilatory response: An individual's Hypoxic Ventilatory Response (HVR) dictates how well he/she will respond to the hypobaric hypoxia of altitude. Those with a brisk HVR will do well, whereas those with a blunted response tend to suffer the symptoms of AMS. A poor ventilatory response leads to hypoxia and hypercapnia (elevated C02 levels). These two combine to augment cerebral blood flow by inducing cerebral vasodilatation.

(2) Fluid retention: The increased sympathetic tone associated with hypobaric hypoxia explains many of the changes seen in the renal system at altitude. Low Renal Blood Flow, increased anti-diuretic hormones, low Glomerular Filtration Rates, and decreased Urine Output all contribute to a Net Fluid Gain. This retained fluid redistributes within the body's fluid spaces. As a result, intravacular and extravascular fluid shifts occur, especially in the brain, lung, and peripheral interstitial tissues.

(3) Increased Intra-cranial Pressure: Increased cerebral blood flow coupled with CNS inflammation resulting in increased vascular permeability leads to increased intra-cranial pressure and associated symptoms.

(4) Altered Pulmonary Mechanics. On ascent to altitude, there is an increase in extravascular lung water. This is thought to be mediated, in part, by a phenomenon known as hypoxic pulmonary vasoconstriction. Blood vessels in the lungs constrict when they are exposed to low oxygen levels, this global vasoconstriction leads to increased right-sided heart pressures. The result of increased right-sided heart pressures causes fluid to leak from the vascular space into the alveoli, which results in impaired gas exchange. This has been demonstrated via Pulmonary Function Testing bronchoalveolar lavage, right sided-heart catheterization and arterial blood gas analysis showing decreased Vital Capacity, decreased Peak Expiratory Flows, pulmonary exudates rich in proteins and Increased Alveolar-Arterial oxygen Gradients. One study showed a 54% increase in estimated lung mass. It is not uncommon to hear crackles in all lung fields when auscultating asymptomatic individuals. The presence of intra-alveolar fluid inhibits the diffusion of oxygen

across the alveolar membrane, thereby reducing oxygenation across the pulmonary capillary membrane.

Signs and Symptoms of AMS. Symptoms tend to occur within the first 1- 3 days of exposure to altitude. Symptoms should resolve spontaneously within 15 - 24 hours, if ascent is arrested. The three most prominent symptoms are Headache, Nausea and Vomiting. (FMST.07.13c) Although nonspecific, headache must be present to make a diagnosis of AMS. 95% of AMS victims will complain of one or more of these three. Other common complaints include dizziness, anorexia, drowsiness, malaise, weakness, and insomnia. It is possible that findings of dyspnea on exertion and mental status changes along with ataxic gait may accompany the classic AMS signs and symptoms. These are clear indications for descent, as these are the hallmark signs of progression to HACE and the development of HAPE. Again, headache is the most common and prominent symptom of AMS. The Lake Louise Consensus criteria was created to give clinicians formal guidance in diagnosing AMS. The criteria require the presence of headache plus one or more symptoms from the above list. (i.e. nausea, vomiting, anorexia, insomnia, dizziness, fatigue, and dyspnea on exertion.)

c. Field Management (FMST.07.13d)

(1) Stop any further ascent. Light duty. No tobacco.

(2) Symptomatic Treatment. Mild analgesics such as Aspirin or Tylenol for the headache, Compazine or Phenergan for nausea/vomiting.

(3) Other Meds: Acetazolamide (Diamox), Dexamethasone (Decadron), and oxygen. The effects of Diamox and Decadron are additive when used together and are a very effective means for treating AMS.

(4) Hyperventilate: It is unrealistic to try to consciously control one's respiratory rate over any appreciable period of time. It is nearly impossible to maintain conscious hyperventilation while functioning in the field.

(5) Descend. If none of these methods help, then individuals should descend 1,000-3,000 feet. This will usually result in marked relief of symptoms. Most cases, however, improve in 1-2 days with symptomatic treatment.

(6) ALL PATIENTS WITH AMS MUST BE EVALUATED FOR HAPE AND HACE.

HOW?

- Field sobriety test.
- Mini-mental status exam.
- Auscultation of lung fields.

d. Preventive Measures: (FMST.07.13e)

(1) The best approach to high-altitude travel is staged, gradual ascent, when moving at altitudes greater than 3000 meters. Gradual ascent means no faster than 3000 ft/day up to 14,000 and no greater than 1,000 ft/day over 14,000 ft.

(2) Avoid any alcohol and/or sedatives during the first two nights.

(3) Maintain adequate fluid intake.

(4) Work high - Sleep low.

(5) High Carbohydrate Diet (approximately 70% of diet).

(6) Avoid overexertion, but remember that mild exercise assists acclimatization.

(7) No smoking.

(8) Pharmacological options: Prevention of AMS is possible with Acetazolamide (Diamox). It dramatically lowers the incidence of AMS when taken prophylactically. The usual dose is 125 mg PO BID starting 24-48 hours before ascent and should be continued for 3-4 days after the start of the mission. It is also effective when used as a treatment especially when combined with Decadron. In individuals with contraindications or allergies to Diamox use Decadron (dosed PO/IM/IV 4 mg Q6).

3. **HIGH ALTITUDE CEREBRAL EDEMA (HACE).**

a. Definition. HACE is a high-altitude illness characterized by swelling of the brain. (FMST.07.13f)

(1) HACE can occur as low as 8,000 feet, but typically occurs at more than 12,000 feet.

(2) HACE, considered by most, to be a progression of AMS, occurs as a failure to heed the warnings of AMS signs and symptoms and continue ascent.

b. Pathophysiology. Increased cerebral blood flow along with the fluid shifts mentioned above lead to vasogenic edema (leaky blood vessels) inside the confined space of the skull. It is also thought that inflammation plays a key role in the development of HACE and occurs as follows. Hypoxia triggers the release of inflammatory mediators (likely Transforming Growth Factor Beta, TGF-β and Vascular Endothelial Growth Factor, VEGF.) These mediators cause increased vascular permeability that result in fluid accumulation within the brain – edema and increased intracranial pressure. Once this pressure nears arterial pressure, cerebral blood flow will be impaired. Eventually, intra-cranial pressure will rise further with a concomitant increase in the risk of brain herniation leading to marked alterations in mental status, ataxia and even cranial nerve palsies. Increased CSF production further compounds the problem. (FMST.07.13g)

c. Differential Diagnosis: Other illnesses are possible in individuals ascending to altitude. Your differential should include the following when neurologic deficits are involved:

(1) Cerebrovascular spasm and Transient Ischemic Attacks.

(2) Cerebral vascular thromboembolic events.

(3) Intra-cranial hemorrhage (aneurysm, AVM).

(3) Hypothermia.

(4) Meningitis

(5) Encephalitis

(6) "Hangover"

d. Signs and Symptoms:

(1) Early signs and symptoms are those found in AMS. (i.e. H/A, NV)

(2) When the above warning signs go unheeded the true signs and symptoms of HACE can quickly follow: (FMST.07.13h)

(a) Ataxia (loss of muscle coordination leading to difficulty maintaining balance), especially prominent in the victim's gait.

(b) Mental Status Changes, poor judgement, personality change which may progress to stupor, coma and death relatively rapidly without proper treatment.

(c) Lassitude, confusion, hallucinations, convulsions, behavioral aberrations.

(d) Paralysis of one or more extremities (see Differential Diagnosis).

(3) The most important impediment to early recognition of HACE is its insidious onset. Early signs and symptoms frequently go unrecognized or are ignored by patients, as well as their companions, who also may be suffering to some degree from the effects of altitude.

e. Field Treatment (FMST.07.13i)

(1) Early recognition is the KEY. Look for Ataxia and Mental Status Changes.

(2) Treatment should be immediate since fatalities can occur within a few hours in severe cases.

(3) Once diagnosed, the patient should be placed in the most comfortable position possible, descent should be directed immediately and administration of high concentration O2 should be given, in route, if available.

(4) Decadron (early), 10mg PO/IM/IV followed by 4 mg Q6. Diamox can also be used 125-250 mg BID.

(5) Consider diuretics – use carefully.

(6) If loss of consciousness, think airway management and ABC's.

(7) Casevac to an advanced medical facility ASAP!

(8) A Gamow Bag can be a lifesaver if decent is not immediately possible. Perform 4-6 compressions per day each treatment should last 40-60 minutes.

NOTE: The Gamow Bag is a one-man portable hyperbaric chamber. It weighs 14.5 lbs., is operated by a foot-pump, and is made of nylon with an airtight zipper. With the HAPE or HACE victim inside, the bag is pressurized with the foot-pump to an internal pressure of 2 psi. This is the equivalent to a 5000 - 6000-foot descent in regards to the partial pressure of oxygen inside the bag. Air is circulated within the bag via the foot-pump. This can be a life-saving resource in the event a victim is caught at altitude and cannot descend due to weather or other uncontrollable variables.

f. Prevention

(1) Preventive measures are the same as those discussed for AMS.

(2) Decadron can be used as a prophylactic medication along with Diamox, in cases of unavoidable rapid ascent to extreme altitude. However one must continue this regimen, while at altitude, or face rapid onset of symptoms once the medication is stopped.

4. <u>**HIGH ALTITUDE PULMONARY EDEMA (HAPE)**</u>.

a. Definition. HAPE is a high-altitude illness that is characterized by filling of the lungs with edema fluid. (FMST.07.13j)

(1) HAPE rarely occurs below 12,000 feet.

(2) HAPE victims in the past were commonly described, after autopsy, as victims of "pneumonia" and "Inflammation of the lungs". Since 1960, medical science has had the technology to investigate the complexities of the physiological changes associated with hypobaric hypoxia.

(3) Incidence: Occurs in 1-2 % of people brought rapidly to 12,000ft.

(4) Commonly occurs within 2-4 days of ascent.

(5) Risk for developing HAPE remain the same as above:

(a) Rate of ascent
(b) Sleeping altitude
(c) Level of exertion
(d) Gender (M>F)
(f) Individual physiology

b. Pathophysiology of HAPE:

(1) Hypoxic Pulmonary Vasoconstriction

(a) Normal part of the acclimatization process (elevated PAP).
(b) Adaptive at sea level (i.e. lobar pneumonia).
(c) Maladaptive at altitude (global hypoxia).
(d) Tends to be uneven, leading to high pressure/high flow area.

(2) Overperfusion

(a) Localized areas exposed to high pressure and high flow.
(b) Shearing forces and stress failure at the microvascular level.
(c) Endothelial damage to capillaries leads to inflammatory response and increased vascular permeability.
(d) Elevated hydrostatic pressure combined with increased vascular permeability leads to leakage of protein rich fluids into the alveoli and decreased oxygenation

(3) Nitric Oxide deficiency

Individuals prone to HAPE may have a deficiency in the amount of Nitric Oxide produced in the vascular endothelium resulting in more profound hypoxic pulmonary vasoconstriction.

c. Pathologic Findings of HAPE: Postmortem studies of victims of HAPE have yielded some interesting results:

(1) Grossly, the lungs are congested and swollen, as seen with other causes of pulmonary edema. Average lung weight on autopsy is 2-4 times normal. Histologically, one sees pulmonary edema with a protein rich exudate filling the alveoli. It is also common to find the alveolar spaces filled with hyaline membranes.

(2) The left heart is completely normal in appearance. However, the right ventricle, right atrium, and pulmonary arteries are distended and dilated.

(3) Greater than 50% of HAPE victims are found to have evidence of HACE.

d. Signs and Symptoms. These tend to occur within 2-4 days of arrival at altitude. Usually the symptoms of AMS are present before or occur with the symptoms of HAPE. (FMST.07.13k)

(1) Early signs:

(a) Dry cough, frequently occurring at night.

(b) Dyspnea on Exertion (DOE), especially with ambulation uphill.

(c) Mild chest pain - usually perceived as an ache beneath the sternum.

(d) Decreased work performance and increased recovery time between events.

(e) Peripheral or Central Cyanosis.

(2) Later signs:

(a) Dyspnea at Rest.

(b) Productive cough which yields large amounts of pink, frothy sputum.

(c) Rapid pulse and respiratory rates.

(d) Audible crackles on auscultation.

(e) **Mental status changes**, **Ataxia**, Loss of Consciousness.

(f) CXR findings:

- Prominent pulmonary arteries.
- Normal heart size.
- Patchy infiltrates.

<u>ECHO findings:</u>

-Non-cardiogenic pulmonary edema -increased PAP
-Tricuspid regurgitation.
-Normal left ventricle
-Prominent pulmonary arteries -enlarged right atrium
-Enlarged right ventricle.

NOTE: Crackles heard on lung exam, as an isolated finding is NOT an indication for descent or grounds for the diagnosis of HAPE. It is common (up to 15% of climbers on Mt. Rainier) to find asymptomatic rales in climbers at altitudes of 10,000-12,000ft. Concern is heightened when crackles occur in constellation with other signs and symptoms.

e. Field Treatment (FMST.07.131)

(1) The most important emergency care measure is immediate descent to a lower altitude, since fatalities can occur within 6-12 hours in severe cases.

(a) A descent of at least 2,000-3,000 feet can be a definitive treatment if the condition is caught early enough.

(2) The patient should be placed in the most comfortable position (usually sitting) and given **high-flow O2 if available**. It is important to minimize activity, as exertion increases pulmonary artery pressures. Remember oxygen is Lasix for the lungs.

(3) Medication regimens include:

(a) Nifedipine (10mg PO initially then 30mg SR Q12) – calcium channel blocker - pulmonary artery/arteriole dilator. SR=sustained relief.

(b) Diamox and Decadron may be appropriate regimens, depending on your medication availability.

(c) Morphine (5mg IV initially decreased PAP) as a last resort.

(4) Oxygen is always appropriate, however, it is not always available.

(5) Gamow Bag can be a lifesaver if decent is not immediately possible.

f. Prevention Measures. Preventive measures are the same as those discussed for AMS.

(1) Acetazolamide (Diamox) has been proven to be successful in the prevention of HAPE. 125 - 250 mg PO bid helps prevent HAPE in individuals with recurrent episodes.

(2) Nifedipine (30 mg SRq12h) - also effective in individuals with recurrent HAPE.

(3) It should be noted that a history of HAPE is NOT a contraindication to the individual participating in high-altitude activities. Each case should be evaluated individually.

NOTE: Remember decent is the most important treatment for both HACE and HAPE. (FMST.07.13m)

5. **SYSTEMIC EDEMA AT HIGH ALTITUDES.**

Occasionally, certain individuals, especially women, will develop edema of the face, hands, and feet at altitude. This edema usually tends to occur in the absence of other symptoms. In susceptible persons, repeat episodes are common. The edema goes away upon return to lower altitude. There is no need for descent in these cases because the patient is usually able to tolerate the edema and continues to function well. However, they should be examined carefully to rule out the presence of HAPE or HACE. Diuretics are usually not warranted. It is also probably reasonable to restrict salt intake. The only indication for descent is if the edema is causing functional disability, i.e. inability to don one's boots, or facial edema severe enough to restrict vision.

6. **SICKLE CRISIS.**

Persons with sickle cell trait are at risk for developing sickling attacks when exposed to hypoxic stress, as might occur at high altitudes. However, sickling can occur at lower elevations. Individuals with Mediterranean family ties should be considered at risk for having sickle cell trait. The trait is more likely in black individuals as well. Any patient complaining of Left Upper Quadrant pain, SOB, and/or arthralgia at altitude should have sickling crisis included in the differential diagnosis regardless of race. It has been known to occur in white females as well. It has been recommended by some that individuals with sickle cell trait should be restricted to an altitude of 8,000 feet and below. However, more liberal recommendations exist, suggesting that the individual and their symptoms dictate ultimate altitude.

7. **ALTITUDE THROAT.**

At higher altitudes, mouth breathing is often increased during exertion. Inhalation of this cold, dry air can result in drying of the mucous membranes in the throat and upper airway. This condition responds to humidified air, salt-water gargles or lozenges. If exudate or edema is present, throat culture and/or antibiotics may be warranted.

8. **HIGH ALTITUDE FLATUS EXPULSION HAFE).**

Ascent to altitude can result in rapid expansion of intestinal gas, due to the decreased pressure of altitude. This produces abdominal discomfort and the passage of colonic gas. This condition is not associated with serious consequences; however, it may create tension amongst tent-team members. Treatment consists of oral administration of anti-flatulence such as simethicone, Mylanta II, or Gaviscon. Know and avoid your gas producing foods!

9. **HIGH ALTITUDE RETINAL HEMORRHAGE.**

Up to 60% of individuals ascending to altitudes greater than 18,000 feet will develop asymptomatic retinal hemorrhages. On fundoscopic exam, there will be hyperemia of the disc and dilation and tortuosity of the vessels. "Flame" hemorrhages and "cotton wool spots" can be seen on exam. The only indication for descent in an individual with retinal hemorrhages is loss of visual acuity. This usually results when a large hemorrhage involves the macular region of the retina. Remember that Carbon Monoxide poisoning can present with retinal hemorrhage as well.

10. **THROMBOEMBOLIC EVENTS.**

Thromboembolic disease occurs with increased frequency at higher altitudes. It is believed that the common circumstances of volume depletion due to dehydration and Polycythemia due to altitude exposure predispose individuals to coagulopathic events. In addition, extended periods of inactivity due to "going to ground" during storms contribute to this predisposition. There does not seem to be any direct influence of hypobaric hypoxia on the body's normal coagulation system.

11. **ACETAZOLAMIDE.**

Acetazolamide (Diamox) is a Carbonic Anhydrate Inhibitor, which is especially active in the brain, the lung, and the kidney.

a. Kidney:

(1) Uptake of bicarbonate ion is inhibited.

(2) Bicarbonate diuresis is induced.

(3) Metabolic acidosis ensues within one hour of medication.

(4) Sequence mimics and accelerates the natural acclimatization process.

(5) Enhanced HVR is the end result.

(6) Diuresis addresses problems associated with fluid retention.

b. Brain:

(1) Direct effects at the level of the Central Respiratory Center.

(2) Diminishes periodic breathing during sleep.

(3) Fewer episodes of apnea and extreme hypoxemia.

(4) CSF production is inhibited and CSF absorption is enhanced.

c. Side Effects:

(1) Polyuria.

(2) Flattens taste of carbonated beverages.

(3) Peripheral paraesthesias, myopia, impotence, nausea, and drowsiness.

(4) Contraindicated for Sulfa Allergic and G6PD-deficient individuals (reversible when medication is discontinued.)

UNITED STATES MARINE CORPS
Mountain Warfare Training Center
Bridgeport, California 93517-5001

FMST.07.35
03/13/02

STUDENT HANDOUT

HEAT ILLNESSES

TERMINAL LEARNING OBJECTIVE: Given a casualty in a wilderness environment, and necessary equipment and supplies, manage common heat related injuries to prevent death or further injury per the reference (FMST.07.35)

ENABLING LEARNING OBJECTIVES:

(1) Without the aid of references, given a list of heat illnesses match the definition to the correct condition, in accordance with the references. (FMST.07.35a)

(2) Without the aid of references, select from a given list the correct field treatment for heat cramps, in accordance with the references. (FMST.07.35b)

(3) Without the aid of references, select from a given list the correct field treatment for heat exhaustion, in accordance with the references. (FMST.07.35c)

(4) Without the aid of references, select from a given list the correct field treatment for heat stroke, in accordance with the references. (FMST.07.35d)

1. **CAUSES OF HEAT ILLNESSES**. Heat illness is caused by one of or a combination of the following factors:

a. Excessive environmental heat and humidity.

b. Faulty heat dissipation mechanisms within the body.

c. Stressed heat dissipation mechanisms.

d. Prolonged strenuous physical activity with inadequate fluid and electrolyte replacement.

2. **THERMOREGULATION**. The body is constantly adjusting itself in order to maintain its optimal temperature. This is called thermoregulation. A basic physics principle applies, in that heat is transferred from a higher to a lower temperature. What this means is that when the environmental temperature is higher than the skin temperature, the body gains heat, but when the environmental temperature is lower, the body loses heat. As a result, elevated environmental temperature adds to the heat burden of the body and can interfere with heat dissipation.

a. Heat Transfer Mechanisms. Keeping in mind the above principle we will now discuss the four methods of heat transfer.

1) Conduction: This is heat exchange between two surfaces in direct contact. An example would be lying on a hot or cold surface.

2) Convection: This is heat exchange from a surface to a gas or liquid, usually air or water. The amount of heat transferred depends on many variables, such as density, flow rate and surface area exposed.

3) Radiation: This is heat transfer between the body and its surroundings. All matter either absorbs heat or emits thermal radiation. The sun emits heat and the body absorbs it. A interesting fact is that highly pigmented skin while protective against UV radiation absorbs about 20% more heat than lightly pigmented skin.

4) Evaporation: This is the heat flow from the body surface to the environment by sweating.

b. If the body is unable to maintain thermal equilibrium by radiation, convection, and conduction and core temperature rises, sweating must occur to permit heat loss by vaporization of water. Since gastric emptying of water is about 1L/hr and sweat rates can exceed 1L/hr, a healthy diet along with constant replenishment of water lost is imperative to prevent illness.

3. **PREDISPOSING FACTORS**.

a. An underlying illness such as infection.

b. The temperature and humidity of previous days effect the likelihood of becoming a heat casualty.

c. Most heat illnesses occur during the morning hours.

d. Thirst is not an accurate indication of hydration; therefore you must drink even when you don't feel thirsty.

e. People usually do not voluntarily drink as much water as they lose.

f. All persons in the heat should be considered dehydrated, unless they have been forced to drink more than they desired.

4. **PREVENTIVE MEASURES**.

a. Drink 6-8 qts every day to replace fluids lost by sweating or from other bodily functions.

b. Consumption of a solute rich meal or a fluid replacement beverage about 2 hours before exercise can potentially counteract the sweat/salt losses.

5. **EARLY SIGNS OF HEAT RELATED ILLENSSES**. Changes in cognitive functions appear to occur frequently prior to the physical symptoms of heat illnesses. It is therefore imperative that medical personnel and leaders watch for confusion and impaired ability to function or work in the members of their group.

6. **SPECIFIC HEAT RELATED ILLNESSES**.

a. Heat Induced Syncope. (FMST.07.35a) This event is caused by the gravitational displacement of blood into the legs of an upright and stationary person combined with hemodynamic displacement of blood into peripheral circulation to support heat transfer at the body surface.

1) Prodromal symptoms.

(A) Restlessness

(B) Sighing

(C) Dysphoria

(D) Nausea

(E) Yawning

2) Treatment.

(A) The fainting is usually brief and responds to horizontal positioning.

(B) Rest in a cool, shaded area.

(C) Oral fluids.

(D) Persons should be cautioned against standing in hot environments and advised to flex leg muscles repeatedly while standing still to enhance venous return and warned to sit or lie down if warning signs reappear.

b. Heat Cramps. These are painful spasms of muscles, caused by sweating resulting in excessive loss of body salt. Generally, the spasms involve voluntary skeletal muscle subjected to prior exertion. This can occur even if a person is replacing water loss, but not salt losses.

1) Precipitating Factors: Cramps are preceded by several hours of sustained effort, combining heavy sweating in hot surroundings, with the ingestion of large volumes of water.

2) Symptoms: Painful muscle cramps in the arms, legs, and abdomen.

3) Treatment: (FMST.07.35b)

(A) Rest in a cool, shaded area.

(B) Gentle massage of affected muscles (may relieve spasms) or place the affected muscle in a stretch.

(C) Generally, heat cramps respond quickly to salt solutions.

1.) 1/4 to 1/2 teaspoon of salt in a quart of water (sip).

2.) In severe cases IV isotonic saline 0.9% NaCL.

NOTE: Salt tablets are not recommended due to the gastric irritation they cause.

c. Heat Exhaustion. This occurs when body salt losses and dehydration from sweating are so severe that a person can no longer maintain adequate blood pressure. It tends to develop over several days, that present ample opportunity exists for electrolyte and water imbalances to occur. Some 20% of cases are found to also have some form of gastroenteritis. Additionally, 60% of heat exhaustion casualties will also have symptoms of heat cramps.

1) Problems in recognition.

(A) Early signs may go unrecognized or be mistaken for malingering.

(B) Headache and visual disturbances may mimic prodromal of migraine headache.

(C) Dizziness and syncope may lead to an evaluation for arrythmia.

(D) Chills, ^ temperature, and sweating may lead to an evaluation for infection.

(E) Altered mental status or ataxia may lead to an evaluation for Meningitis or high altitude medical problem.

2. Symptoms:

(A) Headache

(B) Fatigue

(C) Piloerection

(D) Nausea/Vomiting

(E) Profuse sweating

(F) Dizziness

(G) Hyperirritability, Anxiety

(H) Chills

(I) Heat cramps

3. Clinical Findings:

(A) Tachycardia

(B) Hypotension

(C) Core temperature may be normal or elevated. Spontaneous cooling may occur, usually with profuse sweating. This will not happen with Heat stroke.

(D) Hyperventilation

(E) Syncope

(4) Treatment is based on symptoms and focuses on reduction of body temperature and rehydration. (FMST.07.35c)

(A) Lay down in a cool, shaded area.

(B) If body temperature is elevated, lower by any means available.

(C) Sprinkling with water and fanning.

(D) Ice bags over superficial great vessels and fanning.

(E) Periodic toweling of the skin renews the evaporative process.

(F) If patient is alert, orally rehydrate with cool water or commercially available rehydration mixture. Goal is 2L over 1st 2 hours.

(G) If patient is not alert, rehydrate via IV with LR.

(H) Victim must rest and continue to be rehydrated over next 24 to 36 hours.

d. Heatstroke (Hyperthermia). " A true medical emergency". This is a failure of the body's cooling mechanisms that rid the body of excessive heat build-up. Prior to 1950 the mortality rate was 40 to 75%. Long term survival of victims is directly related to the speed of effective treatment measures. Traditionally, the diagnosis of Heatstroke required the following 3 signs: a core temperature > 104°F(41°C), altered mental status, and cessation of sweating. Although, this symptom complex does represent full-blown Heatstroke adherence to this strict criteria may delay critical interventions. Unless an alternative etiology is obvious, the previously healthy person who collapses after physical exertion in hot weather should be considered to have exertional Heatstroke. Treat for Heatstroke whenever you suspect it and always suspect it in the heat.

1. Symptoms:

(A) Headache
(B) Shortness of breath
(C) Nausea/ vomiting
(D) Low BP
(E) Skin may be wet or dry
(F) Constricted pupils
(G) Seizures
(H) Dry mouth
(I) Core temp > 104°F (41°C)
(J) Weakness/ dizziness
(K) Hot, red skin
(L) Tachycardia
(M) Confusion, Aggression, Combative
(N) Coma

2. Clinical Findings.

(A) CNS: altered mental status, agitation, ataxia, delirium, hallucinations, convulsions, seizures, coma, pupillary constriction.

(B) Cardiovascular: hypotension (a late and ominous finding), shock, sinus tachycardia with ST segment & T wave abnormalities.

(C) Pulmonary: hyperventilation, pulmonary edema.

(D) Renal: acute renal failure, hematuria, pyuria, proteinuria, UA casts.

(E) Gastrointestinal: nausea, vomiting, +/- disseminated intravascular coagulation, hematemesis, melena.

(F) Hematologic: WBC 20-30 thousand, < platelets, < clotting factors.

(G) Hepatic: increased liver function tests.

3. Emergency Treatment. (FMST.07.35d)

(A) Rapid reduction of body core temperature by any means available until core temperature is 102°F.

1) Ice water bath (controversial).
2) Cool water bath.
3) Pouring or hosing cool water over victim and fanning.
4) Placing ice bags at neck, armpits, groin, forehead, chest etc..

(B) Maintain ABC's.

(C) Monitor core temperature.

(D) IV access infuses RL or normal saline 500cc/hr x 4hrs. (watch for pulmonary edema).

(E) Control seizures with valium.

(F) Transport STAT. Cool enroute.

UNITED STATES MARINE CORPS
Mountain Warfare Training Center
Bridgeport, California 93517-5001

FMST.07.15
04/01/02

STUDENT HANDOUT

COMBAT CASUALTY CARE

TERMINAL LEARNING OBJECTIVE: Given a simulated casualty in a simulated combat environment, treat the casualty in accordance with the references. (FMST.07.15)

ENABLING LEARNING OBJECTIVES:

(1) Without the aid of references, choose from a given list the three correct phases of care of tactical casualty management, in accordance with the references. (FMST.07.15a)

(2) Without the aid of references, given a simulated casualty and the necessary equipment, conduct combat casualty care, in accordance with the references. (FMST.07.15b)

OUTLINE

1. Summary of Findings and Recommendations.

a. Pre-hospital care of combat casualties is divided into three phases: (FMST.07.15a)
   1. Care under Fire.
   2. Tactical Field Care
   3. Combat Casualty Evacuation (CASEVAC) Care.

b. Suppression of hostile fire may take temporary precedence over the rendering of care in the Care under Fire phase.

c. Cervical spine immobilization is not required for victims of penetrating head or neck trauma.

d. CPR should not be attempted on the battlefield for victims of blast or penetrating trauma who have no pulse, respirations, or other signs of life.

e. The use of tourniquets is encouraged in the Care under Fire phase. The tourniquet is removed and bleeding controlled with direct pressure in subsequent phases when feasible.

6

f. The nasopharyngeal airway is the airway of first choice for unconscious patients until the CASEVAC phase.

g. For patients with airway obstruction in the Tactical Field Care phase unrelieved by a nasopharyneal airway, the next airway maneuver is cricothyroidotomy.

h. Endotracheal intubation is not the standard of care until the CASEVAC phase.

i. Evaluation of the laryngeal mask airway and the Combitube for use by SOF corpsmen and medics is recommended.

j. Progressive, severe respiratory distress in the setting of unilateral blunt or penetrating chest trauma on the battlefield should result in a presumptive diagnosis of tension pneumothorax and decompression of that hemithorax with needle thoracostomy.

k. Chest tube insertion is not indicated until the CASEVAC phase of care.

l. Oxygen is not expected to be carried onto the battlefield, but CASEVAC assets should have it.

m. No IV fluids are to be given until the CASEVAC phase unless the patient is in hypovolemic shock from a bleeding site which has now been controlled. Fluid resuscitation is not necessary for patients who are not in shock and not appropriate for patients in hypovolemic shock from uncontrolled hemorrhage (penetrating wounds of the chest and abdomen).

n. Saline locks may be used instead of IV's if fluid resuscitation is not required.

o. 18 gauge catheters may be used instead of 14 or 16 gauge.

p. Hespan (1000cc) is the fluid of choice for initial resuscitation of patients with controlled hemorrhagic shock.

q. Morphine is to be used IV (5mg) instead of IM.

r. IV cefoxitin is to be used as soon as possible for patients with penetrating abdominal trauma, grossly contaminated wounds, massive soft tissue trauma, open fractures, or any patient in whom a long delay until definitive treatment is expected.

s. Casualties should not be completely undressed for a secondary survey in the field. Removal of clothing should be limited to that necessary to expose known or suspected wounds.

t. More specific combat casualty care planning should be carried out and based on specific mission-oriented scenarios.

u. The establishment of Combat Casualty Transport Teams and their use on CASEVAC assets is recommended.

v. Electronic monitoring should be routinely used on CASEVAC assets unless conditions prohibit.

w. MAST trousers are not recommended as a standard of care in any phase.

2. Phases of Care:

a. Phase One: Care Under Fire

1. Return fire as directed or required.

2. Keep yourself from getting shot.

3. Try to keep the casualty from sustaining additional wounds.

4. Stop any massive external hemorrhage with a tourniquet.

5. Take the casualty with you when you leave.

b. Phase Two: Tactical Field Care

1. Airway management

- Unconscious casualty without airway obstruction: > Nasopharyngeal airway.

- Unconscious casualty with airway obstruction: > Cricothyroidotomy.

- Cervical spine immobilization is not necessary for casualties with penetrating head or neck trauma.

2. Breathing

- Consider tension pneumothorax and decompress with needle thoracostomy if a casualty has unilateral penetrating chest trauma and progressive respiratory disstress.

3. Bleeding

- Control any remaining bleeding with a tourniquet or direct pressure.

4. IV

- Start an 18 gauge IV or saline lock.

5. Fluid Resuscitation

- Controlled hemorrhage without shock: > No fluids necessary.

- Controlled hemorrhage with shock: > Hespan 1000cc

- Uncontrolled (infra-abdominal or thoracic) hemorrhage: > No IV fluid resuscitation.

6. Inspect and dress wound.

7. Check for additional wounds.

8. Analgesia as necessary: > Morphine (5mg) IV, wait 5minutes, repeat as necessary.

9. Splint fractures and recheck pulses.

10. Antibiotics

Cefoxitin 2gm slow IV push (over 3-5 minutes) for penetrating abdominal trauma, grossly contaminated wounds, massive soft tissue trauma, open fractures, or any patient in whom a long delay until CASEVAC or definitive treatment is expected.

11. Cardiopulmonary Resuscitation

CPR on the battlefield for victims of blast or penetrating trauma who have no pulse, no respirations, and no other signs of life will not be successful if attempted.

c. Phase Three: Combat Casualty Evacuation (CASEVAC) Care

1. Airway management

Unconscious casualty without airway obstruction: > Nasopharyngeal airway, endotracheal intubation, Combitube or laryngeal mask airway.

Unconscious casualty with airway obstruction: > Cricothyroidotomy if endotracheal intubation and/or other airway devices are unsuccessful.

2. Breathing

Consider tension pneumothorax or hemothorax and decompress with needle thoracostomy if a casualty has unilateral penetrating chest trauma and progressive respiratory distress.

Consider chest tube insertion if a suspected tension pneumothorax is not relieved by needle thoracostomy

Oxygen.

3. Bleeding

Consider removing tourniquets and using direct pressure to control bleeding if possible.

4. IV

Start an 18 gauge IV or saline lock if not already done.

5. Fluid resuscitation

No hemorrhage or controlled hemorrhage without shock: > Lactated Ringer's at 250cc/hr.

Controlled hemorrhage with shock: > Hespan 1000cc initially.

Uncontrolled (infra-abdominal or thoracic) hemorrhage: > No IV fluid resuscitation.

Head Wound patient: > Hespan at KVO unless there is concurrent controlled hemorrhagic shock.

UNITED STATES MARINE CORPS
Mountain Warfare Training Center
Bridgeport, California 93517-5001

FMST.07.37
04/01/02

STUDENT HANDOUT

BURN MANAGEMENT

TERMINAL LEARNING OBJECTIVE. Given a casualty in a wilderness environment, and necessary equipment and supplies, treat burn injuries to prevent death or further injury per the reference. (FMST.07.37)

ENABLING LEARNING OBJECTIVES

1. Without the aid of references, given a list of the depths and appearances of burns, match them to the type of burn, in accordance with the references. (FMST.07.37a)

2. Without the aid of references, given a list choose the major burns that require rapid casualty evacuation, in accordance with the references. (FMST.07.37b)

3. Without the aid of references, given a list choose the first step in the treatment of burn injuries, in accordance with the references. (FMST.07.37c)

OUTLINE

1. **GENERAL:** Each year there are about 2 million individuals burned severely enough to seek medical attention. Around 70,000 require hospitalization. Deaths are on the decrease largely due to the increase in the use of smoke detectors. It has been estimated that 90% of all burn injuries are preventable. Children and young adults are frequently affected by burn injuries.

2. **PHYSIOLOGY:** For non-chemical burns, the primary events of injury occur during the time of heat contact. Coagulation necrosis takes place within cells and denaturation of collagen in the dermis. Blood vessels are completely destroyed or endothelium damage is severe enough to cause clotting which leads to ischemic necrosis of remaining viable cells. Burn wounds are not static. Surrounding the "coagulation zone" is a zone of capillary and small vessel stasis. Circulation becomes stagnant from a strip formed by red blood cells, and aggregation of platelets and white blood cells. The fate of the burn wound depends on the progression of this zone of stasis.

7

3. In patients with burns less than 10% of the total body surface area (TBSA), the inflammatory process is generally limited to the burn site itself. As burns approach 20% TBSA, the local

response becomes systemic. Burns of 30% TBSA or greater, often lead to multiple organ system failure, which in a patient with a severe burn almost invariably leads to a fatal outcome, often due to acute renal failure.

Cardiovascular changes begin immediately after a burn. The degree of these changes depends initially on the size of the burn and to a lesser extent the depth of the burn. Burns less than 15% TBSA can be fluid resuscitated with crystalloid fluids. As burns pass 20% TBSA, massive shifts of fluid and electrolytes occurs. This results in massive fluid shift from the intravascular to the extravascular space. Unless intravascular volume is repleted, classic hypovolemic shock occurs.

3. TYPES OF BURNS

a. Scald Burns: Scalding is the most common cause of burns, usually resulting from hot water. Water at 140°F creates a deep partial-thickness or full-thickness burn in three seconds. At 156°F the same burn occurs in one second. Freshly brewed coffee is generally about 180°F. Boiling water always causes a deep burn. Soups and sauces are thicker, remaining in contact with the skin longer. Exposed areas tend to burn less deeply than areas covered with a thin layer of clothes. Clothing retains heat and keeps liquid in contact with the skin longer. Immersion scalds are deep and cause severe burns resulting from longer contact time. These burns are more common in children and elderly patients because of their thin skin. Burns from hot oils are generally deep partial-thickness or full-thickness. Temperatures for cooking oils reach around 400°F, and tar can be up to 500°F.

b. Flame Burns: The next most common burn injuries are flame burns. Although household injuries have gone down with the advent of smoke detectors, a high number of burn injuries still occur from careless smoking, improper use of flammable liquids, automobile accidents, and clothing ignited from stoves or space heaters.

c. Flash Burns: Flash burns result from intense heat for a brief time caused by explosions of natural gas, propane, gasoline, and other flammable liquids.

d. Contact Burns: These burns usually result from direct contact of hot metals, plastics, glass, or coals. Contact burns are usually limited in size, but are deep wounds. With the increase of wood stove usage, this type of burn injury is on the rise in children. Contact burns, dealing with an unconscious patient or involving molten materials, are usually fourth degree. The most common contact burn in the wilderness is from hot coals.

e. Electrical Burns: Electrical burns are actually thermal burns from very high-intensity heat. When electricity encounters resistance from the body it turns into heat. Severity of the burn depends on the portion of the body affected relative to the amperage. The smaller the body part, the more intense the heat is, and the part is less able to dissipate the heat. Consequently the fingers and toes are almost always destroyed with severe damage to the foreams. Larger areas can dissipate the current enough to prevent extensive damage. Arc burns take the most direct route, not the path of least resistance. These are deep and destuctive burns which mostly occur at the joints.

Electrical burns may have other associated injuries that should not be overlooked on the initial assessment. Fall related injuries are common. The intense muscle contraction resulting from electrocution can cause fractures of the lumbar vertebrae, humerus, or femur, and also may cause dislocation of the hips and shoulders. The casualty may have cardiac symptoms of a myocardial contusion or infarction. There can be a conduction system failure or an actual rupture of the heart wall. If there are no signs present of cardiac problems intially after a shock of 110 or 220 volts there is a minute chance they will appear later. Nervous system damage is possible whenever a current passes from one side of the body to the other more than likely affecting the spinal cord.

f. Chemical Burns: Usually caused by strong acids or alkalis and are most often the result of industrial accidents, home use of drain cleaners, assault, and improper use of harsh solvents. Chemical burns continue to burn until they are inactivated by reaction with tissue or dilution by flushing with water. Acids tend to tan the skin, creating an impermeable barrier that limits futher penetration. Alkalis combine with cutaneous lipids and saponify[1] the skin until they are neutralized. The appearance of the wound can be misleading. Unless the observer can be 100% sure, all chemical burns should be considered full thickness.

g. NBC Burns: Nuclear weapons cause burns in two ways: by direct absorption of thermal radiation through exposed surfaces (flash burns); or by the indirect action of fires caused by the weapon (flash burns). The first step in treatment is to prevent contamination to the rescuers and other patients.Vesicants such as mustard gas cause skin burns with edema and blister formation. Burns resulting from vesicants should be flushed with copious amounts of water. Absorbent powder (flour, talcum powder, Fuller's earth) can be used if water is scarce. The powders should be wiped away with a moist towel. The military uses M258A1 kits for skin decontamination. White Phosphorus ignites spontaneously when exposed to air. Remove the air with an air-tight seal; mud is optional. A brief rinse with 1% copper sulfate[2] will impead further oxidation.

4. **CLINICAL PRESENTATION:** The severity of the burn injury is related to the size of the burn, the depth of the burn, and the part of the body that is burned.

a. Burn Size: Burns are the only quantifiable form of trauma. The burn size is the single most important factor in predicting mortality, need for specialized care, and the complications expected from the burn. We can measure burn size with the "rule of nines". The "rule of nines" is where areas of the body represent approximately 9% of a person's total body surface area (TBSA), each leg 18% (the front of the leg 9%, and the back of the leg 9%), the front of the torso represents 18%, the back of the trunk is 18%, the head represents 9%, and the perineum 1%. For infants and small children under four years of age, the head represents a larger percentage 18%, and the legs a smaller percentage 13%, each arm 9%, front troso 13%, back troso 13%, and the perineum 1%. The percentage of the burn area and that of the unburned area need to add up to equal 100%. For smaller burn areas, use the Rule of Palmar Surface: the patient's palmar surface equals about 1% TBSA.

[1] Saponify: to convert fat into soap by treating it with an alkali.
[2] Navy uses sodium bicarbonate instead of copper sulfate

b. Burn Depth: The depth of the burn is either described in degree: first, second, third or fourth, or by the depth of the injury. The following describes burns in terms of partial-thickness or full-thickness. While these descriptions appear to separate burns into defined categories, many burns have a mixture of characteristics making a precise diagnosis difficult.

(1) First-Degree: These burns involve the epidermis only and do not blister. They are most commonly caused by ultraviolet light. An example would be a sunburn. The burned skin is painful and red. It should heal in 7 days without scarring.

(2) Superficial Second-Degree, or Superficial Partial-Thickness: These burns include the epidermis and upper layers of the dermis. They characteristically form blisters, under which the skin is red and moist, these burns are painful to the touch. Blisters can take up to two days to appear. These burns are usually caused by hot liquids. Wounds should heal in 14 to 21 days; they may scar depending on the extent of the burn.

(3) Deep Second-Degree, or Deep Partial-Thickness: Burns that involve deeper layers of the dermis. There is damage to hair follicles and sweat glands. They are characterized with a molted pink and white color with blisters forming immediately. These wounds may be less sensitive to touch than the surrounding normal skin, or maybe tender to touch. The patient often complains of discomfort rather than pain. Capillary refill may be slow or absent, when pressure is applied to the wound. These burns are caused by hot liquids, oil, steam, or flame. They may be difficult to distinguish from Third-Degree/Full-Thickness burns. Healing takes 3-9 weeks. Scarring is probable, the degree is related to the amount and the depth of dermal injury. Surgical grafting may be required.

(4) Third-Degree, or Full-Thickness: Involves the entire thickness of the skin, epidermis through the dermis, down to subcutaneous fat. All structures of the epidermis and dermis are destroyed. The wound is classically described as leathery, firm, and depressed when compared to normal skin. The tissue is charred, pale, and insensitive to light and touch. It is often misdiagnosed as a Deep Second-Degree/Deep Partial-Thickness due to the similar clinical findings. This burn will not spontaneously heal. Surgical repair and/or skin grafts are necessary. There is significant scarring.

(5) Fourth-Degree: These burns involve all layers of the skin, subcutaneous fat, muscle, and bone. These are devastating, life threatening injuries. They almost always have a charred appearance, and often only the cause of the burn gives a clue to the amount of underlying tissue destruction. (FMST.07.37a)

c. Burn Classification: The American Burn Association has devised a classification of burns, dividing them into major, moderate, and minor burns. Patients are placed in groups related to their risk. Low-risk patients are between the ages of 10 and 50 years old. High-risk patients are less than 10 and greater than 50 years old. Poor-risk are patients with underlying medical illnesses such as heart disease, diabetes, or chronic pulmonary problems.

(1) Minor Burns: Imply outpatient treatment. Minor burns involve TBSA of less than 15% in the low-risk group, or less than 10% TBSA in the high-risk group; full-thickness burns of less than 2% TBSA in anyone; and no other injuries.

(2) Moderate Burns: Are partial-thickness burns of 15-25% TBSA in the low-risk group, 10-20% TBSA in the high-risk group; full-thickness burns of equal to or less than 10% TBSA in anyone; burns not involving the hands, face, feet, perineum, or circumferential limbs.

(3) Major Burns: Defined as partial-thickness burns greater than 25% TBSA in the low-risk group, or greater than 20% TBSA in the high risk group; full-thickness burns of greater than 10% TBSA in anyone; burns involving the hands, face, feet, or perineum; burns crossing major joints; circumferential limb burns; burns complicated by inhalation injury; electrical burns; burns complicated by fractures or other forms of trauma; and burns in poor-risk patients. These burns require rapid casualty evacuation. (FMST.07.37b)

5. **TREATMENT:**

a. Initial Care:

(1) The first step in the treatment of burn injuries is to stop the burning process. (FMST.07.37c) Remove patients from heat, remove all their clothes that are affected by hot liquids, shut off electricity (very important), remove chemicals by flushing the affected area for 10 to 15 minutes, and decontamination of NBC. Heat can continue to injure tissue for a surprisingly long time. No first aid will be effective until the burning process has stopped.

(2) Manage the ABC's. Once an airway has been established give oxygen. If qualified, consider intubation earlier rather than later.

(3) Assess for associated injuries such as fractures or lacerations and inhalation injury.

(4) Remove clothing and jewelry from the burn area. This will help prevent constriction from swelling of the burn tissue. Do not try to remove anything that has adhered to the wound.

(5) Evaluate the Burn (depth, extent, pain).

b. General Treatment for the Patient:

(1) Gently wash the burn with tepid water and mild soap to remove any debris and to clean the skin surface around the burn site. Pat dry. Remove the skin from blisters that have popped open (do not open blisters unless necessary for function of hands or feet).

(2) Dress burn with a thin layer of antibiotic ointment.

(3) Cover the burn with Second Skin (if the burn is small enough) or cover with a thin layer of gauze, or clean, dry clothing. Burns on the face, neck and hands may be left open to the air after applying silvadene ointment. Covering the wound reduces pain and evaporative losses, but do not use an occlusive dressing.

(4) When evacuation is imminent, do not redress or reexamine the injury. If evacuation is prolonged, redress once a day. Remove old dressings, reclean (removing the old ointment), and apply fresh ointment and a clean dry covering. (Note: if stuck, soak off old dressings with clean, tepid water.)

(5) Do not pack wounds, which are larger than 20% of the body surface area, in ice. Do not leave wet coverings on burns for more than two hours at a time to reduce the risk of hypothermia.

(6) Stabilize the body temperature. When skin is lost, so is the patient's ability to thermoregulate.

(7) Have the patient drink as much fluid as he or she can tolerate without vomiting. Include some salt in the oral fluids, but do not make these solutions stronger than . 9%. This is the equivalent of a pinch of salt per 8 ounce glass.

(8) IV therapy as follows, Baxters (Parkland) Formula: In the first 24 hours give Ringers Lactate, 4ml times the weight in kilograms times the percent TBSA burned. Half of the solution is given in the first eight hours. The second half is given over the next 16 hours. The second 24 hours should be given fluids to maintain blood pressure, colloids (usually in the form of albumin) are the preferred choice.

4 x wt (kg) x TBSA= amount LR given in the first 24 hours

Example a 70kg patient with 50% TBSA this is the formula for IV therapy for the first 24 hours.

4 x 70kg x 50% = 14,000ml in 1ST 24 hours
7,000ml hours 1-8
3,500ml hours 9-16
3,500ml hours 17-24

c. General Management of the Burn: Caring for the wound itself is often the least important aspect of burn care. All burn wounds are sterile for the first 24 to 48 hours. Burn management is aimed at keeping the wound clean and reducing the pain.

(1) Elevate burned extremities to minimize swelling. Swelling retards healing and encourages infection. Have the patient gently and regularly move burned areas as much as possible.

(2) Burns are tetanus-prone wounds. Check on the last known tetanus shot. The patient may need a renewal.

(3) Ibuprofen is probably the best over the counter analgesic for burn pain (including sunburn).

(4) If you have no ointment or dressings, leave the burn alone. The burn's surface will dry into a scab like covering that provides a significant amount of protection.

6. **INHALATION INJURIES:** Half of all fire-related deaths are from smoke inhalation. Smoke inhalation doubles the mortality rate for any burn. Smoke inhalation is a general term for carbon monoxide poisoning, thermal airway injury, and smoke poisoning.

a. Carbon Monoxide Poisoning: Carbon monoxide (CO) is a colorless, odorless, tasteless gas that has an affinity for hemoglobin 200 times greater than that of oxygen. Carboxyhemoglobin (COHb) levels are measured as the percentage of hemoglobin bound to carbon monoxide. The following COHb levels correlate with clinical symptoms:

<10%	No symptoms
20%	Headache, nausea, vomiting, loss of dexterity
30%	Confusion, lethargy, ECG ST depression
40%	Coma
50%	Death

The half-life of COHb is 4 to 5 hours at room air, 90 minutes with 100% oxygen, and 20 to 25 minutes at 3 ata in a hyperbaric chamber. Treatment for less than 40% COHb is a 100% oxygen. At 40% or greater the hyperbaric chamber should be used. If patient is still in a coma after COHb levels are normal, prognosis is poor, and it has been experience that they rarely awaken.

b. Thermal Airway Injury: Injury to the respiratory tract from inhalation of hot gas or steam. The lower respiratory tract is usually not damaged, unless there was an inhalation of flame or steam. Any patients who were in an explosion, with burns of the hands, face, and upper torso are at risk. Signs of carbon such as soot in the pharnyx should alert the caregiver to airway injury. Maintaining the airway is the important issue here.

c. Smoke Poisoning: Is the inhalation of noxious gases that are the products of combustion. Cyanide poisoning is known to occur in victims of smoke inhalation. Other toxic gases released are sulfur dioxide, hydrogen chloride, phosgene, and ammonia. These gases cause damage to the respiratory tract similar to chemical burns. Physical signs include facial burns, intraoral or pharyngeal burns, singed nasal hairs, soot in mouth or nose, hoarseness, carbonaceous sputum, and expiratory wheezing.

UNITED STATES MARINE CORPS
Mountain Warfare Training Center
Bridgeport, California 93517-5001

WMC
04/09/02

LESSON PLAN

NOMENCLATURE AND CARE OF MOUNTAINEERING EQUIPMENT

LESSON PURPOSE. The purpose of this period of instruction is to introduce you to the types of equipment used here and how to care for it so as to prevent its untimely failure. This lesson relates to all climbing and installation work that you perform here.

1. **ROPES.** All ropes used in the military must meet UIAA standards or U. S. Federal Test Standard 191A. Most ropes have a 5-year shelf life and maximum 2-year service life.

a. Static. Black in color.

(1) Construction. Kernmantle

(2) Minimum tensile strength. 7500 lbs.

(3) Maximum elongation. 1.5%

(4) Diameter. 11mm

(5) Sizes.

(a) 165 ft ± 5 ft

(b) 300 ft ± 10 ft

(6) Usage. Rescue operations and bridging where a low amount of elongation is desirable under a working load.

b. Dynamic. Olive Drab in color.

(1) Construction. Water-resistant treated Kernmantle to reduce friction.

(2) Minimum tensile strength. 6500 lbs.

(3) Maximum elongation. 6%

(4) Diameter. 10.5mm and 11mm

(5) Size. 165 ft ± 5 ft.

(6) Usage. For lead climbing/party climbing.

c. Maxim Dry Rope. Olive Drab or Multi-Colored.

(1) Construction. Water-repellent treated Kernmantle.

(2) Minimum tensile strength. 3472 lbs.

(3) Maximum elongation. 6%

(4) Diameter. 9mm

(5) Size. 150 ft ± 5 ft

(6) Usage. For glacier travel/ice climbing

d. Gold Line II.

(1) Construction. Eight strand braided nylon plymor.

(2) Minimum tensile strength. 4500 lbs.

(3) Maximum elongation. 20%

(4) Diameter. 11 mm

(5) Size. 300 ft or 600 ft spools

(6) Usage. Sling Ropes and litters only.

Three strand twisted or laid rope.

Static, kernmantle rope showing core (kern) and sheath (mantle).

Dynamic, kernmantle rope used for rock climbing.

Double braid rope showing the braided sheath and braided core.

<u>INSIDES AND OUTSIDE SHEATHS OF VARIOUS TYPES OF ROPES</u>

NOTE: Sling ropes are made from 15 foot lengths of plymor or dynamic rope ONLY. Twenty-five foot practice coils should be constructed with static rope, but dynamic rope can be used.

2. (5 Min) **<u>ADVANTAGES/DISADVANTAGES</u>**

   a. <u>Advantages of Nylon Rope</u>

      (1) High strength to weight ratio.

      (2) Good energy absorption in dynamic ropes.

      (3) Flexible.

      (4) Rot resistant, not affected by frost.

   b. <u>Disadvantages of Nylon Rope</u>

(1) Low melting point. Nylon fuses at 400°F and melts at 480°F.

(2) Susceptible to abrasions and cuts.

(3) Affected by chemicals and light.

c. Advantages of Manila Rope

(1) Easily gripped.

(2) Hard wearing.

(3) Does not deteriorate in heat.

d. Disadvantages of Manila Rode

(1) Heavy, kinks, especially when wet. Absorbs water and swells.

(2) Burns at +300°F.

(3) Edible by rodents.

3. (3 Min) **GENERAL INFORMATION**

a. Nylon rope stretches under tension and will rupture at between 30% and 70% elongation, depending on construction.

b. Nylon rope loses as much as 30% strength when wet.

c. Temperatures as low as 250°F will damage a nylon rope.

4. (5 Min) **NYLON WEBBING**

a. The type of nylon webbing available is tubular. Tubular nylon webbing is very strong and flexible. All rules that apply to nylon rope apply to tubular nylon webbing. The size of nylon webbing used is:

(1) 1 inch tubular nylon. Tensile strength approximately 4,000 - 4,500 lbs., depending on the manufacturer.

b. Pre-sewn Spectra Runners. Tensile strength approximately 5,500 lbs.

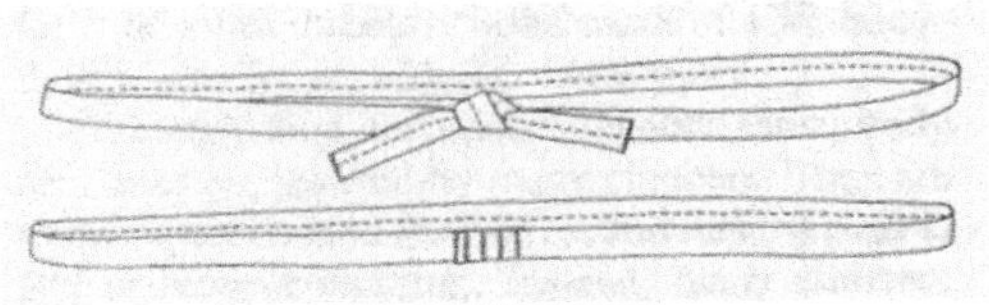

NYLON WEBBING

NOTE: These are minimum strengths. Some manufactures make even stronger webbing.

TRANSITION: We have just discussed general information in nylon webbing, are there any questions? Not only do we use ropes, but we also use carabiners in our installations, we will discuss the types of carabiners used:

5. (5 Min) **CARABINERS**. Also commonly known as snaplinks. Both locking and non-locking are used.

a. Purpose. Carabiners are used for the following purposes:

(1) To attach ropes or runners to pieces of protection.

(2) To attach the rappel rope to the rappel seat for seat-hip rappels or for crossing rope bridges.

(3) To attach the individuals safety rope to a safety line on a rope installation.

(4) To form field expedient pulley systems.

b. Nomenclature of a non-locking carabiner

(1) Gate

(2) Gate pivot pin.

(3) Locking pin.

(4) Body.

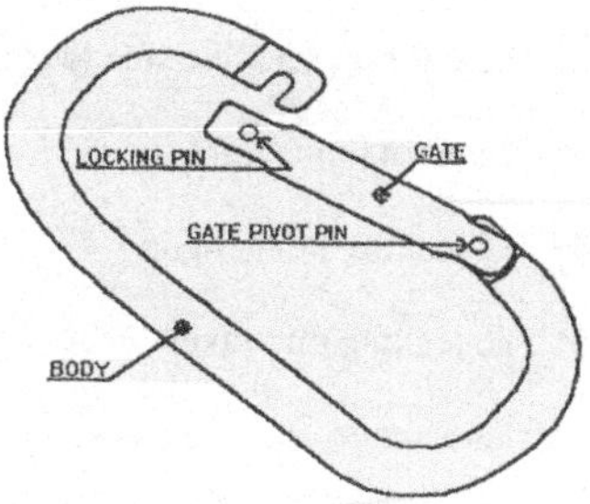

c. <u>Nomenclature of a locking carabiner</u>

(1) Gate.

(2) Gate pivot pin.

(3) Locking notch.

(4) Locking nut.

(5) Body.

NON-LOCKING ALUMINUM CARABINER

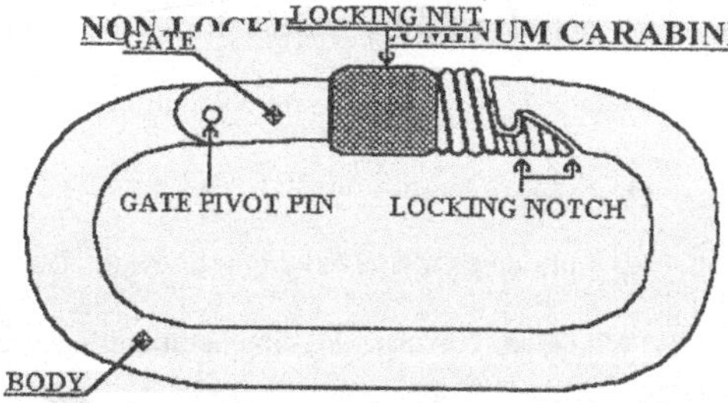

STEEL LOCKING CARABINER

d. There are two types of carabiners used. The two types and their characteristics are:

(1) <u>Steel locking carabiners</u>

(a) Large steel locking "D" (various manufacturers): Minimum tensile strength of 5,500 lbs.

(b) Steel-locking oval Stubai 82 is not in the MAC Kit and obsolete. However, it is being used at MWTC to save money, even though they are beyond the service life. Tensile strength of only 3,300 lbs.

(2) <u>Aluminum non-locking carabiners</u>

(a) Aluminum non-locking oval (various manufacturers): minimum tensile strength of 4,200 lbs.

e. Serviceability Check for a Carabiner. The following steps are used for you to check a carabiner for serviceability:

(1) The gate snaps shut with no friction and with no gap between the locking pin and locking notch.

(2) There is no excessive side to side movement of the gate.

(3) The pivot pin is tight.

(4) The locking pin is tight.

(5) The locking nut travels freely and locks securely.

(6) There are no cracks or flaws in the metal.

NOTE: The weakest part of a carabiner is the gate. If an engraver is used to mark a carabiner, it should be applied to the gate and not the load bearing side.

f. Preventive Maintenance for a Carabiner.

(1) Remove all dirt, moisture and grime.

(2) Lubricate with tri-flow graphite and clean off thoroughly.

NOTE: Whenever you use a locking carabiner ensure that the locking nut is always locked down (tightened).

6. (2 Min) **CARE OF THE CARABINER**. Do not drop the carabiner as this may result in either actual damage to the carabiner or in dirt getting into the workings of the carabiner and damaging it.

UNITED STATES MARINE CORPS
Mountain Warfare Training Center
Bridgeport, California 93517-5001

FMST.07.11
05/01/01

STUDENT HANDOUT

HYPOTHERMIA/REWARMING

TERMINAL LEARNING OBJECTIVE: Given a hypothermia casualty in a cold weather environment and necessary equipment and supplies, manage a hypothermia casualty to prevent death or further injury per the reference. (FMST.07.11)

ENABLING LEARNING OBJECTIVES:

(1) Without the aid of references, select from a given list the definition of hypothermia, in accordance with the references. (FMST.07.11a)

(2) Without the aid of references, select from a given list three of the five mechanisms by which the body loses heat, in accordance with the references. (FMST.07.11b)

(3) Without the aid of references, select from a given list the reason why gentle handling of a hypothermia victim is important, in accordance with the references. (FMST.07.11c)

(4) Without the aid of references, select from a given list the four rewarming techniques used to rewarm a hypothermia victim, in accordance with the references. (FMST.07.11d)

(5) Without the aid of references, select from a given list the proper steps of the sleeping bag rewarming method of rewarming a hypothermia victim, in accordance with the references. (FMST.07.11e)

(6) Without the aid of references, select from a given list the temperature of the water when rewarming a hypothermia casualty in a warm water bath, in accordance with the references. (FMST.07.11f)

8

1. **HYPOTHERMIA**:

a. Definition: A fall in core body temperature to 95°F (35°C) or less. (FMST.07.11a)

b. Hypothermia is potentially lethal. The mortality rate is greater than 50% in severe cases and cases complicated by trauma where the mortality can approach 100%. The most common cause of death in hypothermia is ventricular fibrillation.

c. When speaking of hypothermia, core temperature values must be used. In a controlled environment of an MTF or ER the gold standard is a core temperature taken by rectal or esophageal thermometry. The term "exposure" has often been used to speak of hypothermia but this term is inexact and has numerous other connotations. It should not be used as a synonym for hypothermia.

d. Incidence.

(1) Classically, hypothermia has been thought to occur in extremely cold environments. However, it can occur in temperate climates and at Southern latitudes. In fact, greater than 50% of cases of hypothermia occur at ambient temperatures of 50°F or greater. Contrary to popular belief, the highest incidence of hypothermia is seen in urban areas.

(2) Hypothermia tends tends to occur more frequently in males with a male to female ratio of 2.8:1.0. It's also more common in the elderly, those with drug and alcohol dependency and those with psychiatric diagnoses.

Many standard clinical thermometers don't register a low enough temperature to accurately diagnose hypothermia. Currently, a low-reading rectal thermometer is available through the Federal Stock System (NSN 6515-00-139-4593). All medical personnel attached to cold weather units or deploying to cold weather environments should carry this thermometer. In a high altitude cold weather environment medical providers should be adept at correlating signs and symptoms with specific temperature ranges so as not to unduly expose the patients to the elements. The more accurate, and invasive temperature determination should only be performed when the patient can be adequately protected from the elements.

2. **PATHOPHYSIOLOGY**:

a. Heat Conservation/Loss.

(1) The body is a furnace that generates heat. The ability of the body to thermo regulate depends on the ratio of heat production to heat loss. An imbalance of these two processes can lead to hypothermia. About 75% of food energy consumed goes toward generating body heat while the other 25% goes toward maintaining basic metabolic function.

(2) Thermogenesis can be divided into two main categories:

(A) Shivering thermogenesis creates heat by muscle activity with subsequent breakdown of glycogen. Shivering usually begins at a core temperature of around 97°F (36°C) and finally ceases at core temperatures of 85°- 90°F (30°- 34°C).

(B) Non-shivering thermogenesis creates heat by hormonal interaction with target organs such as the heart. Epinephrine, norepinephrine, growth hormone and thyroid hormone have all been implicated in playing a role in non-shivering thermogenesis.

(3) Changes in hormone secretion patterns are extremely important in acclimatization to cold weather.

(4) Once heat is generated it is retained by three main mechanisms: Peripheral vasoconstriction, abolition of sweating and piloerection. Peripheral vasoconstriction is by far the most important and efficient mechanism of heat conservation. The main internal temperature sensors in the body are located in the hypothalamus in the brain. When these areas sense a drop in core temperature they initiate peripheral vasoconstriction at the surface of the skin to prevent a loss of core body heat.

(5) The body loses heat by five main mechanisms: (FMST.07.11b)

(A) **Radiation** is heat transferred to the surrounding environment via infrared radiation.

(B) **Convection** is the loss of heat via air/water current.

(C) **Conduction** is the direct contact-transfer of heat from a hot to a cold object.

(D) **Evaporation** is the conversion of water from a liquid phase to a gaseous phase (perspiration).

(E) **Respiratory heat loss** takes place when warm humidified air from the lungs is expelled into the atmosphere.

(F) The mechanism accounting for the greatest percentage of heat loss is highly variable and depends on ambient conditions, i.e., a nude man standing in a cold room will lose approximately 65% of his heat by radiation. The same man standing in calm cold water will lose most of his heat by conduction and at a rate 25 times faster.

**NOTE: The liver and brain are the most thermogenic organs on a per weight basis. However, due to their greater mass, skeletal muscle and skin produce the largest percentage of body heat.

(6) Numerous factors predispose to the development of hypothermia. These include but are not limited to:

(A) Malnutrition.
(B) Lack of adaptation.
(C) Inactivity.
(D) Sedatives.
(E) Physical exhaustion.
(F) Burns.
(G) Extremes of age.
(G) Dehydration.
(H) Comorbid illness.

3. **SIGNS AND SYMPTOMS**:

a. Clinical manifestations of hypothermia reflect CNS, and cardiorespiratory involvement.

(1) 98°-95°F (37°-35.6°C): Mild shivering, cold sensation and impaired fine motor coordination.

(2) 95°- 92°F (35°-32.8°C): Violent shivering, difficulty speaking, sluggish thinking, and amnesia. Large muscle in coordination.

(3) 92°-86°F (32.2°-30°C): Shivering is replaced by muscular rigidity. Exposed skin is blue or puffy. Movements are jerky. Dulled sensorium, but victim may be able to maintain posture and the appearance of being in contact with surroundings. Possible atrial arrhythmias.

(4) 85°-81°F (29.4°-27.2°C): Coma, lack of reflexes, possible ventricular arrhythmias. Cessation of shivering below 86oF.

(5) Below 78°F (25.6°C): Failure of cardiac and respiratory centers, pulmonary edema, ventricular fibrillation, and asystole.

b. Some patients with hypothermia will exhibit a phenomenon that is called, "paradoxical undressing."

(1) In these instances the hypothermia victim is found nude with their cloths beside them. If the victims are women they are often misdiagnosed as sexual assault cases. This phenomena results from loss of hypothalamic control over peripheral vasoconstriction which leads to flushing of the periphery with relatively warm blood causing the individual to feel hot. Subsequently the individual undresses to cool down.

c. Again it must be stated that the only way to correctly diagnose hypothermia is with a core temperature reading.

4. **FIELD TREATMENT**:

a. ABC's, C-Spine precautions, remove wet clothing, replace with dry clothes and insulate with a vapor barrier system. (Wrapping the victim in two plastic garbage bags or bubble wrap is ideal. Further insulation may be necessary). Handle the victim carefully to prevent ventricular fibrillation caused by unnecessary jarring and rough handling. (FMST.07.11c) CASEVAC.

b. Victims with altered consciousness should be NPO. Under no circumstances should ETOH be given. This is a peripheral vasodilator and will only enhance heat loss.

c. If CASEVAC is not feasible for a long period of time, or if the victim is only mildly hypothermic, it may be necessary to rewarm in the field by using a proper sleeping bag rewarming technique. Additionally, insulated hot packs placed at the high heat loss areas will help.

d. The hypothermic victim often has no discernible vital signs.

(1) Before beginning CPR thoroughly assess pulses for 30-45 seconds to detect viable cardiovascular activity.
(2) Severely hypothermic victims in ventricular fibrillation and asystole have been successfully resuscitated even after periods as long as 4 hours.

5. **REWARMING**: The re-warming of a hypothermia casualty in a BAS setting is less than optimal since the AMAL for a BAS does not include monitoring devices, laboratory facilities, etc., but rewarming of a casualty can be successfully conducted even in these fairly primitive conditions. The following are three basic techniques of re-warming:

- Passive external re-warming: a sleeping bag or blanket that is not pre-warmed.
- Active external re-warming: a pre-warmed sleeping bag, bear hugger or warm water bath.
- Active internal or core re-warming: IV fluids, warmed/humidified O2, lavage, extracorporal re-warming.

a. Four ideal re-warming strategies that may be available in the field include: (FMST.07.11d)

(1) Sleeping bag rewarming.

(2) Water bath/counter Current rewarming.

(3) Warm/humidified airway rewarming.

(3) Warmed I.V. solutions. NS is preferable as a cold liver does not like to metabolize Lactate.

6. **REWARMING TECHNIQUES (BAS)**:

a. Sleeping Bag Rewarming. (FMST.07.11e) This is the simplest, but least effective method.

(1) Make the diagnosis. Take a rectal temperature to determine if the casualty is clinically hypothermic (95°F or less). Use this temperature as a baseline to determine stabilization.

(2) Warm sleeping bag (pre-warm sleeping bag with two volunteers). Placing a hypothermic casualty into a cold sleeping bag will cause further heat loss by conduction.

(3) Strip casualty. Remove all wet clothing avoiding unnecessary handling. Muscular movement will pump cold blood to the core.

(a) Place the casualty between the two volunteers, if space permits. Their body heat is transferred from the two volunteers to the cold casualty.

(4) Monitor core temperature frequently throughout transport to ensure that the casualty's core temperature is not continuing to drop.

(5) Adequate insulation is required to prevent further heat loss to the environment.

(6) Augment heat by placing insulated heating pads in the high heat loss areas: head, axilla, groin, popliteal region and antecubital fossa. (Be careful not to cause burns)

**NOTE: It should be noted that an evacuation bag can be used; this will provide more room for the casualty and volunteers.

b. Water Bath Re-warming. (FMST.07.11f) This method has a long history of success in rewarming even severely hypothermic casualties. The casualty is placed in a portable/field expedient bathtub or a life raft. The extremities and head are kept out of the water to avoid significant afterdrop. The water bath temperature should be 104°-108°F. An anesthesia temperature probe should be used to constantly monitor the core temperature. Since the casualty will cool the bath water around him, it will be necessary to continue to add warm water to maintain the proper water bath temperature. Keep in mind however that this technique has definite drawbacks. In the event of ventricular fibrillation caused by re-warming shock, unsynchronized cardioversion can be very dangerous in a wet environment. Furthermore, if the casualty in the water is not on a firm surface CPR, if needed, can be very difficult to perform ineffective.

**NOTE: Individual in heat distress can be rapidly cooled using this method with "cool" water instead.

c. Warm Airway Re-warming. A warmed and humidified air/oxygen mixture is used to provide a warming media within the lungs. These devices consist of a method of generating warm humidified air and are usually portable. The warm humidified air modestly increases the amount of heat that can be delivered to core. There should be a thermometer in the airway tubing to monitor the temperature. To prevent injury to the bronchi, the temperature should not exceed 115°F (108° to 115°F is ideal). Warm airway rewarming should be used in conjunction with other rewarming methods. This method, by itself, will probably not deliver sufficient heat to rewarm a severely hypothermic casualty by itself and may in fact blunt the shivering response robbing the body of an important endogenous mechanism for producing heat.

Caveat: When possible, active external re-warming should always be combined with active core re-warming techniques so as to minimize the effects of afterdrop. This occurs as a result of peripheral vasodilatation causing warm core blood to move to the surface while relatively cold surface blood moves to the core and may lower the fibrillation threshold.

**NOTE: Hypothermic submersion incident casualties cannot tolerate humidified air for any length of time.

d. Warmed I.V. Solutions. Hypothermia is a common response to I.V. therapy. Solutions that have been prewarmed have been shown to prevent this complication. There are various methods for warming I.V. solutions:

(1) Crystalloid solutions can be warmed by warm water bath and microwaved with no adverse changes to its integrity.

(2) Fresh Frozen Plasma (FFP; -20°C): Studies have shown that microwaving for 30 seconds followed by gentle manipulation of the bag for 10 seconds, repeated five times will result in only 2.6% cell destruction.

(3) Packed Red Blood Cells (PRBC 4°C): Using conventional rewarming methods, studies have shown that the outer five millimeters of the bag can have isolated areas of hemolysis. Until a device is made that can uniformly shake up PRBC during warming, we do not recommend microwaving of blood. Rewarming of blood by dilution with warm, calcium-free crystalloid solution is useful.

WIND CHILL CHART

WIND	AMBIENT TEMPERATURE (FAHRENHEIT/CENTIGRADE)													
CALM F	35	30	25	20	15	10	5	0	-5	-10	-15	-20	-25	-30
CALM C	27	-1	-4	-7	-9	-12	-15	-18	-21	-23	-26	-29	-32	-34
EQUIVALENT WIND CHILL FACTOR (FAHRENHEIT/CENTIGRADE)														
5 MPH F	33	27	21	16	12	7	1	-6	-11	-15	-20	-26	-31	-35
8 KPH C	1	-3	-6	-9	-11	-14	-17	-21	-24	-26	-29	-32	-35	-37
10 MPH F	21	16	9	2	-2	-9	-15	-22	-27	-31	-38	-45	-52	-58
16 KPH C	-6	-9	-13	-17	-19	-23	-26	-30	-33	-35	-39	-43	-47	-50
15 MPH F	16	11	1	-6	-11	-18	-25	-33	-40	-45	-51	-60	-65	-70
23 KPH C	-9	-12	-17	-21	-24	-28	-32	-36	-40	-43	-46	-51	-54	-57
20 MPH F	12	3	-4	-9	-17	-24	-32	-40	-46	-52	-60	-68	-76	-81

32 KPH C	-11	-16	-20	-23	-27	-31	-36	-40	-43	-47	-51	-56	-60	-63
25 MPH F	7	0	-7	-15	-22	-29	-37	-45	-52	-58	-67	-75	-83	-89
40 KPH C	-14	-18	-22	-26	-30	-34	-38	-43	-47	-50	-55	-59	-64	-67
30 MPH F	5	-2	-11	-18	-26	-33	-41	-49	-56	-63	-70	-78	-87	-94
48 KPH C	-15	-19	-24	-28	-32	-36	-41	-45	-51	-53	-57	-61	-66	-70
35 MPH F	3	-4	-13	-20	-27	-35	-43	-52	-60	-67	-72	-83	-90	-98
56 KPH C	-16	-20	-25	-31	-33	-37	-42	-47	-51	-55	-58	-64	-68	-72
40 MPH F	1	-4	-15	-22	-29	-36	-45	-54	-62	-69	-76	-87	-94	-101
64 KPH C	-17	-20	-26	-30	-34	-38	-43	-48	-52	-56	-60	-66	-70	-74

7 REWARMING TECHNIQUES (HOSPITAL):

a. Optimal re-warming techniques for higher echelon care facilities still tend to employ rather simple therapeutic modalities. The three safest and most effective are listed below:

- Warmed IVF.
- Warmed/humidified oxygen.
- Bear Hugger.

b. Many texts advocate the use of extracorporal blood re-warming. However, this technique has some disadvantages. It requires a high level of technical and medical support, which are often not available. Since patients undergoing bypass must be anti-coagulated, patients with multiple traumatic injuries may not be good candidates. Lastly, research indicates that anticoagulation with heparin may actually worsen the severity of frostbite that may occur in patients with hypothermia.

c. Lavage: peritoneal, thoracic and mediastinal lavage have all been suggested for use in severe hypothermia. However, the use of these techniques in inexperienced hands tends to carry a much higher complication rate. If performed correctly, peritoneal lavage can be very beneficial with a lower complication rate than thoracic lavage. Gastric, colonic and bladder lavage are indeed safer but are of limited efficacy, as they do not offer access to areas with significant surface area.

d. ACLS: At core temperatures below 86°F attempt three progressive unsynchronized shocks at 200J, 300J, and 360Js. If no response continue CPR and REWARM THE VICTIM. Re-attempt counter-shocks when core temperature rises above 86°F. Above 86°F, perform counter shocks in 2° increments above 86°F, i.e.88°F, 90°F and 92°F. Employ ACLS drug protocols above 93°F along with conventional normothermic ACLS protocol.

UNITED STATES MARINE CORPS
Mountain Warfare Training Center
Bridgeport, California 93517-5001

FMST.07.12
03/21/02

STUDENT HANDOUT

SUBMERSION INCIDENTS

TERMINAL LEARNING OBJECTIVE: Given a casualty in a cold weather environment and the necessary equipment and supplies, manage a submersion incident casualty to prevent death or further injury per the reference. (FMST.07.12)

ENABLING LEARNING OBJECTIVES:

(1) Without the aid of references, given a list of the events, which take place to a submersion victim, select the correct sequence of events, in accordance with the references. (FMST.07.12a)

(2) Without the aid of references, select from a given list, three of the five methods that can be used to increase survival time during cold water immersion, in accordance with the references. (FMST.07.12b)

(3) Without the aid of references, select from a given list, three of the six favorable prognostic signs post submersion incident, in accordance with the references. (FMST.07.12c)

(4) Without the aid of references, select from a given list, three of the four unfavorable prognostic signs post submersion incident, in accordance with the references. (FMST.07.12d)

(5) Without the aid of references, select from a given list the correct field management priorities of a submersion victim, in accordance with the references. (FMST.07.12e)

10

OUTLINE.

1. **TERMINOLOGY**.

a. Definition: A submersion incident is that medical problem that occurs after a casualty has been submersed under water.

(1) Drowning refers to death by suffocation following submersion in water.

(2) Near drowning refers to submersion in water with at least temporary survival (greater than 24 hours).

NOTE: An effort is being made to standardize the terminology used regarding submersion incidents, in order to clarify communication between care-givers. Terms like secondary drowning and other definitions are being discouraged.

2. **PATHOPHYSIOLOGY**.

a. Sequence of events: (FMST.07.12a)

(1) Panic initially, followed by a violent struggle.

(2) Gulping and swallowing air and water to avoid aspiration.

(3) Breath holding until hypoxia leads to unconsciousness.

(4) Once consciousness is lost, the gag reflex relaxes and passive influx of water into the lungs occurs. A small percentage of victims (10-15%) have significant laryngospasm, which prevents any appreciable volume of water to enter the lungs. This is referred to as "dry drowning".

(5) Aspiration is more likely in submersion victims who enter very cold water (10°C). An involuntary gasp reflex occurs, with a marked increase in respiratory rate and depth. This individual is at increased risk of aspiration, and subsequent drowning.

b. Submersion in very cold water will quickly results in hypothermia.

(1) Rapid induction of total body hypothermia (core temperatures of 89.6°F or less) may be protective in submersion incidents. Significant decreases in oxygen requirement result, especially in the case of the brain itself. Marked decreases in cerebral blood flow and oxygen demand have been demonstrated (see Hypothermia lesson). This protective effect of hypothermia has been used to explain the many cases of complete recovery after extended periods of submersion, which have only occurred in cold water.

(2) The "Mammalian Diving Reflex" slows the heart rate, shunts blood to the brain and closes the airway when more primitive marine mammals are exposed to extremely cold water. General consensus holds that this reflex is not a significant response in human beings. However, it has been found to be more prominent in new-born human infants. It has also been speculated that this "Reflex" may be more significant in those individuals who have survived extended submersion times.

c. Fatalities in submersion victims associated with hypothermia are primarily due to three known mechanisms:

(1) Immersion syndrome. Sudden death due to cardiac arrest secondary to massive vagal stimulation. Usually associated with sudden total body exposure to frigid water.

(2) Excessive fatigue and confusion leading to fatal errors of judgment.

(3) Direct hypothermic effects. Thermal conductivity in water is approximately 25 times that of air. Despite the evidence showing hypothermia's protective effects, the complications associated with hypothermia (cardiac arrhythmias, CNS depression, loss of motor control) can lead to submersion and drowning.

d. Some authorities emphasize distinguishing between "wet" and "dry" drowning. In wet drowning, aspiration of water into the lung occurs. (85-90% of casualties). In dry drowning, laryngospasm prevents aspiration of water (10-15% of cases). Ultimately, the end result is the same: obstruction to respiration, oxygen depletion, and asphyxiation. The morbidity and mortality associated with the two types of drowning are not significantly different.

e. Fresh-water and salt-water drowning are somewhat different.

(1) In fresh-water aspiration, the osmotic difference between hypotonic fresh-water and the relatively hypertonic blood leads to net water movement into the vasculature.

This flow across the alveolar membrane also leads to "washout" of surfactant, which leads to collapse of the alveolar spaces. Hypoxemia follows.

(2) In sea-water aspiration, the osmotic shift is reversed, leading to fluid outpouring from the blood stream flooding the alveoli and the interstitium. This leads to decreased lung compliance as well as direct damage to pulmonary capillaries. Again, hypoxemia results.

(3) Hypoxemia is the end result in both types of drowning. No significant difference in morbidity in survivors has been shown between fresh-water and salt-water drowning. The only bodies of water shown to have a worsening effect on prognosis are the hyper-saline bodies, like the Dead Sea and the Great Salt Lake. Otherwise, prognosis is similar.

f. Hypoxemia is the primary insult after submersion. Efforts to correct this problem should be vigorous and sustained. All clinically dead submersion casualties should receive CPR, or ACLS if available, immediately upon removal from the water, regardless of water temperature or duration of submersion. Since the overwhelming majority of naturally occurring water bodies in the world have temperatures below 92°F, all submersion victims should be treated aggressively as hypothermic casualties. Survivals after submersion times up to 66 minutes have been documented. Remember: No one is dead until they are warm and dead.

g. In the case of prolonged submersion or significant aspiration, pulmonary derangements occur rapidly. In lesser insults, the onset of symptoms may be delayed for as long as 24 hours. Physiological changes include:

(1) Cardiovascular. Casualties develop respiratory acidosis and hypoxemia, which may trigger dysrhythmias. Electrolyte changes are not usually significant in survivors of submersion incidents and should correct once spontaneous respirations are re-established.

(2) CNS. C-Spine injuries are common in submersion incidents, most commonly a result of shallow-water diving. Injury patterns include wedge, tear-drop and burst fractures of the upper C-Spine, frequently resulting in tetraplegia. Concussive head injuries also occur. Drug and alcohol use should always be considered.

(3) Metabolic. Respiratory acidosis is the initial problem. Acid/base derangement should correct after resuscitation of the casualty without specific intervention.

3. **SURVIVAL TECHNIQUES IN COLD WATER.**

a. Swimming motion increases heat loss more than it increases heat production. The direct hypothermic effects of cold-water immersion make it impossible to swim for long periods of time. The average swimmer is unlikely to swim more than a kilometer in 50°F water. Swimming is advised only when a rescue vehicle is nearby, or the distance to shore is less than ½ mile.

b. Treading Water and Drown Proofing.

(1) Treading water results in reduced survival time due to increased heat loss secondary to increased activity.

(2) "Drown Proofing" is a technique whereby the swimmer relaxes in the water, floating facedown, arms spread, allowing his face to submerge between breaths. "Drown proofing" is the least desirable survival technique in cold water since the head and neck (high heat loss areas) are maximally exposed.

c. Personal Floatation Device (PFD). Without the aid of the PFD, the swimmer caught in cold water is forced to use less desirable methods to stay afloat.

(1) If wearing a PFD, the individual can utilize the "Heat Escape Lessening Posture" (HELP). This position minimizes the exposure of the swimmer's high heat loss areas because the arms are folded across the chest and pressed to the sides, the knees are drawn up to the chest and the legs are crossed at the ankles. Unfortunately, this position is fatiguing and becomes difficult to hold.

(2) If there is more than one person in the water, the "Huddle" position should be used. Similar to a football huddle, the chest, abdomen and groin area should be pressed together. Children or injured individuals should be placed in the middle of the huddle, since they become hypothermic more quickly.

d. Increasing Survival Time: (FMST.07.12b)

(1) Try to enter the water in a lifeboat or raft.

(2) Locate and don a PFD as quickly as possible.

(3) Wear several layers of clothes, especially to protect high heat loss areas.

(4) If you must enter the water, enter gradually rather than jumping or diving.

(5) If afloat in the water, avoid unnecessary movement. Assume the HELP or Huddle position.

(6) Don't try to swim.

(7) Remember the situation is not hopeless. Have the will to SURVIVE!

4. **RESCUE TECHNIQUES**.

a. Reaching the Casualty. The basic order for water rescue is: Reach and Pull, Throw, Tow and, as a last resort, Go.

(1) Reach and pull. When the casualty is responsive and close to the shore, begin by holding out an object for him to grab then pull him from the water. Floating line is considered the best choice, but if it is not available, a tree branch, fishing pole, oar or other object will do as well.

(2) Throw. Should the casualty be alert, but too far away for you to reach and pull, throw an object that will float to them. A PFD or ring buoy is best, if available. The primary course of action is to throw anything that will float, and to do it as quickly and as accurately as possible.

(3) Tow. Once the conscious casualty has an object to hold onto, try to find a way to tow him to shore. From a secure position, tow the casualty ashore.

(4) Go. Swim to the casualty as a last resort. You should be a strong swimmer, trained in water rescue, in order to avoid the dangers associated with this form of rescue. If the casualty is unconscious or unresponsive, you must go to the casualty.

5. **SIGNS AND SYMPTOMS**.

a. History of Incident. Several questions need to be answered, as they will help determine the approach to the casualty.

(1) When a person is found floating face down on the water surface or is brought to the surface by a rescuer, there usually is little question about the cause of unconsciousness.

(2) When near-drowning is unwitnessed or occurs after a dive or other water activity (surfing), a head injury and a C-spine injury should be presumed, and C-Spine precautions should be instituted.

(3) Determine how long the casualty was submerged. This helps determine prognosis.

(4) How cold was the water? Very cold water (less than 70°F/21°C) may be protective.

(5) What was the casualty's health prior to submersion? Drowning could have been secondary to alcohol ingestion, cardiac arrest or other illness/injury.

b. Casualties may be symptomatic or asymptomatic at the scene:

(1) Respiratory signs and symptoms may include dyspnea, tachypnea (SOB), sore throat, substemal burning, pleuritic chest pain, persistent cough, frothy sputum, cyanosis, rales, rhonchi and wheezes.

(2) Cardiovascular signs may include tachycardia and hypotension.

(3) Neurological findings may include anxiety, restlessness, lethargy, confusion, convulsions, incontinence, hyporeflexia and coma.

(4) Abdominal distention is common as is fever and hypothermia.

c. Favorable prognostic signs: (FMST.07.12c)

(1) Casualty is alert.

(2) Basic or advanced life support is received at the scene.

(3) Good response to initial resuscitation.

(4) Hypothermic.

(5) Older child or adult.

(6) Brief submersion time.

d. Unfavorable prognostic signs: (FMST.07.12d)

(1) Age under three years.

(2) Fixed dilated pupils.

(3) Submersed five minutes or longer.

(4) No resuscitation attempts for the first ten minutes.

(5) PH less than 7.10.

(6) Cardio-pulmonary arrest as initial presentation.

(7) Resuscitation time greater than 30 minutes.

6. **FIELD MANAGEMENT**.

a. The scene of a submersion incident is frequently chaotic and first consideration should be the safety of the rescuers, then retrieval of the casualty from the water.

b. Priorities: (FMST.07.12e)

(1) Establish an airway and check breathing and circulation (ABC's). Rescue breathing, can be performed by a trained rescuer, while the casualty is still in the water.

(2) Observe C-Spine precautions, since cervical spine and head injuries are common, especially in diving accidents.

(3) Begin CPR as soon as possible, if necessary, when the casualty reaches the shore, or suitable rescue platform.

(4) The symptomatic patient should be given high-flow 02 and IV access should be established, if available.

(5) Medevac ASAP.

(6) It is recommended that all victims of submersion be observed in a medical facility for a minimum of 6-8 hours. It is not uncommon for asymptomatic near-drowning victims to refuse further care. All that can be done is attempt to get a contact number in order to follow-up with the victim via telephone 8 hours later.

c. Vomiting is common and should be anticipated by turning the patient's head and having suction ready, if available. Frequently, the cause of vomiting is gastric distention.

(1) The American Heart Association recommends that gastric distention be relieved only when it interferes with CPR.

(2) Gastric decompression can be achieved by "breaking" the patient. This maneuver is performed by placing the patient in the prone position and lifting upwards at the abdomen.

(3) Another method is to turn the patient to a lateral recumbent position and compress the abdomen with the palm of one hand.

(4) The preferred method of gastric drainage is to pass an NG tube and relieve distention with gentle suction.

(5) Gastric drainage should not be attempted without suction equipment available, due to the risk of aspirating vomitus.

7. **PROGNOSTIC PREDICTORS**.

a. 0 - 2 "Unfavorable Prognostic Indicators" yield a greater than 90% survival rate (see above).

b. 3 or more "Unfavorable Prognostic Indicators" yields a less than 5% survival rate.

c. Corm's Criteria: Based on the status of the patient 1 hour after initial resuscitation.

(1) A - Awake	All survive (99+%)
(2) B - Blunted	Most survive (95+%)
(3) C – Comatose	
C1 - Flexor Response	Good survival with ICU care (80+%)
C2 - Extensor Response	Good survival with ICU care (80+%)
03 - Flaccid	Poor survival
C4 - Arrest	Dismal survival

UNITED STATES MARINE CORPS
Mountain Warfare Training Center
Bridgeport, California 93517-5001

FMST.07.14

04/02/02

STUDENT HANDOUT

WILDERNESS ORTHOPAEDIC/ TRAUMA INJURIES

TERMINAL LEARNING OBJECTIVE: Given a casualty in a mountainous enviornment and, necessary equipment and supplies, manage common orthopedic injuries in a mountainous enviornment to prevent death or further injury per the reference. (FMST.07.14)

EABLING LEARNING OBJECTIVES:

(1) Without the aid of references, select from a given list the phases of clearing a cervical spine in the field in accordance with the references. (FMST.07.14a)

(2) Without the aid of references, select from a given list, the correct treatment for an open book pelvic fracture in accordance with the references. (FMST.07.14b)

(3) Without the aid of references, select from a given list, six considerations for orthopedic patient evacuation in accordance with the references. (FMST.07.14c)

OUTLINE

1. INTRODUCTION.

Orthopedic injuries occuring in the wilderness are not significantly different from those suffered in garrison. Bones break, ligaments are torn and strained, and joints are dislocated in both settings. The key to preparing for and managing wilderness orthopedic injuries is understanding that the provider will have significant limitations with regard to diagnostic tools and initial management resources at his/her disposal. The personal risks each expedition member is exposed to can rise dramatically if the team is needed to transport a victim out of the back-country. Determination of a casualty's mobility becomes a major consideration for the wilderness medical provider. Assessment of the terrain, creative use of available natural resources, and effective leadership are all necessary components of caring for orthopedic injuries in the wilderness.

10

←

2. GENERAL: Making the diagnosis.

The wilderness "doc", whatever his/her training background, needs to have good physical exam skills. NO X-RAY and NO LABS will be available in the back-country. These basic tools are not available in the field. The provider will need to rely on his/her observation, palpation, and auscultation skills to make a diagnosis. This person will also need to use the patient, if possible, to uncover the likely injury by taking a thorough history which includes careful review of the mechanism of injury.

A. HISTORY: Critical information can be gained from talking to your patient.

- Mechanism of Injury - Determine:
    1. High vs. low-velocity accident
    2. Direction of force
    3. Nature of activity
    4. Loss of consciousness

- Ask the casualty:
    1. Did they hear a pop, snap, crack, or breaking sound?
    2. How long ago did the injury occur?
    3. Can they move the injured body part?
    4. Was there swelling within 15 minutes of injury or did it take longer?
    5. Was there immediate disability or was there a gradual loss of function?
    6. Was there immediate discoloration of a joint?
    7. Can they bear weight on the injured extremity?

- Other important information:
    1. Evaluate the environment. (i.e. Hot/Cold, Wet/Dry)
    2. Allergies.
    3. Medications currently being taken.
    4. Previous injuries/surgeries to the site in question.

B. PHYSICAL EXAMINATION

Your powers of observation become crucial.

- Look for:
    1. Swelling.
    2. Discoloration and bruising.
    3. Obvious deformity/angulation.
    4. Open wounds. (with or without protruding bone fragments)
    5. Differences right vs. left.

6. Muscle spasm surrounding injured site.
7. Point tenderness.
8. Crepitus.

- Beware:

Consider the environment while assessing your patient
- Don't undress the casualty in the cold!
- Protect your patient from the elements. (Sun, Rain, Cold)
- Talor your physical exam to meet the constraints of the environment.
  * palpate under clothing.
  * visualize one region at a time, then re-dress.
  * set up temporary shelter from the elements.

3. Head Injuries:

a. A blow to the head may lead to increased intracranial pressure (ICP) or intracranial bleeding neither of which are managable in the wilderness. The job of the clinician is to diferentiate a serious life threatening injury from a minor one.

1. Minor injury: No loss of consiousness (LOC), or LOC of less than 15 seconds with immediate return to full alertness. The casualty can not be on medications which increase risk of bleeding or have a history of bleeding disorders. Patient may be monitored every two hours for mental status changes, lethargy, irritibility, persistent nausea and vomiting, changes in speech or visual changes.

2. Serious Injury: LOC greater than 15 seconds, and/or persisitent confusion or memory loss; signs or symptoms of increased ICP: Debilitating headache, mental status changes, persistent nausea and vomiting, appearance of clear fluid in external auditory canal, Battle sign, raccoon eyes, or seizures.

3. Field treatment of serious head injury: Suspect injury of C-spine, manage airway-be able to clear vomit, elevate head 30 degrees. Evacuate to treatment facility ASAP.

4. Spinal Injuries.

A. Cervical Spine

1. High Risk Activities:

C-spine injuries in the wilderness usually occur after either a fall from a significant height or high-velocity accident from skiing. Common winter activities predisposing participants to C-spine injuries are skiing, snow-boarding, and snow-mobiling.

2. Anatomy:

The cervical spine consists of seven cervical vertebrae interposed between the base of the skull and the thoracic spine. The cervical spine has a great deal of mobility to allow maximal range of motion for the skull. This increased mobility comes at a high price; this portion of the spine is less stable than the rest of the vertebral column, and hence more subject to injury. The C-Spine is stabilized by three longitudinal ligaments: The anterior longitudinal ligament (ALL) runs longitudinally along the anterior surface of the vertebral bodies. It is broad and very strong and helps to prevent hyper-extension of the head and neck. The posterior longitudinal ligament (PLL) also runs longitudinally, this time along the posterior surface of the vertebral bodies, within the vertebral canal itself. This ligament is relatively narrow and somewhat weaker than the ALL, and helps to prevent hyper-flexion at the neck. Third is the supraspinous ligament, connecting the spinous processes. This, too, is a strong ligament, acting to inhibit hyper-flexion.

The peripheral nerves innervating the muscles and the sensory nerves of the upper extremities originate from the cervical portion of the spinal cord and these are some of the nerves likely to be affected in the event of a C-Spine injury. Hence, it is important to assess the neurologic status of the upper extremities if a cervical injury is suspected.

3. C-Spine Injury Statistics:

-Most common injury: Flexion injury at C5/6
(so look for a deficit in the C6 distribution)

-28% of C-Spine injuries have another spinal fracture associated with it.
(so examine entire spine)

-10% of Head injuries/Facial fractures also have a C-Spine injury, especially if there was LOC.

4. Remember:

a. Head injury -> Assume C-Spine injury.
b. Perform complete Neurovascular exam
-Motor, Sensory, Reflexes, Pulses, Babinski reflex.

B. Clear a C-Spine. (FMST.07.14a)

When faced with a possible C-Spine injury in the Wilderness, clearing it without X-ray should be considered only in extreme situations. However, there are situations which demand that every effort be made to make a casualty ambulatory, because the movement of the individual would demand herculean effort from the rest of the team, potentially placing more individuals at risk of injury or prolonging exposure to extreme environmental conditions.

Clearing a C-Spine in the field should follow a three-phased assessment. If the patient fails the assessment at any one of the phases, then the team is obligated to maintain spinal precautions and transport the patient. The phases are outlined below, and expanded upon in the lecture.

PHASE ONE:

-ensure no alteration of mental status.
-maintain in-line traction.
-perform complete neurovascular exam (focusing on the upper extremities)
-palpate spinous processes one-by-one while asking casualty whether palpation causes pain (supraspinous ligament)
-distinguish between pain over spinous process and muscle soreness associated with paraspinal musculature.
-only if palpation was completely pain-free can you move on to ...

PHASE TWO:

-continue to maintain in-line traction
-palpate anteriorly to the left and right of trachea. along the anterior vertebral bodies (anterior longitudinal ligament).
-Assess for bony tenderness under the angle of the jaw to the clavicle.
-Distinguish between tenderness of the sternocleidomastoid muscle and actual vertebral pain.
-only if palpation was pain-free can you progress to...

PHASE THREE:

-while loosely maintaining in-line traction, have patient move head through active range of motion

*first flexion/extension
*followed by rotational movement

-stop test at the first indication of pain with movement

Note: If you are able to move the patient's head through a full range of motion without pain, you have effectively cleared his C-Spine. Have the patient resume activity slowly and cautiously; evaluate any other complaints noted.

B. Thoracic and Lumbar Spine:

-Look for a T-L Spine injury with calcaneal fractures. Approximtely 10% of patients with a calcaneal fracture will also have an associated lumbar fracture.

C. Pelvis:

-Think HEMORRHAGE/SHOCK if faced with a Pelvic Fracture
Up to 6 liters of blood loss possible – (internal iliac areteries)

Place gentle constricting wrap around pelvis if open book fracture is suspected. (FMST.07.14b)

4. FRACTURES.

Some common wilderness fractures:

a. Metacarpal/Phalangeal	-the hands are your means of interfacing with the environment, sometimes it's a forceful meeting.
b. Distal Radius	-Very common wilderness fracture.
c. Scaphoid	-check for snuffbox tenderness, pain with axial load to thumb.
d. Lunate	-also common, pain dorsally at base of 3rd MC.
f. Clavicle	-need good NV exam, check for pneumothorax.
g. Long bone	-Radius/Ulna: if proximal, check elbow -Humerus: radial nerve runs in spiral groove -Tib/Fib: think compartment syndrome- Fib fx may ambulate with cane -Femur: think Hemorrhage, think traction.
f. Ankle	-commonly a fracture/dislocation, check NV
g. Hip	- Leg will typically be externally rotated

REMEMBER:

- Always think HEMORRHAGE with Long Bone fractures
- Always perform NEUROVASCULAR Exam before and after treatment
- Always SPLINT AND PAD for stability, comfort, function and transport.
- Indications for REDUCTION are: NV deficit , severe angulation, severe pain, if angulation prediposes to open fx or makes transport difficult.
- OPEN FRACTURES: Have 6-8 hours to get to surgery. Field Tx- Gently wash off with betadine/ iodine solution to get off obvious dirt- don't scrub. Wrap in sterile gauze. Don't place exposed bone back under skin. Give antibiotics: Augmentin, 2nd or 3rd generation cephalosporin, a quinolone or tetracycline.

THE REASONS WE PLACE FEMUR FRACTURES UNDER TRACTION.

1. Re-establishing length tightens fascia and tamponades bleeding.
2. Dramatic pain relief.
3. Helps prevent open fractures.

4. Helps reduce secondary soft tissue damage.
5. Increases ease and safety of transport.

SOME BASIC RULES OF EXTREMITY SPLINTING:

- Splint all fractures before moving the casualty, unless in imminent danger.
- Splint all fractures as they are found, unless severe angulation complicates transport or is causing neurovascular deficits.
- Splint to include the joint above and below.
- Construct splints using uninjured extremity, then splint in the position of function, comfort or stability.

5. DISLOCATIONS.

a. Rapid Diagnosis and Reduction is imperative, if evacuation time is > 1-2 hours.

- Easier to reduce immediately after injury. (muscle spasm)
- Makes transport much easier. (increased patient comfort)
- Dramatic pain relief.
- Early reduction reduces risk of Neurovascular injury.
- Reduction could make difference between an ambulatory vs. litter patient
  the safety of the entire expedition could be placed at risk

b. Signs and Symptoms: Remember, NO X-RAYS!

- Decreased Range of Motion.
- Obvious Deformity. (compare right vs. left)
- Usually NO Crepitus.
- Typical Posture.

c. Reduction Techniques:

- Phalanges	Linear traction, buddy tape. Can not reduce volar displacement of 1st phalanyx.
- Shoulder	Bedsheet method- counter traction. Prone with weights method. External rotation. Snowbird technique.
- Patella	Extension with medially directed pressure.
- Ankle	Linear traction with knee flexion.

- Hip	Anchor pelvis, hip flexed at 30 degrees, upward traction.
- Sternoclavicular	Roll between scapulae, and apply forceful downward pressure.
-Knee	Should only be reduced as an absolute last resort. Apply anterior traction gently.

REMEMBER: Analgesia and Sedation should be utilized when available before attempting joint or fracture reduction, unless multiple traumatic injuries are present.

6. EVACUATION CRITERIA:

Conditions demanding expeditious evacuation: (FMST.07.14c)

1. Suspected cervical, thoracic, and/or lumbar spine injuries.
2. Pelvic injury with instability and/or significant blood loss.
3. Any open fractures. (6-8 hour window, antibiotics early)
4. Compartment syndromes. (pain with passive motion, pallor, pulseless, paresthesia)
5. Hip or Knee dislocations.
6. Any other injury the medical provider feels he/she is not prepared to manage.

7. OTHER TRAUMA.

Heavy bleeding:	Usually can be treated by direct pressure, if a tourniquet is required,release pressure every five minutes to reassess.
Contusion:	Apply ice 1st 48 hours which will provide pain relief and limit expansion of hematoma.
Lacerations and avulsions:	High pressure irrigation with clean water, may remove debris with flame treated forceps. Use steri-strips for minor clean lacs. Keep dirty or puncture wounds open and give antibiotics.
Animal Bites:	Copious irrigation, and antibiotic, do not close. Consider rabies if bite was unprovoked and animal was acting strange. (bats, raccoons, skunks, canines, and felines) Has never been seen in livestock, rabbits, squirrels, rats or mice. Most common zoonotic source in North America is the skunk.

Severed part:	Can be reattached up to 24 hours after injury if the cut is clean. Gently clean part, wrap in gauze and keep cool without freezing.

UNITED STATES MARINE CORPS
Mountain Warfare Training Center
Bridgeport, California 93517-5001

FMST.07.38
04/02/02

STUDENT HANDOUT

REPTILE AND ARTHROPOD ENVENOMATION BITES AND STINGS

TERMINAL LEARNING OBJECTIVE: Given a simulated casualty, in a wilderness environment, treat reptile and arthropod envenomation, in accordance with the references. (FMST.07.38)

ENABLING LEARNING OBJECTIVES:

1) Without the aid of references, select from a given list the first aid measures used for pit viper envenomation, in accordance with the references. (FMST.07.38a)

2) Without the aid of references, select from a given list the first aid measures used for coral snake envenomation, in accordance with the references. (FMST.07.38b)

3) Without the aid of references, select from a given list the first aid measures used for widow spiders (lactordectus species) bites, in accordance with the references. (FMST.07.38c)

4) Without the aid of references, select from a given list the first aid measures used for brown recluse spider (loxosceles species) bites, in accordance with the references. (FMST.07.38d)

5) Without the aid of references, select from a given list choose the first aid measures used for bee, wasp, and hornet stings, in accordance with the references. (FMST.07.38e)

12

OUTLINE.

1. **PIT VIPERS**. Crotalidae have a triangular head, cat-like vertical pupils, hinged fangs and a heat-sensitive "pit" on each side of the head between the tip of the nose and the eye. Rattlesnakes have a variable number of rattles depending upon age and number of molts. They sometimes strike without rattling. About 60% of this country's venomous bites are attributed to rattlesnakes. Cottonmouths (water moccasins) and copperheads are the other two commonly encountered North American pit vipers. Copperhead and cottonmouth venom are quite similar but weaker than most rattlesnake venom.

a. Anatomy of Pit Vipers.

(1) Venom apparatus: Venom gland; Compressor glandular muscle; Primary duct; Accessory gland; Secondary duct; Fang sheath; and fangs.

(2) Facial pit organs: Highly sensitive paired receptors of infrared radiation that can detect temperature changes of less than 0.2°C; used to detect warm-blooded prey/predators and aim strike; also pit organs may have a role in determining volume of venom injected.

(3) Rattles or buttons: Interlocking keratin rings at tip of tail in rattlesnakes; a new rattle is added with each molting cycle (every 50-400 days); some are lost due to trauma.

(4) The strike: Aimed primarily by facial pits; rarely strike farther than 1/2 their body length; speed - approximately 8 feet/second.

(a) Rattlesnakes generally inject 25-75% of their venom when they bite humans.

(b) Can bite without injecting any venom ("dry bite").

1. 20% - no envenomation.
2. 20-30% - mild envenomation.
3. 30-50% - moderate to severe envenomation.

b. Signs and Symptoms of Snake Venom Poisoning:

(1) Pit Vipers:

(a) Local:

1. Puncture wounds/scratches:

A. Pattern can be misleading (venomous vs. nonvenomous).

B. Must differentiate from other animal/insect bites or plant puncture wounds- usually venomous snakebite wounds are larger, with more bleeding.

C. May be one or many fang marks present; range from a few mm to 4cm apart.

2. Pain, edema, erythema:

A. Usually presents within 30 minutes if envenomation.

B. More severe envenomation generally yields more rapid progression and more severe pain.

C. Pain is usually characterized as burning and immediate in onset.

D. Pain usually confined to bite site.

E. Edema limited to subcutaneous tissues (with no increase in intra-compartmental pressures).

F. Usually starts within 5 minutes (if none in 10 minutes, probably no significant envenomation unless dealing with Mojave rattlesnake).

G. Gradually spreads up extremity for 36 hours.

H. Danger of airway obstruction in bites to head and/or neck.

3. Ecchymosis:

A. Ecchymosis starts within several hours.

B. May involve entire extremity.

4. Lymphangitis:

A. Venom absorbed rapidly through lymphatic system results in lymphatic inflammation.

5. Petechiae, vesicles, hemorrhagic bullae:

A. Will occur in untreated rattlesnake envenomation.

B. Rarely seen when treated early with adequate antivenin.

C. Usually occurs 6-36 hr after bite.

6. Necrosis, tissue destruction:

A. Due to direct venom effects.

B, Can be prevented if adequate antivenin is given within 2 hours of bite.

(b) Systemic:

1. Nausea, vomiting:

A. Common (early onset may indicate severe envenomation).

2. Weakness, diaphoresis, chills, dizziness/vertigo.

A. Syncope is common with all pit viper bites.

B. Frequency is proportional to severity of bite.

3. Change in taste:

A. May complain within minutes of rubbery or metallic taste.

4. Increased salivation.

5. Fever.

6. Tingling, numbness in scalp/face/fingers/toes:

A. Can occur within 10 minutes and indicates moderate to severe envenomation.

7. Fasciculations in face/neck/back/ or other involved extremity:

A. Can occur early and indicates severe envenomation.

8. Visual disturbances:

A. Blurred vision, yellowing of vision and blindness.

9. Tachycardia, bradycardia.

10. Hemorrhage, hemolysis, thrombosis and DIC:

A. Bleeding from wounds, gingival bleeding, epistaxis, Hematuria, hematemesis, Melena, lower GI bleeding, hemoptysis, peritoneal hemorrhage and cerebral hemorrhage.

B. Can occur as early as 6 hours; more common 12-72 hours after bite.

C. Systemic bleeding only occurs in moderate to severe envenomation.

D. Most coagulopathic effects secondary to proteases acting at various sites in the coagulation cascade.

11. Hypotension, shock:

A. Can occur rapidly in severe envenomation.

B. Early, due to pooling of blood.

C. Late may be due to loss of volume.

D. Generally little effect from decreased cardiac contractility.

12. Pulmonary edema:

A. Common in severe envenomation.

B. Secondary to toxin and inflammatory injury of pulmonary capillary membranes and pooling of blood in major vessels and capillary beds of the lung.

C. Compounded by direct cardiodepressant factor in some venoms.

13. Oliguria, anuria:

A. Hypotension with resultant decreased GFR. (#1).

B. Hemoglobinuria and myoglobinuria with renal tubular obstruction.

C. Direct toxic effect of venom on kidneys.

D. Acute tubular necrosis from rhabdomyolysis.

14. Paresis, Paralysis:

A. Seen with Eastern Diamondback and Mojave rattlesnakes.

B. Uncommon with other rattlesnakes.

15. Respiratory failure:

A. Uncommon with pit viper bites (except Mojave).

B. Complicated by cardiovascular failure.

16. Seizures:

A. Probably due to hypertension and hypoxia.

17. Coma:

A. Usually sensorium is clear.

B. When coma occurs, it is secondary to cerebral anoxia.

18. Death:

A. Generally occurs in 6 to 48 hours in untreated severe envenomation.

B. Can occur sooner with IV envenomation.

c. First Aid Measures for Pit Viper Envenomation.
(FMST.07.38a)

(1) Do not try to kill the snake and bring it in for identification.

(a) Only vital to identify the snake when Mojave rattlesnake or coral snake bites are a possibility, as management is altered.

(b) Risk of a second bite to the victim or rescuer.

(c) Should never delay transportation of the victim.

(2) Remove any jewelry, which could become a tourniquet with progression of edema.

(3) Calm and reassure patient to decrease heart rate and circulation of venom.

(4) Constriction band.

(a) Most widely accepted first aid measure in the literature.

(b) Should be more than 1/2 inch wide.

(c) Apply to proximal end of extremity affected and wrap to approximately 5 cm above swelling (site of envenomation). Then wrap back towards proximal end of extremity from where the wrap started.

(d) Apply only tight enough to occlude lymphatic and superficial venous return.

(e) Probably no benefit when applied 30 min or more after bite.

(f) Maintain until antivenin is started.

(5) Incision and suction.

(a) Very controversial; should only be used by experienced medical personnel.

(b) Do not incise wound.

(c) Suction best applied by mechanical means (Sawyer extractor).

(d) Do not use mouth (unknown oral lesions may expose rescuer to risk).

(e) Should be started within 30 minutes of envenomation.

(f) Continue suction for 30 to 60 min.

(6) Rest and immobilization.

(a) Put victim at rest and splint extremity in position of function, at or just below heart level.

(b) Allow room for swelling in splint.

(7) Ice.

(a) Avoid any method of cooling.

(8) Watch for adverse reactions.

(a) Rare documented cases of anaphylaxis caused by snake venom in patients previously bitten or otherwise exposed to snake venom.

(b) Some evidence that venom can activate the alternate-complement system directly and cause a similar reaction.

(9) Rapid transport to a medical facility.

(a) The most important measure in first aid.

(b) No measures should delay this; antivenin is the only proven definitive therapy.

2. **CORAL SNAKES**. The elapidae species of coral snakes are brightly colored, with black noses and alternating red-yellow-red-black bands around their bodies (remember "red on yellow, kill a fellow"). They have relatively small mouths with fixed fangs. From southern Mexico through tropical South America, the rules for distinguishing coral snakes are highly unreliable. Unless you are a knowledgeable herpetologist, it is best not to pick up colorful snakes in the tropics.

a. Signs and Symptoms of Coral Snake Envenomation:

(1) Fang marks may be hard to see.

(2) Frequent delay in onset of symptoms, followed by extremely rapid progression.

(3) Little or no pain at bite site.

(4) No local necrosis.

(5) Earliest evidence may be drowsiness or euphoria.

(6) N/V, increased salivation, paraesthesias at bite site.

(7) Bulbar paralysis progressing to peripheral paralysis.

(8) Paraesthesias, and fasciculation.

(9) Occasional seizures.

(10) Mild Hypotension.

(11) Death is due to respiratory and cardiovascular failure.

b. First Aid Measures For Coral Snake Envenomation. (FMST.07.38b)

(1) None of proven benefit except rest, reassurance and rapid transportation to medical facility for antivenom administration.

3. **GILA MONSTERS**. No direct injection mechanism for their venom, but have powerful jaws that allow them to chew and tear at their victims. Drooling venom produces a substantial amount of pain and possible complications resembling pit viper envenomation. Follow the recommendations for pit viper envenomation. Fatal Gila monster encounters are extremely rare.

4. **VENOMOUS ARTHROPODS**. Arthropods represent the largest phylum in the animal kingdom. It contains spiders, scorpions, insects, ticks, kissing bugs, water

bugs, caterpillars, moths, butterflies, grasshoppers, centipedes, and millipedes, among others. Some arthropods sting (bees, ants, scorpions, etc.), others bite (spiders, centipedes, kissing bugs, etc.), while still others discharge a secretion that is toxic (millipedes, caterpillars, etc.). Although the number of arthropods that are sufficiently venomous to endanger humans is not known, these animals are implicated in far more bites and stings than all other phyla combined. Almost all of the 20,000 species of spiders are venomous, but luckily for, man, only relatively small numbers have fangs long and strong enough to penetrate the human skin. There are some 500 species of scorpions and all are venomous, although only a small number pose a significant danger to humans.

a. The number of deaths from arthropod stings and bites is not known, nor do most countries keep records of the incidence of such injuries. In Mexico, parts of Central and South America, and in North Africa, deaths from scorpion stings may exceed several thousand per year. Spider bites probably account for no more than 200 deaths per year, worldwide. The number of deaths from arthropod bites or stings in the temperate countries is far greater than the number of deaths from snakebite. Almost all of these deaths, however, are from anaphylactic reactions.

5. **SPIDERS**. -There are at least 200 species of spiders that have been implicated in significant bites to humans (Russell, 1988a). Spiders got the misplaced blame for the bites seen in one large series of case studies. Of some 600 suspected spider bites, 80% were found to be caused by other arthropods. The arthropods most frequently involved in these mis-diagnoses were ticks (including their embedded mouth parts), kissing bugs, mites, bedbugs, fleas, flies, beetles, water bugs and various other stinging arthropods.

a. Lactordectus species (Widow spiders): It is most commonly referred to as the Black Widow, Brown Widow or Frog-legged spider depending on the species. There are many other commons names such as: hourglass, poison lady, deadly spider, red-bottom spider, T-spider, Gray Lady spider, or shoe-button spider. Although both male and female widow spiders are venomous, only the latter have fangs large and strong enough to penetrate the human skin. Mature females range in body length from 10 to 18mm, whereas males are from 3 to 5mm.

(1) Signs and Symptoms of spider envenomation.

(a) Sharp, pinprick-like bite, but in some cases the incident is so minor that it goes unnoticed.

(b) Initial bite-pain is often followed by dull-numbing pain in the affected extremity, and cramps in one of the large muscle masses.

(c) Muscle fasciculations can be seen within 30 minutes of the bite.

(d) Sweating is common.

(e) Complaints of weakness and pain in the regional lymph nodes.

(f) Pain in the lower back, thighs, or abdomen is common and rigidity of the abdominal muscles is seen in severe envenomations.

(g) Severe paroxysmal muscle cramps may occur accompanied by arthralgias.

(h) Tissue reaction around the bite may vary, depending on the species involved. In most cases in the United States, there is no reaction at the bite site and the puncture wounds may be impossible to find. There may be an immediate 2 to 4 mm blanched area around the puncture wounds, surrounded by a slightly erythematous area, which in time becomes pallid. This may persist for 20 to 30 minutes.

(i) Hypertension is a characteristic finding in moderate to severe poisonings. It usually appears 2 to 3 hours post envenomation.

(j) Elevation of body temperature.

(k) Increased salivation.

(1) Anorexia.

(2) Treatment for Widow Spider envenomation.
(FMST.07.38c)

(a) First aid measures are supportive in nature. If the pain is intense, ice can be placed over the wound until the patient arrives at a medical facility.

(b) Antivenin is given in severe cases at medical facilities.

b. Loxosceles species (Violin spiders): Are known in the United States as fiddle-back, or brown recluse spiders. There are over 100 species of Loxosceles. The abdomen of these spiders varies in color from gray to orange or reddish-brown to dark brown. The violin on the cephalothorax is brown to blackish and distinct from the pale yellow to reddish-brown background of the cephalothorax. This spider has three pairs of eyes, forming a curved row. Females average 8 to 15 mm in body length whereas males average 6 to 10 mm. Both are venomous.

(1) Signs and Symptoms of Violin spider envenomation.

(a) Pain; about the same degree of a bee sting but patients may be completely unaware of the bite.

(b) Localized burning sensation around the site of injury.

(c) Pruritus is often present and the area begins to appear red, with a small-blanched area around the immediate bite site.

(d) The reddened area enlarges during the subsequent 1- 8 hours. It often becomes irregular in shape and, as time passes, hemorrhages may develop throughout the area.

(e) A small bleb or vesicle forms at the bite site and increases in size.

(f) Vesicle subsequently ruptures and a pustule forms.

(g) The whole area becomes swollen and painful.

(h) Lymphadenopathy may develop.

(i) Necrosis can develop from superficial to deep layers.

(j) Rare systemic signs and symptoms include:

1. Malaise.
2. Stomach cramps.
3. Nausea and vomiting.
4. Jaundice.
5. Spleen enlargement.
6. Hemolysis.
7. Hematuria.
8. Thrombocytopenia.

(2) Treatment for Loxosceles spider envenomation: (FMST.07.38d)

(a) There are no first-aid measures of value. In fact, all first-aid measures should be avoided, because the natural appearance of the lesion is most important in determining the diagnosis.

(b) An ice pack may be placed over the wound if the pain is severe.

(c) Excision of the bite site is no longer recommended as this has been demonstrated to cause worsening of necrosis and tissue damage. Definitive therapy includes use of dapsone, antibiotics, tetanus prophylaxis and possible steroid use.

c. There are many other species of spiders that include Steatoda species (cobweb spiders), Phidippus species (jumping spiders), Chiracanthium species (running spiders), Lycosa species (wolf spiders), and the newest member to the United States is the Hobo spider. There have been several reported deaths in the Northwestern United States. Originally a European species of brown spider, the Hobo spider was introduced into the U.S. in 1936, and little research has been done on it to date.

6. **SCORPIONS**. Approximately 75 of the 650 species of scorpions can be considered of sufficient importance to warrant medical attention. In the United States, members of the genera Hadrurus and Vejovis are capable of inflicting a painful and erythematous, but non-lethal lesion. However, species of the genus Centruroides are sufficiently dangerous to warrant definitive medical care. Most stings are inflicted by the Vejovis species.

a. Centruroides species. There are approximately 30 species of this genus confined to the New World. Of these, eight are of considerable medical importance, and most of these species are found in Mexico. In the United States, they are commonly referred to as "bark scorpions", because of the preference for hiding under the loose bark of trees, in dead trees, or in logs. Their general color is straw to yellowish-brown or reddish-brown, and they are often easily distinguishable from other scorpions in the same habitat by their long thin tail with its "telson" on the end, the pedipalps, or pincer-like claws. They can reach lengths up to 55 mm.

(1) Signs and symptoms of envenomation.

(a) Initial pain, rarely severe.

(b) Area becomes sensitive to touch (tap test).

(c) Tachycardia may ensue after one hour.

(d) Respiratory rate increases.

(e) Fasciculations may be seen over the face or large muscle masses.

(f) Complaints of generalized muscle weakness.

(g) Respiratory distress may progress to paralysis.

(h) Slurring of speech may be present.

(i) Convulsions have been reported.

(2) Treatment for scorpion envenomation.

(a) There is no first-aid measure of value for scorpion stings.

(b) Ice may be placed over the wound to reduce pain.

(c) In severe cases, antivenin may be recommended.

b. Vejovis species is closely related to the Urectonus. Both genera are ground scorpions. Whereas Urectonus usually inhabits mountain habitats from southern California to Oregon, Bejovis has a wide distribution from the southern portions of Canada, south through Wyoming and Colorado to Texas and west to California. Some species measure up to 85mm, but most are from 30 to 35mm. Vejovis spinigerus, a common western U.S. species, is often called the striped-tail scorpion.

c. Hadrurus species is native to North America. These are the longest and most stout-bodied of our scorpions, generally referred to as the giant hairy scorpions because of their size and the conspicuous bristles on their legs, pedipalps and caudal segments. Adults may can measure up to 135 mm. These are burrowing species and may be found as deep as 2 feet in sandy soil. They are native to Arizona, California and parts of Utah, Nevada and Idaho, as well as Mexico.

7. **BEES, WASPS, HORNETS AND ANTS**. While it may take over 100 bees to inflict a lethal dose of venom in most adults, one sting can cause a fatal anaphylactic reaction in a hypersensitive person. There are 3 to 4 times more deaths in the United States from bees than from snakebites. In the few fatalities that have resulted from multiple bee stings, death has been attributed to acute cardiovascular collapse.

a. Treatment for bee stings: (FMST.07.38e)

(1) The stingers of many bees may remain in the skin and should be removed by teasing or scraping rather than pulling, to avoid even more venom being injected into the patient.

(2) Ice may be applied to the sting site.

(3) Persons with known hypersensitivity to stings should carry a Bee-Sting Kit, containing an antihistamine and epinephrine when in endemic areas.

(4) Desensitization can be carried out using whole-body antigens or, preferably, whole-venom antigens.

8. **TICKS AND MITES**. Carry many diseases. In North America, some species of Dermacentor and Amblyomma cause tick paralysis.

a. Signs and symptoms of Tick Paralysis.

(1) Restlessness and irritability.

(2) Lethargy and anorexia.

(3) Generalized muscle weakness.

(4) Incoordination and ataxia.

(5) Nystagmus.

(6) Ascending flaccid paralysis with loss of deep tendon reflexes.

(7) Bulbar or respiratory paralysis may develop.

b. Treatment for Tick Paralysis:

(1) Remove the tick.

(2) Treat symptomatically (ventilator support may be necessary).

(3) Lesions should be cleaned.

(4) Antivenin and/or corticosteroids are used in severe cases.

9. **LEPIDOPTERA (CATERPILLARS)**. Caterpillars commonly cause envenomation in humans. There are at least ten families of caterpillars that are venomous.

a. MEGALOPYGE OPERCULARIS (Puss Caterpillar or Woolly Slug) The most commonly seen venomous caterpillar in the United States.

(1) Description: Hairy, flat, ovoid shaped: 30-35 millimeters in length.

(2) Location: Texas north to Maryland and Missouri.

(3) Stinging apparatus: Consists of spines intermingled with the hairs of the body:

b. SIGNS AND SYMPTOMS.

(1) Needling pain (Can be intense).

(2) Redness.

(3) Swelling.

(4) Nausea.

(5) Headache.

(6) Fever.

(7) Vomiting.

(8) Lymphadenopathy.

(9) Hypotension.

(10) Shock.

(11) Bleeding of the mucus membranes.

a. Treatment

(1) Apply contact tape to affected area, and remove tape. This will effectively remove spines from affected area.

(2) Group I corticosteroid cream and ointment.

(3) Codeine or Demerol may be required for pain.

(4) Antiemitics may be needed to control vomiting.

Note: Anaphylactic reactions have been reported, but are rare.

UNITED STATES MARINE CORPS
Mountain Warfare Training Center
Bridgeport, California 93517-5001

FMST.07.10

04/02/02

STUDENT HANDOUT

FREEZING AND NEAR FREEZING TISSUE INJURIES

TERMINAL LEARNING OBJECTIVE: Given a casualty in a cold weather environment, and necessary equipment and supplies manage common cold weather injuries in a cold weather environment to prevent death or further injury per the reference. (FMST.07.10)

ENABLING LEARNING OBJECTIVES.

1) Without the aid of references, from a given list select the correct definition of frostbite, in accordance with the references. (FMST.07.10a)

2) Without the aid of references, from a given list select the correct two mechanisms of injury for frostbite, in accordance with the references. (FMST.07.10b)

3) Without the aid of references, from a given list select the correct signs and symptoms of frostbite, in accordance with the references. (FMST.07.10c)

4) Without the aid of references, from a given list select the correct favorable post thaw signs of a frostbite injury, in accordance with the references. (FMST.07.10d)

5) Without the aid of references, from a given list select the correct unfavorable post thaw signs of a frostbite injury, in accordance with the references. (FMST.07.10e)

6) Without the aid of references, from a given list select the correct field treatment for frostbite, in accordance with the references. (FMST.07.10f)

12

7) Without the aid of references, from a given list select the correct primary field treatment consideration when treating frostbite in the field, in accordance with the references. (FMST.07.10g)

8) Without the aid of references, from a given list select the correct four methods to reduce the chance of becoming a frostbite casualty, in accordance with the references. (FMST.07.10h)

9) Without the aid of references, from a given list select the correct definition of immersion foot, in accordance with the references. (FMST.07.10i)

10) Without the aid of references, from a given list select the correct mechanism of injury for immersion foot, in accordance with the references. (FMST.07.10j)

11) Without the aid of references, given a list of categories and characteristics match the category of immersion foot to its characteristics, in accordance with the references. (FMST.07.10k)

12) Without the aid of references, from a given list select the correct field treatment of immersion foot, in accordance with the references. (FMST.07.10l)

OUTLINE.

1. **FROSTBITE**

a. Definition: Frostbite is the actual freezing of tissue. (FMST.07.10a)

(1) Frostbite usually occurs as a result of exposure to subfreezing temperatures for a long period of time. However, brief exposure to high wind-chill or very cold temperatures can also cause frostbite. Frostbite usually occurs after more than twelve hours of exposure to subfreezing temperatures. Skin temperature of the extremities is dependent on heat production and is more a function of underlying muscle mass than of circulation. Thus, the areas of the body most prone to frostbite (hands, feet, ears, nose, chin and molar regions) have small amounts of underlying muscle tissue.

(2) Incidence. Frostbite, unlike hypothermia tends to spare the extremes of age. Thus, it is most prevalent in the middle years of life and like hypothermia, the greatest incidence occurs in urban areas. Other factors predisposing a person to frostbite include: poor nutritional status, dehydration, prior cold injury, individuals from the southeastern part of the United States, diabetics, individuals with atherosclerotic disease, and use of tobacco products.

(3) Although the actual numbers are not known, persons born in the south are 3.7 times more susceptible to frostbite. Persons with a history of prior cold injury are 1.7 times more prone to have a second injury. This includes parts of the body other than the originally frostbitten region.

b. Mechanisms of Tissue Injury. Tissue injury resulting from frostbite is produced in two ways: (FMST.07.10b)

(1) *Actual freezing of the tissues.* Ice crystal formation in the extra-cellular space leads to an osmotic gradient across the cell membrane drawing water out of the cell. This process, known as crenation, causes a marked change in intra-cellular electrolyte and enzyme concentrations, which can lead to cell death or recovery depending on the duration of exposure to freezing conditions and post thaw care of the tissue. Intra-cellular ice crystal formation is much more destructive, however, this requires much faster freezing rates.

(2) *Obstruction of blood supply to the tissues.* Cell injury and death caused by ice crystal formation leads to the release of inflammatory mediators, which damage vascular endothelial cells. The net result is increased vascular permeability. Fluid leaks out of the vascular space and into the intercellular space causing tissue edema and increased blood viscosity. Elevated viscosity combined with platelet aggregation, to the walls of damaged blood vessels, lead to vascular obstruction with resultant tissue damage.

c. Signs/Symptoms. (FMST.07.10c) Initially there is discomfort and pain in the affected tissues, which can progress to loss of sensation. The skin appears white and bloodless. There is decreased or no capillary refill. The skin may feel waxy and will not glide freely over a bony prominence. With minor injury, blisters form, usually in the first 24 hours, and edema may also be present. Deep injury is not associated with substantial formation of blisters. There is also little edema seen in deep injury.

(1) Favorable prognostic signs of frostbitten tissue. These signs indicate probably little or no tissue loss, post thaw. (FMST.07.10d)

(A) Warmth of tissue.

(B) Normal tissue color.

(C) Preservation of sensation.

(D) Blister formation. *A* blister extending the full distal length of the injured digit is a good prognostic sign. However, a blister limited to the line of demarcation between normal and frostbitten tissue is a less favorable sign.

(E) Edema persisting more than 24 hours.

(F) Rapid capillary refill.

(2) Unfavorable prognostic signs. These indicate a poor prognosis with probable tissue loss, post thaw: (FMST.07.10e)

(A) Complete absence of edema.

(B) Cyanotic tissue.

(C) Continued loss of sensation.

(D) Affected areas continue to stay cold.

(3) Since it is extremely difficult to predict ultimate tissue loss from initial examination, any classification must be done retrospectively. Classically, frostbite is divided into four degrees of severity:

(A) First degree - Erythema, edema, transient tingling and/or burning.

(B) Second degree - Blisters, edema, anesthesia and/or paresthesias.

(C) Third degree - Involves the entire thickness of skin and extends into subcutaneous tissue.

(D) Fourth degree - Involves the entire thickness of the skin along with muscle and bone.

(F) Frostnip is a completely reversible injury that is technically not considered frostbite. Frostnip classically presents as a white patch and numbness and is readily warmed in the field by placing the affected extremity in one's axilla or groin. No resultant skin changes occur.

d. Field Management (FMST.07.10f)

(1) The correct treatment for frostbite is as follows:

(A) Once an extremity has suffered frostbite, the primary consideration of treatment is not to rewarm the region if there is a chance of refreezing. (FMST.07.10g) This occurs because the amount of tissue damage is greatly increased if the part is thawed then allowed to re-freeze. Generally, the sooner a frostbitten region is thawed the lower the probability and extent of damage. Rapid rewarming is the preferred method of thawing a frostbitten region. This consists of immersion of an affected part in circulating warm water of 104°-108°F until the tissues are soft and pliable, which usually takes about 40-60 minutes.

(B) If slow rewarming methods (such as allowing a part to sit at room temperature) is used, some refreezing of the melted fluid occurs and the new ice crystals are actually larger than those formed during the original freezing causing further tissue destruction.

(C) Out dated methods such as rubbing with snow or rewarming next to an open fire, etc., only enhance tissue damage.

(D) Once thawing has taken place no benefit of continued rapid rewarming can be demonstrated.

(E) Cold injuries that are old and have thawed spontaneously should not be rapidly re-warmed again.

(F) Nicotine use should be prohibited since it is a vasoconstrictor and may enhance tissue loss.

(G) There is no need for ointments, salves, etc.

(H) Blisters should be allowed to remain intact.

(I) Medevac all cases of frostbite, as soon as the tactical situation permits.

(J) One final note; in military tactical situations such as ours, the victim often must walk down below the snow line to reach a BAS or base camp where rapid rewarming can be carried out. In this situation let the victim walk if he has frostbitten feet. Once his feet have thawed, he will be unable to walk and will have to be carried out.

(K) Hospital and long-term care: After re-warming the frostbitten tissue, attention must turn to prevention of infection and functional restoration. The patient's tetanus status should be determined and meticulous care should be taken to keep the injury clean. Daily whirlpool therapy for 30-45 minutes has been shown to be beneficial. The use of post-thaw medications is controversial but there is evidence that topical aloe vera and NSAIDS to control inflammation are probably beneficial. Some studies have demonstrated benefit from the use of allopurinol as a free radical scavenger. Also, animal studies have shown a definite benefit from the use of 90% DMSO gel, however, this use is not FDA approved on humans and it may be difficult or impossible to obtain this medication. The wide scale use of heparin and dextran 40 has been suggested but a definitive recommendation that supports their use still cannot be made. Studies suggesting the use of thrombolytics are very small and not well controlled therefore there is no current recommendation for use of thrombolytics. Lastly, it has been shown that there is a clear benefit for the use of early and aggressive physical therapy to preserve function and prevent contracture.

e. Prevention (FMST.07.10h)

(1) Some general principles for the prevention of frostbite are listed below

(a) Dress to keep comfortably cool. The head radiates 30-50% or more of all body heat and so goes the old adage, "If your feet are cold put on your hat".

(b) Nutrition. Eat the right type and quantity of foods.

(c) Avoid tight fitting and constrictive clothing. This decreases blood flow and increases the chance of frostbite. Rings, jewelry, etc., conduct heat away from the skin and increase the chance of frostbite.

(d) Wear mittens instead of gloves. They are much warmer. For work involving a high level of dexterity use contact gloves.

(e) Do not touch metals with bare hands at extremely low temperatures. Your flesh will stick to the surface and this will result in subsequent frostbite and tissue loss. If you find yourself in such a predicament pour a warm fluid over the metal to thaw your flesh off the metal.

(f) Avoid nicotine.

2. **IMMERSION FOOT**.

a. Definition. (FMST.07.10i) Immersion foot (AKA trenchfoot) is a non-freezing injury of the extremities in which the tissues are damaged.

b. Usually a disease of the feet (but it can occur in the hands), it occurs when the extremity is exposed to a cold-wet environment (external temperatures above freezing and less than

50°F) for long periods of time (usually greater than 12 hours). Prolonged standing enhances the pathologic potential for injury due to wet and cold conditions.

c. Mechanism of Injury. (FMST.07.10j) Even though this injury is called immersion foot, actual immersion of the feet in water is not necessary. This injury can also occur in the hands.

(1) The circulation of blood to the affected extremity is reduced because the victim's extremity is cold, wet and the peripheral blood vessels have constricted to reduce overall heat loss.

(2) Because prolonged cooling damages sensitive tissues, nerve injuries are also involved. This nerve damage is responsible for the severe pain and paresthesias. Damaged skin and subcutaneous tissues result in erythema, edema and possible blister formation.

d. Signs/Symptoms Early signs include numbness and paresthesias followed by complete numbness and sometimes leg/arm cramps. The skin at first appears erythematous but becomes progressively pale and mottled then grayish-blue as the condition progresses. Often there is a large amount of edema and blister formation. In severe cases the arterial pulses are absent. Immersion foot is classified by severity: (FMST.07.10k)

CATEGORY	CHARACTERISTICS
Minimal	Reddening of skin; slight sensory changes
Mild	Edema; sensory changes (reversible)
Moderate	Edema, redness, blisters, intra-cutaneous hemorrhage; irreversible nerve damage
Severe	Severe edema, massive intra-cutaneous hemorrhage; necrosis, gangrene

e. Field Management (FMST.07.10l) Treatment consists of:

(1) Pat drying of extremity.

(2) Gentle rewarming.

(3) Elevation of affected extremity.

(4) Bed rest.

(5) Treatment after this is supportive. For more severe injuries, debridement or resection of a portion of the foot may be necessary. However, this is best accomplished at a higher echelon of care. Do not be fooled by this injury. These patients can get

extremely ill from wet-gangrene which may necessitate amputation of tissue. However, the initial treatment should be as conservative as possible. DO NOT TREAT THESE PATIENT'S WITH WATER IMMERSION THERAPY. If infection is present obtain surgical consult early, and plan on small sequential amputations.

3. **CHILBLAINS (PERNIO)**.

   a. Definition. Chilblains are a non-freezing injury that leads to red inflamed, pruritic skin.

   b. Mechanism of Injury. Pernio is caused by intermittent exposure to above freezing temperatures and high humidity. (More extreme cold is usually associated with lower humidity). Classically, chilblains occur in young individuals and most commonly on the anterior surface of the legs and the dorsum of the hands. The cause of pernio is not clearly known, but it is thought to reflect an abnormal vascular response to the environment.

   c. Signs/Symptoms. Skin is red, tender, and pruritic, edematous and blanches with pressure. Itching may be severe and hyperhydrosis can occur.

   d. Management. The field treatment is massage and gentle heat. This treatment is usually self-limiting and will not recur without re-exposure. Topical steroids are effective for itching and burning.

4. **CORNEAL FROSTBITE**

   a. Definition. Actual freezing of the cornea.

   b. Mechanism of injury. This occurs when large amounts of cold air pass over and cool the cornea, resulting in corneal clouding and loss of vision. This can occur as the result of even short periods of exposure to activities associated with high wind chill i.e. snowmobiling.

   c. Signs/Symptoms. On examination the cornea will be clouded over.

   d. Treatment and Prevention. Antibiotic ointment and DO NOT PATCH THE EYES. Prevention is the key.

UNITED STATES MARINE CORPS
Mountain Warfare Training
Center Bridgeport, California
93517-5001

FMST.07.22
04/03/02

STUDENT HANDOUT

LAND NAVIGATION REVIEW

TERMINAL LEARNING OBJECTIVE In a mountainous environment, navigate in mountainous terrain, in accordance with the references. (FMST.07.22)

ENABLING LEARNING OBJECTIVES

1) Given a map and compass, orientate the map, in accordance with the references. (FMST.07.22a)

2) Given a map and a protractor locate a six-digit grid coordinate, in accordance with the references. (FMST.07.22b)

3) Given a map, compass, and protractor perform a resection to locate an unknown point from three known points, in accordance with the references. (FMST.07.22c)

4) Without the aid of references, select from a given list the correct definition for true north, grid north, and magnetic north, in accordance with the references. (FMST.07.22d)

OUTLINE.

1. **THE MAP**. Before you can properly use a map, there is some basic information you will need to understand.

a. Definition. A map is a reduced or scale drawing of the ground and important things on the ground as seen from the air. Essentially it is a picture of the surface of the earth, as it would appear looking at it from an aircraft.

13

b. Marginal Information. As you take a quick look at the borders, you will notice that there is a large amount of information. When using a map, the instructions are placed around the outer edges of the map and are known as marginal information. All maps are not the same, some having more marginal information than others, and the information being located in different places around the border. It is important that you know how to use this information if you plan to get the most out of your map. The different information on the margins is as follows:

(1) Sheet name. The sheet name is found in two places: the center of the upper margin and lower right margin of the map. The map is usually named after it's outstanding cultural or geographical features. Whenever possible, the name of the largest city is used.

(2) The sheet number. The sheet number is found in the upper right margin of your map and is used as a reference number for the map sheet.

(3) Series name and scale.

(a) The series name is found in the upper left margin of your map. A map series usually comprises a group of similar maps on the same scale and on the same sheet lines or format designed to cover a particular geographical area. It may also be a group of maps designed to serve a common purpose. The name given a series is of the most prominent area.

(b) The scale notation represents the ratio of the maps distance to the corresponding (actual) distance on the earth's surface. For example, the scale notation of 1:50,000 indicates that one unit of measure on the map equals 50,000 of the same units of measure on the actual surface of the ground (1 map inch = 50,000 earth inches).

(4) Series number. The series number appears in the upper right margin and lower left margin. It is a comprehensive reference expressed either as a four digit number or as a letter followed by three or four numbers.

(5) Index to adjoining sheets. The index to adjoining sheets is found in the lower right margin of your map.

(6) Bar scales. The bar scale is located in the center lower margin of your map.

(7) Contour interval. The contour interval is located in the center lower margin of your map, just below the bar scales. It states the vertical distance between adjacent contour lines on the map. For example, every contour line represents an increase in elevation as indicated by the distance of the contour interval.

(8) Northern References. There are 3 different northern references located on the declination diagram. It is located in the right portion of the lower margin of your map.

(a) True North. A line from any position on the earth's surface to the North Pole, which is the exact top of the earth.

(b) Grid North. The north direction established by the vertical grid lines on the map.

(c) Magnetic North. The direction to the North Magnetic Pole.

(9) Legend. Is located in the lower left margin. It illustrates and identifies some of the symbols on the map. There is not enough room on a map sheet to show the true outlines of objects and land features so the mapmaker uses a set of standard symbols to represent them. Sometimes the symbols are not the same, or standard types of symbols typically used. This is because of the type of map, the scale, or the origin. Because of this possible difference, to prevent errors in symbol identification, the legend should be referred to every time a map is used. (The example on the legend may state, "unimproved road, hard surface road, jeep trail," etc. The right side of the symbol will have three symbols; the first being "unimproved road", the second "hard surface road", and the third being "jeep trail".)

(10) Map colors. The colors you see on your map are not there to make the map pretty or enjoyable to the user. They have important meaning. The colors vary with different types of maps, but on a standard, large scale map, there are five basic colors:

(a) Black. Represents most of the man-made features.

(b) Red. Represents major roads.

(c) Blue. Represents water.

(d) Green. Represents vegetation.

(e) Brown. Represents all terrain features, such as hills, draws, fingers, depressions, and saddles by the use of contour lines. These contour lines are shown in brown.

(f) Other Colors. Occasionally other colors are used to show special information. These, as a rule, will be indicated in the marginal information. Currently, there are changes being made to the color system, which may result in changes of the marginal information in regards to map colors.

2. **HORIZONTAL SCALES**.

a. Bar Scale. This special ruler, located at the bottom of the map sheet, enables you to measure a distance between 2 points on your map and change that map distance into the actual ground distance.

(1) The primary scale. The primary scale is the part of the scale that starts at the zero (0) and extends to the right.

(2) The extension scale. The second part of the bar scale is the extension scale. It starts at the zero (0) and extends left.

b. Measuring Straight-Line Distance. If you are given a mission that requires you to move from your present position to a different position some distance away and you plan to travel in a straight line between these two points, the following method is used to determine the distance you will have to travel:

(1) First locate your present position and your destination on your map. Then you will lay a straight edged piece of paper on the map so that the edge of the paper touches the location of your present position and the location of your destination. Make a tick mark on the edge of the paper at each point. Now move the paper down to the bar scale. Place the left tick mark on the zero (0) of the scale so that the other tick mark will fall somewhere on the primary scale.

c. Measuring a Curved Line Distance. If you are given a mission that requires you to move from your present position to a different position and this time you choose to move along a road to reach your destination, the following method is used to work out the ground distance you will have to travel:

(1) Find your present position, your destination, and the road you will follow, on your map. Position a straight edged piece of paper at your location and make a tick mark,. Now, rotate the paper so that it runs along the road until the road turns, and make a tick mark on both the paper and the map. Keeping these 2 tick marks together, rotate the paper again until the road turns, and again mark the paper and the map. Continue this procedure until you have reached your destination. Now using your bar scale as previously stated, find the actual ground distance your squad will be traveling.

3. **ELEVATION AND RELIEF**. In order to begin learning the subject of elevation and relief, let's first look at some definitions.

a. Contour lines. Of the several ways to indicate elevation and relief on maps, the most common is through the use of contour lines. A contour line represents an imaginary line on the ground along which all points on this line are the same elevation above sea level.

b. Characteristics of Contour Lines. There are a few characteristics of a contour line that you should know that will help you understand what they represent on your map. These characteristics are as follows:

(1) Contour lines never cross.

(2) Contour lines connect on themselves.

c. Types of Contour Lines. In addition to knowing the definition, purpose, and characteristics of a contour line, you need to realize that there are different types of contour lines.

(1) Index contour lines. Starting at zero elevation, every 5th contour line is drawn with a heavier line.

(2) Intermediate contour lines. These are the contour lines falling between the Index contour lines.

(3) Supplementary contour lines. These lines are shown as dashed brown lines.

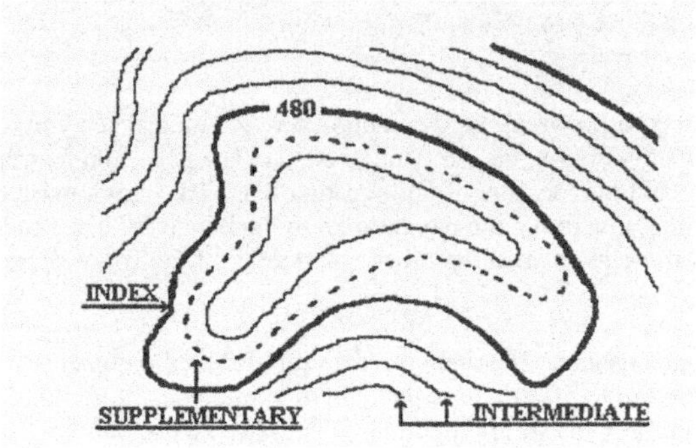

d. Determining Elevation of a Point. Now, you are ready to learn how to determine the elevation of any point of the ground represented on your map.

(1) Given elevation nearest the point, examine the area around your point until you find the nearest point that has a marked elevation. There are a few different ways that contour lines are marked, and they are:

(a) Labeled index contour line. Index contour lines are broken with the elevation of the line printed in the break.

(b) Spot elevations. At various places on your map, usually on prominent landforms such as road junctions or hilltops, a spot elevation will be given. It will be printed in either brown or black.

(c) Bench marks. Similar to Spot elevations, but more accurate. They are marked on your map by a black "x", with the letters "BM", and the elevation.

(d) A point on a contour line. If a point is located on a contour line, its elevation will be the same as that contour line.

(e) A point between contour lines. If a desired point is less than 1/4 the distance between 2 lines, its elevation is considered to be the same as the closest contour line. If the point is from 1/4 to 3/4 the distance, its elevation is considered to be in the middle of the 2 lines.

(f) A point on top of a hill. To estimate the elevation on the top of an unmarked hill, add 1/2 of the contour interval to the elevation of the highest contour line around the hill.)

4. **TERRAIN FEATURES**. There are many types of terrain features that you must be able to identify. By interpreting the arrangement of contour lines, you should be able to get a mental picture of how these terrain features would appear if you were actually out there on the ground looking at them. Some of the features you need to be able to recognize are as follows:

a. Hills. Defined as a point or small area of high ground. A hill is represented on your map by a number of contour lines each of which circles around and connects onto itself. These rough circles get smaller as they progress toward the top of the hill. You need to carefully compare the arrangement of these contour lines to the actual appearance of the hill.

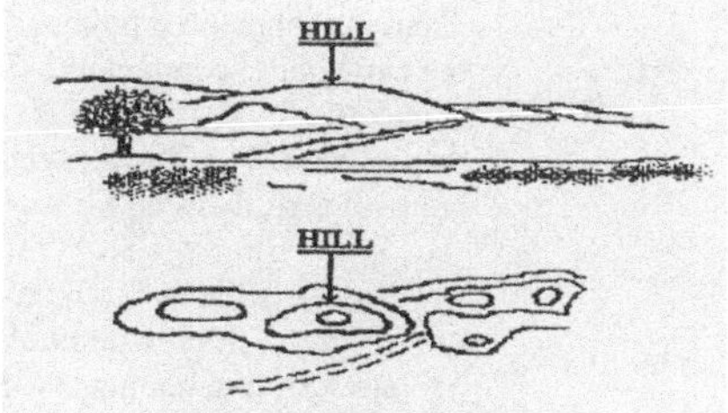

b. Spur. Defined as a line of high ground. The points along the top or crest of the spur are higher than the ground on both sides. A spur is represented on the map by a number of contour lines, each of which form a U-Shaped or V-Shaped design. The end or bottom of the U or V-Shaped lines point down to a lower elevation.

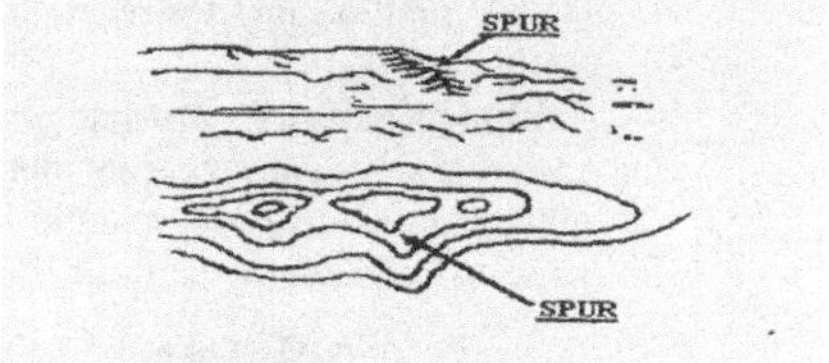

c. Draw. Defined as a line of low ground. The points located in the bottom of a draw are at a lower elevation than the ground to either side. A draw usually runs between 2 spurs. It is sometimes difficult to tell the difference between the contour lines that represent a draw from those that represent a spur because they are both U or V-Shaped lines. Remember that the curved portion representing a spur points down to ground of lower elevation, and the curved portion representing a draw points to ground of higher elevation. Now by looking closely at the area on your map, you will be able to tell which of the 2 you are dealing with. In many cases, draws will contain a stream, while spurs usually never have a stream running down their centers.

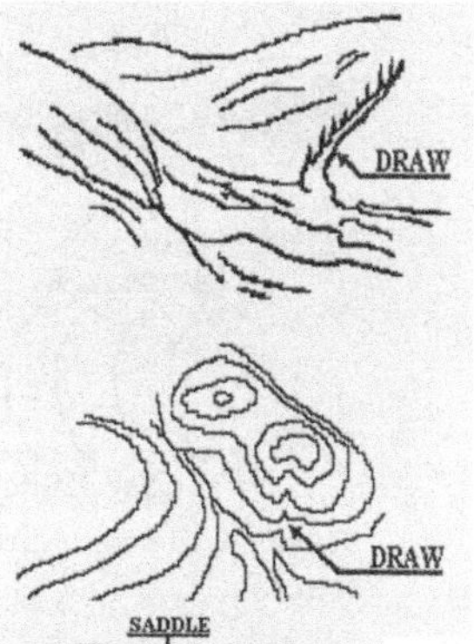

d. Saddle. Defined as the low ground between 2 hilltops, or a dip along the crest of a spur. On your map a saddle is represented by contour lines that form 2 hilltops that are located closely to each other. The area between these 2 hilltops being the location of the saddle.

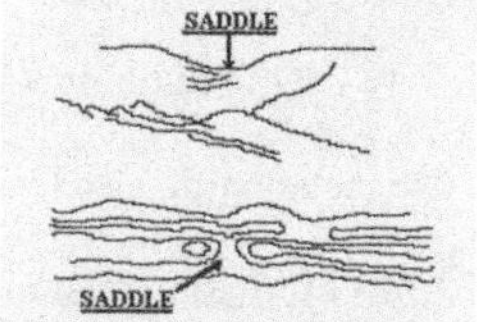

e. Depression. Defined as a point or area of low ground surrounded on all sides by higher ground. A depression is represented by contour lines forming circles within a small area on the map. These rough circles look identical to hilltops except for the presence of hachures. These are short stubby lines that are connected to the contour line at one end and point away from the contour lines at right angles in a downhill direction.

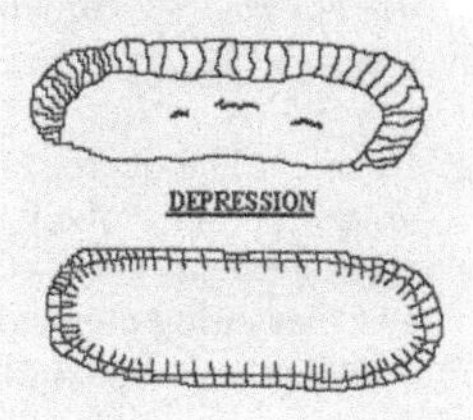

f. Cut. A cut is a man-made feature by which the bed of a road or railroad is leveled by cutting through high ground.

A cut is represented by a sudden ending of contour lines, the ends being connected by a straight line. Hachures are connected to this straight line and point downhill toward the roadbed. This sudden ending of contour lines occurs on both sides of the road bed.

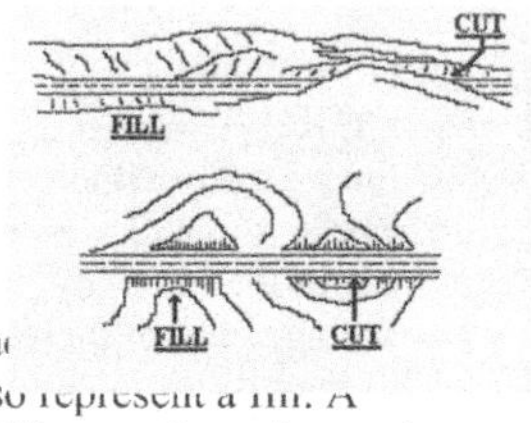

g. Fill. A fill is also a man-made feature by which the bed of a road ... filling in the low area. A sudden ending of contour lines can also represent a fill. A straight line connects these ends with Hachures pointing downhill away from the road bed. Again, this occurs on both sides of the road bed.

5. **SLOPES.** The rate of rise or fall of a ground form is known as its slope. In combat, this factor of slope must be considered when making plans for an operation. You must have the ability to recognize the ground forms in your area of operation by studying your map and the spacing between the contour lines on your map.

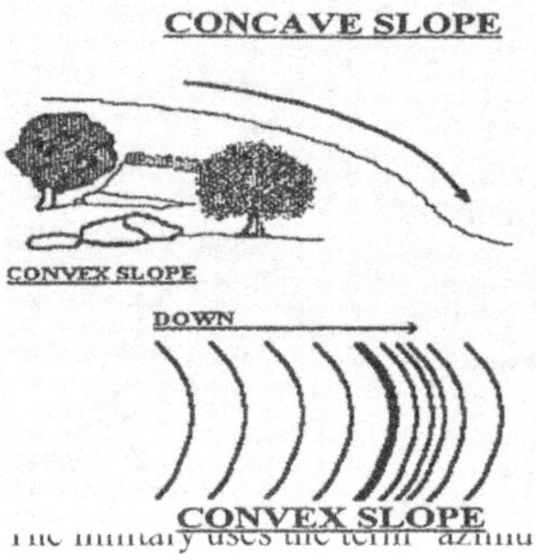

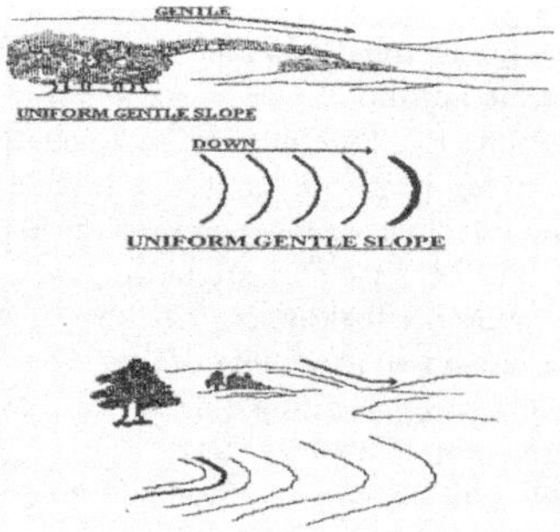

a. The military uses the term "azimuth" to describe the direction from one point to another. There are two base direction lines used to measure the value of an azimuth. They are as follows:

(1) Grid north. It is used when measuring direction from your map.

(2) Magnetic north. Used when measuring direction from your compass.

b. Units of Measurement. Now that you have learned about the two base direction lines that are used to measure direction, we need to learn the units of measurement that are used to measure the angle from one of these direction lines to the direction line.

(1) Degrees. The degree is the unit of measurement that is used most often. Anytime you need an azimuth to move from one point to another, you will use degrees as your unit of measurement. The base direction line, grid north or magnetic north, will be zero (0) degrees and measuring clockwise from this base direction line, the angle increases until you reach 360 degrees which brings you back around in a full circle to the base direction line.

(2) Mils. The mil is also used as a unit of measurement for direction. It is used when working with indirect fire weapons such as mortars and artillery. The azimuth you send will have to

be in mils, not in degrees. Remember, you are still measuring the angle between the base direction line and the line of sight from you to the distant object only you are using a different unit to measure the angle. Moving clockwise from the base direction line, which would be zero (0) mils, the angle back to the direction line. Now you know there are 360 degrees and also 6400 mils in the circle. This means that each degree equals approximately 17.5 mils.

c. Converting Map and Compass Directions. (WMC.5.2d) Earlier you learned that an azimuth was always measured from a base direction line. You also learned that we can use two different base direction lines. Grid north is the base direction line used when you are working with the map and getting a map direction or a grid azimuth. Magnetic north is the base direction line used when you are using the compass to get a compass direction or a magnetic azimuth. Because these two different base direction lines are used and because inmost cases there is a difference of one or more degrees between them, it is necessary for you to learn how to convert from a grid azimuth to a magnetic azimuth or from a magnetic azimuth to a grid azimuth. In order to learn this procedure we must take another look at the declination diagram.

(1) Declination diagram. A diagram that shows the relation between three different norths true north, grid north and magnetic north. It is the difference between grid north and magnetic north that we need to look at. The difference between these two north's is known as the grid-magnetic angle or G-M angle.

(a) Grid-Magnetic angle. The grid-magnetic angle is the angular difference between grid north and magnetic north. As you can see on this diagram a line that connects grid north and magnetic north represents the G-M angle. The size of the angle is stated off to the left side of the diagram in both degrees and mils. The G-M angle on your maps is 18 degrees or 320 mils. It is important to know that the G-M angle will change depending on where you are located on the earth's surface.

(2) Converting. When you receive a mission as a squad leader to move your squad from your present position to some distant point, you take out our map, a protractor and figure out the map direction or the grid azimuth to that distant point. However, before you can use your compass to direct you to your destination, you will have to convert the grid azimuth you have from your map, to a magnetic azimuth that you can set on your compass.

(a) The first step in making this conversion is to examine the declination diagram on your map. The G-M angle on your map is 18 degrees. You know that you will have to add the G-M angle to, or subtract it from the azimuth you measured on your map, or shot on your compass.

(b) The instructions as to whether you will add or subtract this angle are written on the map depending on whether you are going from grid to magnetic or vise-versa.

d. Orientation of a Map. (WMC.6.14a) In order to gain the best use of your map and to avoid confusion while navigating across terrain, you will need to know how to orient your map to the ground. *A* map is oriented when it is lying flat and level with it's north and south pointing in the same directions as north and south on the ground. There are two methods you can use to orient a map.

(1) Inspection. The first method used to orient the map is by inspection. By carefully examining your map and the ground around the area in which you are located to find linear features such as roads, railroads and powerless, you can align the feature on the map with the same feature on the ground. If there is only one linear terrain feature in the area in which you are located, you will have to be careful not to reverse your directions. Even though you have a road running north-south on your map aligned with the same road on the ground, the north end of your map may be pointed in a southerly direction down the road. To avoid this from occurring, find another terrain feature in the area that you can also identify on the map. Once you have aligned your map along the linear feature, check to insure that the second feature is positioned on the same side of the linear feature on your map as it is on the ground.

(2) Compass. This is the second method of orienting a map. You begin by placing your map on the ground so it lies flat and level. Open your compass so that the cover and eyepiece are completely open. Align the graduated straight edge of the compass along one of the north-south grid lines on your map with the cover end of the compass pointing toward the top of the map. This will place the stationary index line parallel to the grid north. Since the north seeking arrow on the dial of your compass will always point to magnetic north, you set your compass up just like a declination diagram. Grid north being the stationary index line and magnetic north being your north seeking arrow. Now rotate your map, insuring that the graduated straight edge of your compass remains aligned with the north-south grid line, until the face of the compass appears exactly as the declination diagram in the lower margin of the map. Your map is now oriented. For the Sonora Pass map sheet you would rotate the map until the north seeking arrow on the compass lies 16.5 degrees to the right of the stationary index line.

7. **GRID SYSTEM.** (WMC.6.14b) In most combat situations it is very important to be able to pinpoint your location, the location of another unit, or the location of an objective. In order to do this, it is necessary to have a location system that will allow you to accurately determine a specific location and state that location in such a way that will allow others to find that location quickly. This system is known as a grid system.

a. Grid Box. This box gives basic instructions on reading grids in determination of specific points on the map.

b. Grid Lines. The straight black lines that run across, and up and down the surface of your maps form the 1000 meter grid squares. These lines are called grid lines. If you look around the border of your map you will see that both the horizontal and vertical grid lines are identified by a two digit number.

c. Grid Square Identification. Each 1000 meter grid square on your map has it's own unique number. This number is called a grid coordinate. A four digit number is the result of combining the two digit number identification of the vertical grid line and the two digit number identification of the horizontal grid line that cross to form the lower left corner of the grid square. To find the grid coordinate of any grid square on your map you will always read your map right and up. You find the two digit number of that vertical grid line and you will have the first 2 digits of your coordinate. Next, find the two digit number of the horizontal grid line and place it after your vertical number and you now have the 4 digit grid coordinate that identifies your grid square.

d. Locating Terrain Features Within a Grid Square. If you are given a 4 digit grid coordinate to find a certain point on the earth's surface, that grid coordinate would put you within 1000 meters of that point. If you are told to find a certain feature within that grid square and you are good at reading your map, a 4-digit coordinate may get you close enough to your point to enable you to find it. However, there are many situations in which a 4 digit number will not get you close enough to a given position to allow you to accomplish your task. It is necessary to be able to locate points within a grid square and using a six-digit grid coordinate does this.

(1) Six Digit Grid Coordinate. The 1000-meter grid square will have to be broken down further into 100-meter grid squares. Because the surface of your map would become too cluttered by all these additional lines, a special instrument has been made to help you divide up a 1000-meter grid square. This instrument is called a coordinate scale or protractor.

8. **RESECTION**. (WMC.6.14c)

a. You will face a situation where you need to pinpoint the location of your position. This may not be possible by associating the terrain in your area with the map because you may have a poor map or the terrain in your immediate area is flat and has no distinguishing features. When you find yourself in this situation you will need to know how to perform a resection.

(1) Examine the terrain around your position and find two terrain features or objects that you can also identify on your map.

(2) Use your compass to find the magnetic azimuth to one of the distant terrain features you have identified.

(3) Convert the magnetic azimuth you found into a grid azimuth.

(4) Change the grid azimuth you found into a back azimuth.

(5) Drawing a line on your map from a distant point you identified in step one back toward your present position. Place a small mark on your map at this point. Using a straight edge, draw a line from where you plotted the distant object to the mark you made by the edge of your protractor. This line should be extended to insure it continues through the area where you are located.

(6) Repeat, finding the magnetic azimuth to the second distant object. That is at least 60 degrees from the first.

(7) When you have drawn your second line on the map, the point where the second line crosses the first line is the location of your position. The process that has just been covered is called a two-point resection.

(8) If your position is located on a linear line feature such as a road, power line, railroad or river, then it is only necessary to perform a one-point resection in order to pinpoint your location. Instead of identifying two distant points, you will only need to select one. Where the single line that you drew on your map crosses the linear feature on which you are located will be the location of your position.

9. **DETOURING AROUND AN OBSTACLE**.

a. There will be occasions when it will become necessary to move off your plotted route in order to avoid an obstacle such as a small lake or swamp. You must be able to move around these obstacles and return to the route you were following before you took your detour. This can be accomplished by using one of three detour methods: Far side landmark method, nearside landmark method, and the 90-degree offset method.

(1) Far side landmark. Used during hours of daylight when you are able to identify some feature on the far side of the obstacle that lies on your line of march and is obvious enough so that you can find it once you have moved around to the far side.

(2) Nearside landmark. Used during hours of daylight when you cannot find a landmark on the far side. However, you can identify a feature on the nearside that is on your line of march and you will still be able to see once you have moved around to the far side. Again, let us use the example of being on an azimuth of 198 degrees, and it brings you to a small lake. From the nearside you cannot locate any features on the far side. There is, however, a small pier on the nearside that is located on your line of march. You can now move around the lake using the easiest and safest route. Move to an area on the far side that is as close to being opposite of the nearside landmark as possible. You then figure out the back azimuth of your route azimuth, which in this case would be 18 degrees. Shoot an azimuth back across the lake to the pier. Move to the right or left until you are in a position where the pier lies at an azimuth of 18 degrees. This puts you back on your line of march and you can continue on following your original azimuth of 198 degrees.

(3) 90 degree offset. Used during hours of reduced visibility or during daylight hours when there are no visible landmarks along the line of march on either side of the obstacle. This time you are moving along a 360 degree azimuth and you encounter a large, impassable area of thick brush. The brush is too high to see any landmarks from either side of the thicket. The following procedure should be taken to detour around the thicket.

(a) Stop the pace count you have been making for movement along your assigned route.

(b) Decide which way you will detour around the thicket. Let's say in this case the right flank appears to be the easiest and safest. Add 90 degrees to your original azimuth of 360 degrees in order to move your patrol to the right.

(c) Move on this 90 degrees azimuth until you have reached a point where you can clear the right flank of the thicket. You will keep a new pace count starting from the point where you made your 90 degrees turn to the right to the point where you can clear the right flank of the thicket.

(d) Make another change of direction. This time you want to return to a 360 degrees azimuth so you will parallel your original course and move by the right flank of the thicket. In order to return to the 360 degrees azimuth you will now subtract 90 degrees.

(e) Move on the 360-degree azimuth until you have by passed the thicket. At the point of your second 90 degree turn, you will pick up the pace count you were making before you began making your detour around the obstacle. Even though you are not on your original line of march, you are moving parallel to your original line of march and therefore must include the distance you cover during this leg of the offset in your pace count for your route.

(f) Once you have bypassed the thicket, you will make another change of direction. This time, you want to move back toward your original line of march. Since that means you will be turning toward the left, you will subtract 90 degrees from the 360 degrees azimuth you have just completed. Your new direction will be 270 degrees.

(g) Pace back toward the original line of march following the 270-degree azimuth, the same number of paces as you made off the original line of march. In doing this you will have returned to your original line of march.

(h) Once you have moved back to your original line of march you will make your last change of direction. By adding 90 degrees to the original azimuth of the last leg of the offset, which in this case was 270 degrees, you will be back on the original azimuth of 360 degrees.

(i) Return to your original line of march pace count.

UNITED STATES MARINE CORPS
Mountain Warfare Training Center
Bridgeport, California 93517-5001

FMST.07.39
04/03/02

STUDENT HANDOUT

SEARCH AND RESCUE MANAGEMENT

TERMINAL LEARNING OBJECTIVE: In a mountainous environment, execute a search and rescue, in accordance with the references. (FMST.07.39)

ENABLING LEARNING OBJECTIVES:

(1) Without the aid of references and from a given list, choose the definition of a search, in accordance with the references. (FMST.07.39a)

(2) Without the aid of references and from a given list, choose the definition of a rescue, in accordance with the references. (FMST.07.39b)

(3) Without the aid of references and from a given list, choose one the four methods used to determine a search area, in accordance with the references. (FMST.07.39c)

(4) Without the aid of references and from a given list, choose one of the five methods for confining a search area, in accordance with the references. (FMST.07.39d)

(5) Given a lost individual scenario in a mountainous environment, plan a search and rescue in accordance with the references. (FMST.07.39e)

OUTLINE.

1. **GENERAL**. When discussing a search and rescue situation we must first be able to differentiate between the two facets of the operation. A "search" means that you are looking for an overdue missing person (FMST.07.39a). A "rescue" 'is the extraction of a known subject from a known location (FMST.07.39b). In either of these situations time is critical for several reasons:

a. Subject may need emergency medical care.

b. Size of the search area is limited.

c. Time and weather destroys clues.

d. Subject may need protection from the environment or from himself.

2. **SEARCH METHODS**. When conducting a search you will be searching for clues, instead of searching for the victim. There are two reasons for this:

a. There are more clues than victims. Anyone moving through an area will leave signs.

b. Clue detection greatly reduces search difficulty by reducing the search area.

c. This determination will assist the search manager in the use of assets, the justification for requesting outside agency assistance, and other management decisions. There are eight points that must be considered in determining the search urgency:

1. Subject profile.
2. Weather profile.
3. Equipment available to subject.
4. Subject's training, experience, and mobility.
5. Terrain hazards.
6. History of incidents in the area.
7. Time.
8. Political sensitivity.

3. **SUBJECT PROFILE**. The subject profile is based on the subject's capabilities and mobility in the terrain and conditions of the search area. The three points are age, medical condition, and number of subjects. Discussing this will assist us is rating the relative urgency:

a. Age: Very young/old, other.

b. Medical condition: Injured, ill, mental problems, healthy, or dead.

c. Number of subjects: Single or group.

d. Subject experience profile: No experience/no knowledge of area, no experience/knows area, experienced/no knowledge of area, or experienced/knows area.

e. Weather profile: Past or current bad weather, bad weather in next 8 hours, bad weather in next 24 hours, or no bad weather predicted. Temperature and visibility.

f. Equipment profile: Inadequate, questionable, or adequate for environment or weather.

g. Terrain/Hazard profile: Known hazardous terrain or, hazards or few or no hazards.

h. History of incidents in the area: Outcome of other searches for individuals with similar profiles in similar situations.

i. Time: Time since the subject went missing and the effects of time on clues.

j. Political sensitivity: External sources; VIPs, politicians, relatives, media, or higher authority in your chain of command.

4. **DETERMINING THE SEARCH AREA**. There are four methods used to determine the search area by rescue managers. The best application is to use a combination of these methods by personnel with search experience. (FMST.07.39c)

a. Theoretical: This is the distance that the victim could have traveled in the elapsed time since he/she was reported missing. (PI x R Squared)

b. Statistical: Historical data on searches for similar individuals in similar conditions.

c. Subjective: Evaluation of limiting factors for this specific incident and location.

d. Deductive reasoning: Sherlock Holmes version, based on analysis of circumstances involving the specific incident.

5. **BEHAVIOR ASPECTS OF A LOST PERSON**. General aspects of lost person behavior will give you an awareness of the actions of the individuals you are searching for by separating them into different categories. Some factors that could effect the search strategy are the general health of the subject, his/her past experiences, physiological effects of the environment, and the victims' mental state or biological cycle. Also, looking at the subject's ability to assist in his/her own survival with simple skills such as fire, shelter, signaling, and land navigation will determine strategy. The psychological impact on the victim should also be taken into account. Will the subject be overcome by Stress? Fear? Panic? Embarrassment? Or will the subject be calm and develop a positive "I'll be OK attitude? A further point to consider is the circumstances under which the subject became lost. Was the subject at a known location such as a campground? Enroute to a specific location or on a trail? Traveling in a wilderness area with no specific location or destination in mind? Or does the subject fall under one of the following categories:

a. Children age 1 to 3: Unaware of concept of being lost, no navigation skills or sense of direction, wander aimlessly, might seek out a convenient place to take a nap. i.e. hollow log, under bush/rock/picnic table.

b. Children age 3 to 6: More mobile, more inquisitive and easily distracted, will attempt to "go home" once they decide that they are lost, taught to avoid strangers and may hide from searchers.

c. Children 6 to 12: Good navigation skills and sense of direction in familiar terrain but become disoriented in unknown surroundings, may have intentionally run away and do not want to be found, may not answer when called, same fears as adults but no experiences to fall back on, once night sets in more likely to respond to searchers and except assistance.

d. Mentally retarded (all ages): Act similar to children 6 to 12, generally will not respond to name, will hide from searchers or seek shelter, once shelter found might remain in place, will do little or nothing to assist in their own survival or rescue.

e. Elderly (above 65): Possibly suffering from senility or Alzheimer's, easily confused or attracted to something that strikes their fancy, may not be oriented to the present, likely to over-extend themselves physically, and often have hearing impairments.

f. Despondent: Often are seeking solitude and will not respond to searchers, usually found near prominent terrain/scenic views, within sight and sound of civilization, rarely found in the underbrush.

g. Hikers: Rely on trails and have a set destination in mind, have poor land navigation once they loose the trail, often get confused at trail intersections or in springtime when trails are covered with snow, hiking parties of different ability levels get separated, will miss trails and head down wrong drainages. Hunters: Concentrate on game more than navigation, often pursue game into drainages/deadfalls/underbrush/deep snow, tend to over-extend themselves, typically not prepared for foul weather.

h. Cone pickers/Berry pickers/Photographers/ rockhounds/etc: Intention to stay or go to a specific location, carry little provisions or survival gear, go in good weather so they don't carry foul weather gear, often are not aware of surrounding as they are intent on pictures or the ground if pickers, these people are an extremely high risk group.

i. Fishermen: Well oriented once they arrive at lake or river, most often late due to accident, these are often recovery operations.

j. Climbers: Generally well equipped and self sufficient, tend to remain on marked routes and leave good plans with friends, primary factors are weather and hazardous conditions such as falling debris or avalanches, once located these searches often become technical rescues. 1. Marines: Subject profile will depend on individual's MOS, mission, training, and abilities learned as a civilian.

6. **BEGINNING THE SEARCH**. Once we have determined that a subject is missing we must determine whether or not he/she is in our search area. The sheriffs department will initiate a "Bastard Search", this is to ensure that the subject is not simply at the local tavern, shacked up with an old friend, or was not in an automobile accident enroute to the area. Next we will begin our search at one of the four following area:

a. Last Known Point (LKP): This is the last place that you have physical evidence to support that the subject was there. i.e. Subject's vehicle at the trailhead, name in a logbook, campsite with gear, etc.

b. Place Last Seen (PLS): The last place he was physically seen. i.e. On the trail by other hikers, at the hunting cabin, at a restaurant or gas station, or back home in San Diego.

c. Departure Point: The place where the subject was supposed to enter the search area. Usually a trailhead or specific point relating to the individual's activity.

d. Clues: As clues are discovered, and you confirm that they belong to the subject, you will update or adjust the search area.

7. **MATTSON**. Once we have determined that we do have an overdue subject and can assume that he/she is in our search area we need to break the search area into manageable sized segments. Depending upon the area to be searched, special training required to search high hazard areas, and assets available, we will divide the area by hazard or by terrain features. The lines of division should follow terrain features so that the searchers on the ground can easily navigate and stay within their zones. We will then determine the POA using the Mattson approach. This is done with each of the five or six members of the management team with the most search experience assigning a percentage to each of the segments using his individual judgment on the likelihood of the subject being in that area. Always include an extra column to represent the "rest of the world". Once all personnel have made their determinations, divide the totals and this will give you your areas to search in order of precedence:

NAME	A	B	C	D	E*	TOTAL
DUNSTON	40%	20%	10%	20%	10%	100%
KEOGH	50%	30%	5%	10%	5%	100%
KOFFEL	40%	25%	15%	10%	10%	100%
AYRES	60%	20%	10%	5%	5%	100%
TOTAL	190%	95%	40%	45%	30%	400%
AVERAGE	47.50%	23.75%	10.00%	11.25%	7.5%	
PRIORITY	1 ST	2ND	4TH	3RD	5TH	

8. **COMMAND SYSTEM**. The Office of Emergency Service (OES), has developed a command system that they use for all emergency situations. This system is called the

Incident Command System (ICS). The breakdown is very similar to the military chain of command and the duties correspond to those of the various "S" shops in the battalion. The personnel in each duty area will complete their assigned duties as you would in a military unit with the Incident Commander (IC) being overall responsible for the entire action. Normally, in California, the IC is the County Sheriff.

a. Incident Commander. CO

b. Operations-Tactics, coordination, insert resources. S-3

c. Plans-Strategy, intelligence, briefing and debriefing teams. S-2

d. Logistics-Beans, Band-Aids, beds, heads, motor-T, Special equipment. S-4

e. Communication-COC and field, landlines, frequency coordinator. S-6

f. Finance-Documentation of all man-hours, damages, legal issues.
S-1

g. Liaisons-inter agency, media, safety, family. PAO/Chaplain

9. **CONFINING A SEARCH AREA**. There are several reasons to confine the search area. The most important is that the longer the subject has free movement, the farther he/she can travel. (Remember the Theoretical Method) Additionally, this will ensure that the subject does not walk out without you being aware of it. It's real dumb to be conducting a search for a subject that is on his/her way home. The use of confinement will not interfere with the other searchers, decreases the chances of a more massive search and if promptly initiated has been shown to be highly effective. There are several methods that we can use. (FMST.07.39d)

a. Sign-cutters - Searchers that hike or ski the perimeter of the entire search area, then the perimeter of the segments. This will let you know if he/she moves from one area to another and allows you to concentrate limited assets in a specific area.

b. Trail-watchers/Road Blocks/Camp-ins - Searchers that camp at trailheads. They can alert you if the subject walks out and interview other individuals concerning possible sightings of the subject.

c. Look-outs - These personnel are already in place in fire towers. Simply notify them of the situation.

d. Track Traps - Clearing areas of the trail that are checked periodically for fresh tracks from the subject.

e. String-Lines - Spools of line with paper arrows attached that point to a safe area of a trail-watcher camp.

10. **USING CLUES**. The use of clues cannot be underestimated. They can tell you if a subject was or was not in the area, by freshness you can tell how long ago the subject was there, by the type of clue you can determine physical and mental condition, and by following clues you can determine a direction and speed of travel. Once you have located a CONCLUSIVE clue, you will re-mattson the area, using the clue as the LKP. The clues we will be looking for fall under four categories:

a. Physical - Footprints, equipment, candy wrappers, cigarette butts, etc.

b. Recorded - Summit logs or trail registers.

c. People - Witnesses who have seen the subject.

d. Event - Mirror flashes, camp fires, whistle signals, etc.

11. **ADDITIONAL ASSETS**. There are a variety of assets available to you in the search and rescue business. Some are volunteers, some charge money and some (Like the military) require higher authority to grant permission for their use. Remember that in addition to active searchers you can also use passive measures such as lights, noise, balloons, fires, smoke generators, sirens, or P.A. systems. Some examples of active searchers are:

a. Scent dogs.

b. Tracking dogs.

c. Human trackers.

d. Hasty teams.

e. Grid teams.

f. Helicopters.

g. Civil Air Patrol.

h. Snowmobile clubs.

i. Horse teams.

UNITED STATES MARINE CORPS
Mountain Warfare Training Center
Bridgeport, California 93517-5001

FMST.07.16
04/03/02

STUDENT HANDOUT

TRIAGE

TERMINAL LEARNING OBJECTIVE: Given simulated casualties, in a wilderness environment, triage casualties, in accordance with the references. (FMST.07.16)

ENABLING LEARNING OBJECTIVES:

1) Without the aid of references and from a given list, choose when actual triage most often begins, in accordance with the references. (FMST.07.16a)

2) Without the aid of references and from a given list, identify one of the four-triage categories used by NATO, in accordance with the references. (FMST.07.16b)

3) Without the aid of references and from a given list, choose the primary cause of death in a mass casualty situation, in accordance with the references. (FMST.07.16c)

OUTLINE.

1. **DEFINITION**: Triage or sorting (often used interchangeably) is the evaluation and classification of casualties for the purpose of establishing the priority for treatment and evacuation.

NOTE: Casualty sorting is one of the most important tasks for the entire medical service. No other job requires more informed judgment, hard work and courage. The Officer and the Corpsman responsible for triaging wounded Sailors and Marines must exercise sound judgment so that casualties receive proper, expedient care. Triage is a continuing process and any individual assigned should be the most capable and experienced available.

a. Decision Factors. The factors concerning casualty triage are based upon the following:

(1) The need for immediate resuscitation.

(2) The need for emergency surgery.

(3) The futility of surgery due to obviously lethal wounds.

(4) Time needed to perform surgery compared with:

(a) Probability of success.

(b) Number of other casualties needing treatment.

b. The actual triage begins most often with the casualty himself. (FMST.07.16a). This usually occurs when the injured individual determines that "Buddy-Aid" or "Self-Aid" will not be adequate.

(1) The corpsman, once presented with the casualty, must make several determinations:

(a) Is the casualty a walking-wounded or a litter case?

(b) If severely wounded, are resuscitative efforts required immediately?

(c) Does the casualty need any other immediate treatment (bleeding, asphyxia or pain control)?

(2) Once the casualty has been transported to a Battalion Aid Station (BAS), triage continues. The dental officer will commonly perform the duties of the Triage Officer at the echelon I or echelon II level, if present.

2. **NATO TRIAGE CATEGORY CODES**. (FMST.07.16b)

a. At each medical treatment facility (MTF) in the area of operations (AO), incoming casualties are classified by level of treatment required. Four triage category groups have been universally adopted for use by both United States and NATO Forces. These categories are listed and defined as they appear in standardization agreement (STANAG) No 2879:

(1) Immediate Treatment (Group T1) (RED Tag). To include those requiring emergency life-saving surgery. These procedures should not be time consuming and should concern only those patients with high chances for survival. Some examples are:

(a) Respiratory obstruction.

(b) Accessible hemorrhage.

(c) Emergency amputation.

(2) Delayed Treatment (Group T2) (YELLOW Tag). To include those badly in need of time-consuming major surgery, but whose general condition permits delay in surgical treatment without unduly endangering life. To mitigate the often, critical effects of delay in surgery, sustaining treatment, such as stabilizing I.V. fluids, splinting, administration of antibiotics, catheterization, gastric decompression, and relief of pain will be required. Examples are:

(a) Large muscle wounds.

(b) Fractures of major bones.

(c) Intra-abdominal and/or thoracic, head or spinal injuries.

(d) Uncomplicated major burns.

(3) Minimal Treatment (Group T3) (GREEN Tag). To include those with relatively minor injuries who can effectively care for themselves or who can be helped by untrained personnel. Some examples are:

(a) Minor lacerations.

(b) Abrasions.

(c) Fractures of small bones.

(d) Minor burns.

(4) Expectant Treatment (Group T4) (BLUE Tag). This group comprises patients who have received serious and often multiple injuries, and whose treatment would be time-consuming and complicated with a low chance of survival. If fully treated, they may make heavy demands on medical manpower and supplies. Until the mass casualty situation is under control, they will receive appropriate supportive treatment. The extent of treatment depends on available supplies and manpower, and may involve the use of large doses of narcotic analgesics. These patients should not be abandoned, and every effort should be devoted to their comfort. The possibility of their survival, despite alarming injuries, must always be kept in mind. Examples are:

(a) Severe multiple injuries.

(b) Severe head or spinal injuries.

(c) Large doses of radiation.

(d) Widespread severe burns.

3. **TREATMENT CONCEPTS**.

a. The triage officer must be familiar with all aspects of combat casualty care (airway management, hemorrhage control, use of anesthesia, surgical procedures, fluid resuscitation and post-operative care, to name a few), but ever-present in his mind should be the following concepts:

(1) Asphyxia and hemorrhage are initially the primary causes of death in any mass casualty situation. (FMST.07.16c)

(2) Once at a definitive care facility, uncontrolled hemorrhage is the leading cause of death.

(3) Life has priority over limb, and preserving function overrides the correction of superficial anatomic defects.

(4) A casualty is in a constant state of change until the wound or injury has been repaired.

(5) Systemic disturbances caused by the wound continue until healing is complete.

(6) Combat conditions will dictate the modifications necessary in triage and care of mass casualties.

b. Priorities of Treatment. With the six concepts of treatment in mind, we can now address three basic priorities for treatment.

(1) First Priority:

(a) Asphyxia:

1. Mechanical obstruction of the airway.

2. Sucking chest wound.

3. Tension pneumothorax.

4. Maxillofacial trauma.

5. Massive hemo-pneumothorax.

(b) Shock:

1. Exsanguinating hemorrhage.

2. Major internal hemorrhage (thoracic, abdominal, and pelvic).

3. Visceral injuries/evisceration.

4. Cardio-pericardial injuries.

5. Massive muscle injury leading to Fat Embolization to the lung.

6. Full-thickness or deep-partial burns of 10 % or greater.

(2) Second Priority:

(a) Visceral injuries:

1. Perforation of G.I. Tract.

2. Pancreatic and Biliary system.

3. Genitourinary tract.

4. Thoracic wounds w/o asphyxia.

(b) Vascular injuries requiring repair:

1. Tourniquet cases.

(c) Closed head injuries:

1. Increasing loss of consciousness.

(d) Major burns:

1. Partial-thickness burns of greater than 25% TBSA in the low risk group, 20% TBSA in the high risk group.

2. Burns involving the poor risk group.

3. Burns involving the face, hands, feet, genitalia/perineum.

4. Circumferential limb burns.

(3) Third Priority:

(a) Brain and spinal injuries - requiring burr hole decompression.

(b) Soft tissue injuries.

(c) Debridement necessary with less than major muscle damage.

(d) Lesser fractures and dislocations.

(e) Eye injuries.

(f) Maxillofacial injuries w/o asphyxia.

(g) Burns under 20% in other locations other than priority 2.

4. **TRIAGE TOOLS**.

a. Trauma Score/Glasgow Coma Scale. In recent years, many methods have been proposed to speed up the classification and sorting of casualties. One method that has gained wide acceptance is the Trauma Score/Glasgow Coma Scale. Several studies have been performed using various parameters, i.e., respiratory rate, chest wall expansion, BP, capillary refill, etc. By assigning a number scale to each specific response, the likelihood of survival as a percentage can be calculated.

1. Glasgow Coma Scale.

a. Eye opening:

1) Spontaneous = 4.
2) To voice = 3.
3) To pain = 2.
4) None = 1.

b. Verbal Response.

1) Oriented = 5.
2) Confused = 4.
3) Inappropriate words = 3.
4) Incomprehensive words = 2.
5) None = 1.

c. Motor Response.

   1) Obeys Command = 6.

   2) Localized Pain = 5.
   3) Withdraw Pain = 4.
   4) Flexion (pain) = 3.Extension (pain) = 2.
   5) None = 1.

d. Total the points up and they equal:

   1) 13-15= 4.
   2) 9-12 = 3.
   3) 6-8 = 2.
   4) 4-5 = 1.
   5) <3 = 0

2. Trauma Scale.

   a. Respiratory Rate:

      1) 12-29 min = 4.
      2) 10-12 min = 3.
      3) 6-9 min = 2.
      4) 1-5 min = 1.
      5) None = 0.

   b. Systolic BP:

      1) >89 mmHg = 4.
      2) 76 mmHg = 3.
      3) 50-75 mmHg = 2.
      4) 1-49 mmHg = 1.
      5) None = 0.

   c. Total the points.

3. Total trauma Scale + GCS = % survival (1-12)

   12 = 99
   10 = 88
   8 = 67
   6 = 63
   4 = 33
   2 = 29

1 = 25
0 = 4

UNITED STATES MARINE CORPS
Mountain Warfare Training Center
Bridgeport, California 93517-5001

FMST.07.08
04/03/02

STUDENT HANDOUT

PREVENTIVE MEDICINE AND WATER PURIFICATION

TERMINAL LEARNING OBJECTIVE. Given a unit in a mountainous environment and the necessary equipment and supplies, perform preventive medicine, in accordance with the references. (FMST.07.08)

ENABLING LEARNING OBJECTIVES.

1) Without the aid of references, from a given list select the five important areas of personal hygiene, in accordance with the references. (FMST.07.08a)

2) Without the aid of references, from a given list select the correct steps taken to ensure proper immunization prior to deployment, in accordance with the references. (FMST.07.08b)

3) Without the aid of references, from a given list select the three methods of water purification, in accordance with the references. (FMST.07.08c)

4) Without the aid of references, from a given list select the four forms of halogens used for water purification, in accordance with the references. (FMST.07.08d)

5) Without the aid of references, from a given list select the definition of giardia, in accordance with the references. (FMST.07.08e)

6) Without the aid of references, from a given list select the most common drug used to treat giardia infection in the United States, in accordance with the references. (FMST.07.08f)

7) Without the aid of references, from a given list select the three important reasons for proper waste disposal, in accordance with the references. (FMST.07.08g)

8) Without the aid of references, from a given list select the correct methods of waste disposal, in accordance with the references. (FMST.07.08h)

16

1. **PERSONAL HYGIENE** Your body's defense mechanisms break down due to the physical demands of the arduous terrain, environment, stress and metabolic changes. Personal

hygiene becomes the key to prevent the spread of communicable diseases. Five important areas are: (WMC.6.17a)

a. Body.

(1) The body should be washed frequently in order to minimize the chances of small cuts and scratches developing into full blown infections and as a defense against parasitic infections.

(2) A daily bath or shower consisting of soap and hot water is ideal. However, when this is not possible, you should:

(a) Give yourself a "sponge bath" using soap and water, making sure particular attention is given to body creases, i.e., armpits, groin area, face ears, hands and feet.

(b) If water is in extremely short supply, you should take an "air bath" by:

(1) Remove all clothing and hang it up to air.

(2) Expose the body for two hours to the sunlight, which is ideal, but the effects will basically be the same if done indoors or during an overcast day. BE CAREFUL NOT TO SUNBURN!

(3) Shaving. Shaving at night before bedtime is preferable because shaving will remove facial oils, which aid in protecting the face from a harsh cold environment during the day.

b. Hair.

(1) Clean frequently.

(2) Inspect at least once a week for parasites. Have your mountain buddy inspect the areas not within your view. This is particularly important during the summer season. Any time the area you are operating in an area that contains wild life there is the possibility of parasites like fleas and ticks.

c. Hands. Hand washing with soap and water will cut down in the transmission of viral and parasitic infections via the person-person and fecal-oral route.

(1) Keep fingernails trimmed and clean to prevent accidentally scratching yourself and to prevent harborage for bacteria.

d. Feet. Are your prime sources of transportation, so take care of them!

(1) Inspect frequently for:

(a) Blisters.

(b) Infections, bacterial and fungal.

(2) Preventive measures:

(a) Keep your feet dry, change socks frequently.

(b) Use antifungal foot powder.

(c) Antiperspirants. Pretreatment for 2 weeks prior to field with an antiperspirant containing Aluminum Chlorhydrate. This will aid in the prevention of athlete's foot.

e. Oral Hygiene. The mouth and teeth should be cleaned at least daily to prevent tooth decay and gum disease.

(1) Ideally with:

(a) Tooth brush.

(b) Tooth paste.

(c) Dental floss.

(2) If these items are not available:

(a) A "chew stick" made from a clean twig about 8" long and finger width. Chew one end until it becomes frayed and brush-like, then use as a toothbrush.

(b) Field expedient dental floss can be pulled from the center lines of paracord.

(c) The gums should be stimulated at least once daily by rubbing them vigorously with a clean index finger.

2. **IMMUNIZATIONS (NAVMEDCOMINST 6230.3)**. In addition to being required, immunizations are your best front line defense against communicable disease.

a. Eight weeks prior to deployment: (WMC.6.17b)

(1) Ensure Basic Series Completed (BSC) on required immunizations.

(2) Determine if additional immunizations/chemophylaxis will be required in that geographical area of operations.

(3) Identify and flag health risk personnel: i.e. TB converters (NAVMEDCOMINST 6224.1), hepatitis (NAVMEDCOMINST 6230.1A), mononucleosis, parasitic infections and anemic personnel.

(4) Review medical intelligence reports:

(a) EPMU, Navy Environmental Preventive Medicine Unit.

(b) G-2, USMC.

(c) NAVDISVECTOECOLCONCEN, Navy Disease Vector Ecology and Control Centers.

(d) CDC - Center For Disease Control.

(e) WHO - World Health Organization.

3. **WATER.** A satisfactory source is one where quantity and quality is enough to supply the needs of all the troops. Water or ice taken from the environment should be considered contaminated and must be purified with available means. Be aware that cold weather slows the chemical reaction time of purification measures.

a. Sources.

(1) Rivers

(2) Streams

(3) Ponds

(4) Lakes

(5) Ice

(6) Snow

b. Water Source Selection. The choice of a water source is influenced by:

(1) Not contaminated by: (Check all to one mile upstream).

(a) Sewage

(b) Enemy pollution

(c) Chemicals

(d) Biologicals

(e) Radiologicals

NOTE: Check condition of vegetation or for dead animals.

(2) Quantity.

(3) Ease of procurement.

(4) Ease of purification.

(5) Freedom from turbidity.

(6) Excessive organic or non-organic contamination.

(7) Should be easily protectable.

4. **PURIFICATION.** Simply consists of removing or destroying enough impurities to make water safe and pleasant to drink.

a. The first step in purifying is to select a treatment method. (All figures are based on one quart of water.) The three different purifying methods are: (WMC.6.17c)

(1) Boiling. The standard recommendation is to bring the water to a rolling boil for 1 minute at sea level for complete sterilization of the water. Recent research showed that additional boiling time for sterilization was not necessary and that boiling for two minutes at any altitude rendered water safe for consumption.

(2) Halogens. The number of tablets and contact time must be considered. All charts based on one-quart canteen. (WMC.6.17d)

(a) Iodine tablets.

TYPE WATER	TABLETS	CONTACT TIME
CLEAR	1	25 MIN
CLOUDY	2	25 MIN
COLD	2	25 MIN

(b) Chlorine bleach. The strength of the solution, number of drops and contact time must be considered.

PERCENT	CLEAR WATER	CLOUDY WATER	COLD WATER
1%	10GTTS/30 MIN	20GTS/30 MIN	10GTTS/60MIN
4-6%	2GTTS/30 MIN	4GTTS/30 MIN	2GTTS/60 MIN
7-10%	1 GTTS/30 MIN	2GTTS/30 MIN	1 GTTS/60 MIN

(c) Iodine solution. The strength, number of drops and contact time must be considered.

PERCENT	CLEAR WATER	CLOUDY WATER	COLD WATER
2%	5GTTS/30 MIN	10GTTS/30 MIN	8GTTS/60 MIN

(d) Betadine solution.

CLEAR WATER	CLOUDY WATER	COLD WATER
8GTTS/30 MIN	16GTTS/30 MIN	8GTTS/60 MIN

(1) Filtration Most filters are not small enough to completely disinfect water. They remove particulate debris, which allows lower halogen dose and improved appearance and taste of "dirty" water. Filters clog quickly if the water is dirty or has a lot of suspended particles. NEVER USE FILTERS AS YOUR SOLE METHOD OF WATER PURIFICATION.

(a) Maximum Filter Pore Size:

1) Parasitic eggs and larvae: 20 micrometers (um).

2) Giardia, E. histolytica: 5 um.

3) Cryptosporidium: 1 um

4) Enteric bacteria: 0.2 um.

5) Viruses: 0.01 um (too small for field water filtration).

5. **GIARDIA**

a. Definition. (WMC.6.17e) An intestinal flagellate protozoa that can cause clinical illness in humans.

(1) Found worldwide in tropical and temperate areas, the infectious cysts are passed by humans and animals and can be transmitted by direct fecal-oral route or waterborne contamination (cysts maintain viability for months in cold water). Due to these factors, Giardia is both a common cause of diarrhea in travelers to foreign countries, and campers in the United States where it has become endemic and is now considered the most common human parasite.

(2) It is found with increasing frequency in formerly "pristine" mountain streams, rivers and lakes and outbreaks have occurred in remote recreational areas and resorts throughout the western mountain states.

(3) Giardiasis is currently a risk for anyone drinking untreated surface water in the western mountains and for communities with water treatment techniques inadequate for eliminating the cysts.

b. Pathophysiology. The organism has two forms: Trophozoite and cyst.

(1) These may be passed in the stool of an infected person. The trophozoites die within one hour outside the body, however the cysts are very hardy, retaining infectivity more than two months in cold water. If the cysts are ingested, excystation is encouraged by stomach acid, and the motile trophozoite attaches by suction disc to the wall of the duodenum and proximal jejunum, where they may cause symptoms of illness.

(2) Water, ice, or food prepared with contaminated water, are the major sources of infection, but direct person-to-person transmission can occur from sexual contact or hand-to-mouth contamination.

(3) Incubation time: 5 - 25 days (7 - 10 medium). Most infections are thought to be asymptomatic carrier states, however, a wide array of clinical syndromes may occur.

c. Symptoms.

(1) A number of people have abrupt onset of explosive diarrhea accompanied by abdominal cramps, foul gas, vomiting, fever and malaise. This commonly lasts three to four days before developing the more common subacute syndrome. In most individuals, the onset is more insidious, and symptoms are persistent or recurrent.

(2) Stools become mushy, greasy and malodorous. Over a period of weeks, watery diarrhea may alternate with soft stools and even constipation. Upper GI symptoms accompany stool changes, mid and upper abdominal cramping, substernal burning, acid indigestion, sulfurous belching, nausea and foul flatus, distention with an early "full" feeling are all typically increased after a meal.

(3) Symptoms of anorexia, fatigue and weight loss are common, but fever and vomiting are infrequent except during an acute onset of illness.

d. Laboratory Parameters.

(1) The "typical" victim should suggest a diagnosis of giardiasis. This individual may have a history of camping, vacationing at a mountain resort, or foreign travel. Symptoms of upper abdominal discomfort, foul gas and soft stools arose one or two weeks after a short trip or during a lengthy one and have fluctuated for weeks. Sometimes improving, the individual held off visiting the doctor, only to have the symptoms recur. Work or school is not interrupted, however, physical discomfort and embarrassment from the gas are concerns.

(2) Laboratory confirmation of giardiasis can be difficult. Stool samples are the first means of diagnosis. Trophozoites may be found in fresh, watery stools, but disintegrate rapidly. Cysts remain in fresh stools for at least 24 hours. Cysts passage, however, is extremely variable and not related to clinical symptoms. Three samples, taken every other day, should be examined for a definite diagnosis.

(3) Average duration of symptoms has varied from three to ten weeks. Acquired resistance has been shown. Immunity is suspected, but not proven and human reinfection definitely occurs.

e. Treatment. (WMC.6.17f) Complete cure of most Giardia infections can be achieved 90% of the time with *Metronidazole* (flagyl) 250 mg three time s a day for 5 days.

f. Prevention There is no safe or effective chemoprophylactic drug and prevention should be practiced by travelers and campers. The purity of American wild waters can no longer be assumed and in developing countries, even tap water should be suspected. In addition to choosing a technique for individual field water disinfection, individuals must become more responsible for prevention of water contamination. This requires attention to personal latrines and camper trailer waste disposal, a subject that most people do not want to touch.

6. **WASTE**. The importance of proper waste disposal cannot be over emphasized.

a. It serves to: (WMC.6.17g)

(1) Eliminate harborage for rodents and vermin.

(2) Prevents attraction of rodents and vermin.

(3) Prevents a pathogenic contamination source.

b. Two basic types of waste are organic and non-organic. The basic ways to dispose of them are burning, burying, or hauling it away. (WMC.6.17h)

(1) Organic wastes:

(a) Human waste - Burn or haul away to a designated waste pit area.

(b) Urine - Use only assigned marked areas.

(c) Edible garbage - Burn. It should not be left exposed for animals, vermin or the enemy.

(d) Paper -Burn.

(e) Contaminated bandages - Sterilize/burn.

(2) Non-organic waste:

(a) Metals - Haul away or bury.

(b) Liquids - Burn or bury.

UNITED STATES MARINE CORPS
Mountain Warfare Training Center
Bridgeport, California 93517-5001

FMST.07.19
04/03/02

STUDENT OUTLINE

REQUIREMENTS FOR SURVIVAL

TERMINAL LEARNING OBJECTIVE Given a survival situation in any type of environmental condition and minimal equipment and resources, apply the requirements for survival to facilitate individual or group survival and recovery per the references. (FMST.07.19)

ENABLING LEARNING OBJECTIVES

1. Without the aid of references and from a given list, choose one of the survival stressors, in accordance with the references. (WMC.A1.a)

2. Without the aid of references and from a given list, choose one of the priorities of work in a survival situation, in accordance with the references. (WMC.A1.b)

1. REQUIREMENTS FOR SURVIVAL

a. This mental "mind-set" is important in many ways. We usually call it the "will to survive" although you might call it "attitude" just as well. This basically means that, if you do not have the right attitude, you may still not survive.

A1

b. A guideline that can assist you is the acronym " SURVIVAL".

(1) <u>S</u>ize up.

(a) Size up the situation.

1. Conceal yourself from the enemy.

2. Use your senses to hear, smell, and see to determine and consider what is developing on the battlefield before you make your survival plan.

(b) Size up your surroundings.

1. Determine the rhythm or pattern of the area.

2. Note animal and bird noises and their movement.

3. Note enemy traffic and civilian movement.

(c) Size up your physical condition.

1. Check your wounds and give yourself first aid.

2. Take care to prevent further bodily harm.

3. Evaluate condition of self and unit prior to developing survival plan.

(d) Size up your equipment.

1. Consider how available equipment may affect survival senses; tailor accordingly.

(2) <u>U</u>ndue haste makes waste.

(a) Plan your moves so that you can move out quickly without endangering yourself if the enemy is near.

(3) <u>R</u>emember where you are.

(a) If you have a map, spot your location and relate it to the surrounding terrain.

(b) Pay close attention to where you are and where you are going. Constantly orient yourself.

(c) Try to determine, at a minimum, how your location relates to the following:

1. The location of enemy units and controlled areas.

2. The location of friendly units and controlled areas.

3. The location of local water sources.

4. Areas that will provide good cover and concealment.

(4) Vanquish fear and panic.

(a) Realistic and challenging training builds self-confidence and confidence for a unit's leadership.

(b) The feeling of fear and panic will be present. The survivor must control these feelings.

(5) Improvise and improve.

(a) Use tools designed for one purpose for other applications.

(b) Use objects around you for different needs. (i.e. use a rock for a hammer)

(6) Value living.

(a) Place a high value on living.

(b) Refuse to give into the problem and obstacles that face you.

(c) Draw strength from individuals that rise to the occasion.

(7) Act like the natives.

(a) Observe the people in the area to determine their daily eating, sleeping, and drinking routines.

(b) Observe animal life in the area to help you find sources of food and water.

NOTE: Remember that animal reactions can reveal your presence to the enemy. Animals cannot serve as an absolute guide to what you can eat and drink.

(8) Live by your wits, **but for now**, learn basic skills.

(a) Practice basic survival skills during all training programs and exercises.

2. **STRESS**. Stress has many positive benefits. Stress provides us with challenges: it gives us chances to learn about our values and strengths. Too much stress leads to distress. While many of these signs may not be self-identified, it remains critical that all survivors remain attentive to each other's signs of distress. Listed are a few common signs of distress found when faced with too much stress:

   a. Difficulty in making decisions (do not confuse this sign for a symptom of hypothermia).

   b. Angry outbursts.

   c. Forgetfulness.

   d. Low energy level.

   e. Constant worrying.

   f. Propensity for mistakes.

   g. Thoughts about death or suicide.

   h. Trouble getting along with others.

   i. Withdrawing from others.

   j. Hiding from responsibilities.

   k. Carelessness.

3. **SURVIVAL STRESSORS**. (FMST.07.19a) **Any** event can lead to stress. Often, stressful events occur simultaneously. These events are not stress, but they produce it and are called "stressors". In response to a stressor, the body prepares either to "fight or flee". Stressors add up. Anticipating stressors and developing strategies to cope with them are the two ingredients in the effective management of stress. It is essential that the survivor be aware of the types of stressors they will encounter.

   a. Injury, Illness, or Death: Injury, illness, and death are real possibilities a survivor has to face. Perhaps nothing is more stressful than being alone in an unfamiliar environment where you could die from hostile action, an accident, or from eating something lethal. Illness and injury can also add to stress by limiting your ability to maneuver, get food and drink, find shelter, and defend yourself.

   b. Uncertainty and Lack of Control: Some people have trouble operating in settings where everything is not clear-cut. The only guarantee in a survival situation is that nothing is guaranteed. This uncertainty and lack of control also add to the stress of being ill, injured, or killed.

c. Environment: A survivor will have to contend with the stressors of weather, terrain, and the variety of creatures inhabiting an area. Heat, cold, rain, winds, snow, mountains, insects, and animals are just a few of the challenges awaiting the Marine working to survive. Depending on how a survivor handles the stress of environment, his surroundings can be either a source of food and protection or a cause of extreme discomfort leading to injury, illness, or death.

d. Hunger and Thirst: Without food and water a person will weaken and eventually die. Getting and preserving food and water takes on increasing importance as the length of time in a survival setting increases. With the increased likelihood of diarrhea, replenishing electrolytes becomes critical. For a Marine used to having his provisions issued, foraging can be a significant source of stress.

e. Fatigue: It is essential that survivors employ all available means to preserve mental and physical strength. While food, water, and other energy builders may be in short supply, maximizing sleep to avoid deprivation is a very controllable factor. Training data collected at MCCDC demonstrates that individuals who lack sleep for 24 hours or more suffer a 25% decrease in effective performance. Further, sleep deprivation directly correlates with increased fear. Effective survival is highly unlikely when fatigue builds to a level where staying awake becomes a stressful evolution.

f. Isolation: Although Marines complain about higher headquarters, we become used to the information and guidance it provides, especially during times of confusion. Being in contact with others provides a greater sense of security and a feeling someone is available to help if problems occur. A significant stress in survival situations is that often a person or team has to rely solely on its own resources.

4. **NATURAL REACTIONS**. Man has been able to survive many shifts in his environment throughout the centuries. His ability to adapt physically and mentally to a changing world kept him alive. The average person will have some psychological reactions in a survival situation. These are some of the major internal reactions you might experience with the survival stresses.

   a. Fear: Fear is our emotional response to dangerous circumstances that we believe have the potential to cause death, injury, or illness. Fear can have a positive function if it encourages us to be cautious in situations where recklessness could result in injury. Unfortunately, fear can also immobilize a person. It can cause us to become so frightened that we fail to perform activities essential for survival.

   b. Anxiety: Anxiety can be an uneasy, apprehensive feeling we get when faced with dangerous situations. When used in a healthy way, anxiety urges us to act to end, or at least master, the dangers that threaten our existence. A survivor reduces his anxiety by performing those tasks that will ensure his coming through the ordeal alive. Anxiety can overwhelm a Marine to the point where he becomes easily confused and has difficulty thinking.

c. Anger and Frustration: Frustration arises when a person is continually thwarted in his attempts to reach a goal. The goal of survival is to stay alive until you can reach help or until help can reach you. To achieve this goal, Marines must complete some tasks with minimal resources. One outgrowth of frustration is anger. Getting lost, damaged or forgotten equipment, the weather, inhospitable terrain, enemy patrols, and physical limitations are just a few sources of frustration and anger. Frustration and anger encourage impulsive reactions, irrational behavior, poorly thought-out decisions, and, in some instances, an "I quit" attitude. If the Marine does not properly focus his angry feelings, he can waste much energy in activities that do little to further either his chances of survival or the chances of those around him.

d. Depression: Depression is closely linked with frustration and anger when faced with the privations of survival. A destructive cycle between anger and frustration continues until the person becomes worn down-physically, emotionally, and mentally. When a person reaches this point, he starts to give up, and his focus shifts from "What can I do" to "There is nothing I can do." If you allow yourself to sink into a depressed state, then it can sap all your energy and, more important, your will to survive.

e. Loneliness and Boredom: Man is a social animal and enjoys the company of others. Loneliness and boredom can be another source of depression. Marines must find ways to keep their minds productively occupied.

f. Guilt: The circumstances leading to your survival setting are sometimes dramatic and tragic. It may be the result of an accident or military mission where there was a loss of life. Perhaps you were the only, or one of a few, survivors. While naturally relieved to be alive, you simultaneously may be mourning the deaths of others that were less fortunate. Do not let guilt feelings prevent you from living.

5. **PRIORITIES OF WORK IN A SURVIVAL SITUATION.** (FMST.07.19b) Each survival situation will have unique aspects that alter the order in which tasks need to be accomplished. A general guideline is to think in blocks of time.

a. First 24 hours: The first 24 hours are critical in a survival situation. You must make an initial estimate of the situation. Enemy, weather, terrain, time of day and available resources will determine which tasks need to be accomplished first. They should be the following:

(1) Shelter.

(2) Fire.

(3) Water.

(4) Signaling.

b. Second 24 hours: After the first 24 hours have passed, you will now know you can survive. This time period needs to be spent on expanding your knowledge of the area. By completing the following tasks, you will be able to gain valuable knowledge.

(1) Tools and weapons: By traveling a short distance from your shelter to locate the necessary resources, you will notice edible food sources and game trails.

(2) Traps and snares: Moving further away from your shelter to employ traps and snares, you will be able to locate your shelter area from various vantagepoints. This will enable you to identify likely avenues of approach into your shelter area.

(3) Pathguards: Knowing the likely avenues of approaches, you can effectively place noise and casualty producing pathguards to ensure the security of your shelter area.

c. Remainder of your survival situation. This time is spent on continuously improving your survival situation until your rescue.

6. **GROUP SURVIVAL.** In-group survival, the group's survival depends largely on its ability to organize activity. An emergency situation does not bring people together for a common goal; rather, the more difficult and disordered the situation, the greater are the disorganized group's problems.

a. Groups Morale: High morale must come from internal cohesiveness and not merely through external pressure. The moods and attitudes can become wildly contagious. Conscious, well-planned organization and leadership on the basis of delegated or shared responsibility often can prevent panic. High group morale has many advantages.

(1) Individual feels strengthened and protected since he realizes that his survival depends on others whom he trusts.

(2) The group can meet failure with greater persistency.

(3) The group can formulate goals to help each other face the future.

b. Factors that Influence Group Survival: There are numerous factors that will influence whether a group can successfully survive.

(1) Organization of Manpower - Organized action is important to keep all members of the group informed; this way the members of the group will know what to do and when to do it, both under ordinary circumstances and in emergencies.

(2) Selective Use of Personnel - In well-organized groups, the person often does the job that most closely fits his personal qualifications.

(3) Acceptance of Suggestion and Criticisms - The senior man must accept responsibility for the final decision, but must be able to take suggestion and criticisms from others.

(4) Consideration of Time - On-the-spot decisions that must be acted upon immediately usually determines survival success.

(5) Check Equipment - Failure to check equipment can result in failure to survive.

(6) Acquiring survival knowledge and skills increases survival Knowledge and Skills - Confidence in one's ability to survive.

UNITED STATES MARINE CORPS
Mountain Warfare Training Center
Bridgeport, California 93517-5001

FMST.
07.20
04/03/02

STUDENT OUTLINE

SURVIVAL KIT

TERMINAL LEARNING OBJECTIVE. In a summer mountainous environment, construct a survival kit, in accordance with the references. (FMST.07.20)

ENABLING LEARNING OBJECTIVES

(1) Without the aid of references, choose from a given list in the criteria for a survival kit, in accordance with the references. (FMST.07.20a)

(2) Without the aid of references, choose from a given list one example of each criteria for a survival kit, in accordance with the references. (FMST.07.20b)

1. COMPONENTS FOR A SURVIVAL KIT

a. The environment is the key to the types of items you will need in your survival kit. How much equipment you put in your kit depends on how you will carry the kit. A kit on your body will have to be much smaller than one carried in a vehicle.

b. Always layer your survival kit, keeping the most important items on your body.

c. In preparing your survival kit, select items that can be used for more than one purpose.

A2

d. Your survival kit does not need to be elaborate. You only need functional items that will meet your needs and a case to hold them. The case might be a first aid case, an ammunition pouch, or another suitable case. This case should be:

(1) Water repellent or waterproof.

(2) Easy to carry or attach to your body.

(3) Suitable to accept various sized items.

(4) Durable.

e. When constructing a survival kit, you should have the following components: (FMST.07.20a)

(1) Fire starting items.

(2) Water procurement items.

(3) Food procurement items.

(4) Signaling items.

(5) First aid items.

(6) Shelter items.

2. **<u>ITEMS CONTAINED WITHIN EACH COMPONENT</u>** (FMST.07.20b)

a. <u>Fire Starting Items</u>.

(1) Matches.

(2) Magnifying glass.

(3) Flint and Steel.

(4) Lighter.

(5) Potassium Permanganate, with a container of sugar or anti-freeze.

(6) Prepackaged Tinder.

(a) Commercially Manufactured

(b) Cotton Balls and Petroleum Jelly

b. Water Procurement Items

(1) Water Disinfecting Chemicals.

(a) Iodine Tablets

(b) Betadine Solution

(c) Iodine Solution

(2) Metal Container. (Serves for boiling water)

(a) Canteen Cup

(b) Survival Kit Container

(c) Any suitable can that contained no petroleum products.

(3) Water Carrying Items.

(a) Canteen

(b) Plastic Bag

(c) Plastic/Metal/Glass Container which contained no petroleum products

c. Food Procurement Items

(1) Fish.

(a) Various sized hooks

(b) Various sized sinkers/weights

(c) Metal leaders and swivels

(d) Small weighted jigs

(e) Fishing line

i. Think about the size of fish for that environment when selecting weights and sizes.

(2) Game.

(a) Snares

i. Commercially Manufactured

ii. Aircraft Cable

iii. Tie Wire

(b) Bait

i. MRE Cheese Spread or Peanut Butter Package

(c) 550 Cord for Gill Net and Trap Construction

i. Engineer/Marking Tape

ii. Sling shot rubber and pouch

d. Signaling Items

(1) Day.

(a) Mirror

(b) Whistle

(c) Pyrotechnics (Smoke, Pen Flares)

(d) Air Panels

(2) Night.

(a) Pyrotechnics (Pen Flares, Star Clusters)

(b) Lights (Flashlight, Strobe, Chemlight)

(c) Whistle

e. Shelter Items

(1) Cordage.

(a) 550 Cord

(b) Wire

(c) Communication wire

(d) Tie wire

(2) Finger Saw.

(3) Sewing Kit with Needles for construction/repair of clothing.

(4) Tentage.

(a) Poncho

(b) Tarp

(c) Space blanket

(d) Plastic trash bags

f. First Aid Items

(1) Band-Aids.

(a) Steristrips

(b) Adhesive Tape

(c) Non-stick pads, 4x4's, Gauze, Battle Dressings

(d) Muslin Bandage

(2) Ointments.

(a) Burn

(b) Anti-septic

(3) Miscellaneous.

(a) Salt

(b) Sugar

(c) Eye Wash

(d) Alcohol prep pads

(e) Suture Kit

(f) Scalpel

(g) Vile of Yarrow

g. Miscellaneous items

(1) Fingernail clippers

(2) Compass

(3) Notebook with pen or pencil

(4) Wood eye screws and nails

(5) Surgical tubing

NOTE: It is assumed that the Marine is always carrying a high quality fixed bladed knife, a multi-tool knife, and a sharpening stone.

UNITED STATES MARINE CORPS
Mountain Warfare Training Center
Bridgeport, California 93517-5001

FMST.07.23
04/05/02

STUDENT OUTLINE

SIGNALLING AND RECOVERY

TERMINAL LEARNING OBJECTIVE Given a survival situation in any type of environmental condition and minimal equipment and resources, signal for aid to facilitate individual or group recovery per the references. (FMST 07.23)

ENABLING LEARNING OBJECTIVES

(1) Without the aid of references and from a given list, choose the audio international distress Signal, in accordance with the references. (FMST.07.23a)

(2) Without the aid of references and from a given list, choose the visual international distress signal, in accordance with the references. (FMST.07.23b)

(3) Without the aid of references and given a list of messages, match in writing the international code symbols to each corresponding message, in accordance with the references. (FMST.07.23c)

OUTLINE

1. **SIGNALING DEVICES**. The equipment listed below are items that may be on your body or items inside an aircraft. Generally, these items are used as signaling devices while on the move. They must be accessible for use at any moment's notice. Additionally, in a summer mountainous environment, Marines may experience areas that are snow covered and must be familiar with the effects that snow will have on specific signaling devices.

    a. Pyrotechnics. Pyrotechnics include star clusters and smoke grenades. When using smoke grenades in snow pack, a platform must be built. Without a platform, the smoke grenade will sink into the snow pack and the snow will absorb all smoke. A rocket parachute flare or a hand flare has been sighted as far away as 35 miles, with an average of 10 miles. Pyrotechnic flares are effective at night, but during daylight their detectability ranges are reduced by 90 percent.

b. M-186 Pen Flare: The M-186 Pen Flare is a signaling device carried in the vest of all crew chiefs and pilots. Remember to cock the gun prior to screwing in the flare.

c. Strobe Light: A strobe light is generally carried in the flight vests of all crew chiefs and pilots. It can be used at night for signaling. Care must be taken because a pilot using goggles may not be able to distinguish a flashing strobe from hostile fire.

d. Flashlight: By using flashlights, a Morse code message can be sent. A SOS distress call consists of sending three dots, three dashes, and three dots. Keep repeating this signal.

e. Whistle: The whistle is used in conjunction with the audio international distress signal. It is used to communicate with forces on the ground.

f. AN/PRC-90 & AN/PRC-112: The AN/ PRC 90 survival radio is a part of the aviator's survival vest. The AN/PRC-112 will eventually replace the AN/PRC-90. Both radios can transmit either tone (beacon) or voice. Frequency for both is **282.8 for voice**, and **243.0 for beacon**. Both of these frequencies are on the UHF Band.

g. Day/Night Flare: The day/night flare is a good peacetime survival signal. The flare is for night signaling while the smoke is for day. A red cap identifies the older version flare with three nubbins while the new generation has three rings around the body for identification during darkness. The flare burns for approximately 20 second while the smoke burns for approximately 60 seconds.

NOTE: Once one end is used up, douse in water to cool and save the other end for future use.

h. Signal Mirror: A mirror or any shiny object can be used as a signaling device. It can be used as many times as needed. Mirror signals have been detected as far away as 45 miles and from as high as 16,000', although the average detection distance is 5 miles. It can be concentrated in one area, making it secure from enemy observation. A mirror is the best signaling device for a survivor; however, it is only as effective as its user. Learn how to use one now, before you find yourself in a survival situation.

(1) Military signal mirrors have instructions on the back showing how to use it. It should be kept covered to prevent accidental flashing that may be seen by the enemy.

(2) Any shiny metallic object can be substituted for a signal mirror.

(3) Haze, ground fog, or a mirage may make it hard for a pilot to spot signals from a flashing object. So, if possible, get to the highest point in your area when flashing. If you can't determine the aircraft's location, flash your signal in the direction of the aircraft noise.

AIMING THE SIGNAL MIRROR

2. METHODS OF COMMUNICATION

a. Audio: Signaling by means of sound may be good, but it does have some disadvantages.

(1) It has limited range unless you use a device that will significantly project the sound.

(2) It may be hard to pinpoint one's location due to echoes or wind.

(3) International Distress Signal. (FMST.07.23a) The survivor will make six blasts in one minute, returned by three blasts in one minute by the rescuer.

b. Visual: Visual signals are generally better than audio signals. They will pinpoint your location and can been seen at a greater distances under good weather conditions.

(1) The visual international distress symbol is recognized by a series of three evenly spaced improvised signaling devices. (FMST.07.23b)

3. **IMPROVISED SIGNALING DEVICES**. Improvised signaling devices are generally static in nature. They must be placed in a position to be seen by rescuers. They are made from any resources available, whether natural or man-made.

a. Smoke Generator: The smoke generator is an excellent improvised visual signaling device. It gives the survivor the flexibility to signal in either day or night conditions. This type of signal has been sighted as far away as 12 miles, with an average distance of 8 miles. Smoke signals are most effective in calm wind conditions or open terrain, but effectiveness is reduced with wind speeds above 10 knots. Build them as soon as time and the situation permits, and protect them until needed.

(1) Construct your fire in a natural clearing or along the edge of streams (or make a clearing). Signal fires under dense foliage will not be seen from the air.

(2) Find two logs, 6 - 10 inches in diameter, and approximately five feet long. Place the two logs parallel to each other with 3 - 4 feet spacing.

(3) Gather enough sticks, approximately two inches in diameter and four feet long, to lie across the first two logs. This serves as a platform for the fire.

(4) Gather enough completely dry branches to build a pyramid fire. The pyramid fire should be 4 feet by 4 feet by 2 feet high.

(5) Place your tinder under the platform.

(6) Gather enough pine boughs to lay on top of the pyramid fire. This serves to protect the fire and the tinder.

(7) To light, remove the pine bough and ignite the tinder. If available, construct a torch to speed up the lighting process, especially for multiple fires.

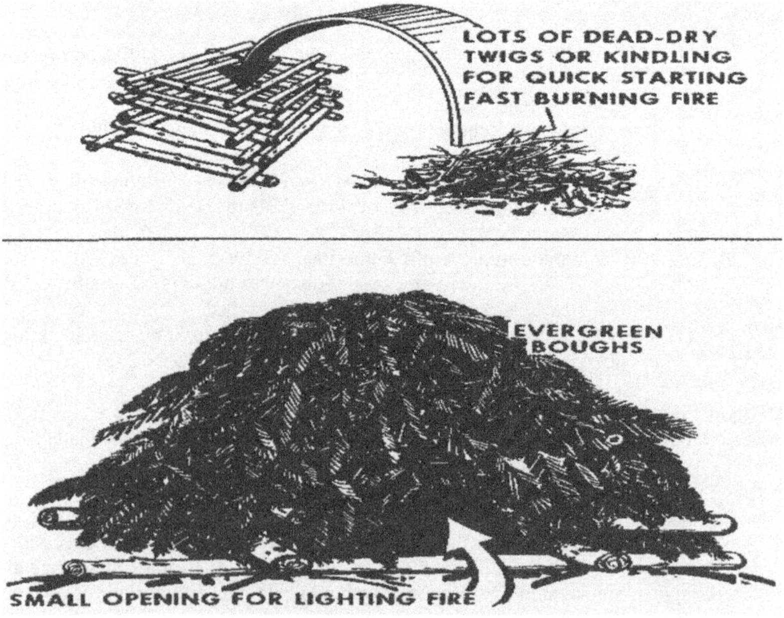

SMOKE GENERATOR

(8) To create a smoke effect during the day light hours, place the pine bough on the ignited fire.

(9) Placing a smoke grenade or colored flare under the platform will change the color of the smoke generated. Remember that you want the fire to draw in the colored smoke, which will create a smoke color that contrasts with the background will increase the chances of success.

b. Arrangement or alteration of natural materials: Such things as twigs or branches, can be tramped into letters or symbols in the snow and filled in with contrasting materials. To attract more attention, ground signals should be arranged in big geometric patterns.

(1) International symbols. The following symbols are internationally known. (FMST.07.23c)

Number	Message	Code symbol
1	REQUIRE ASSISTANCE	V
2	REQUIRE MEDICAL ASSISTANCE	X
3	NO OR NEGATIVE	N
4	YES OR AFFIRMATIVE	Y
5	PROCEED IN THIS DIRECTION	↑

INTERNATIONAL SYMBOLS

(2) Shadows: If no other means are available, you may have to construct mounds that will use the sun to cast shadows. These mounds should be constructed in one of the International Distress Patterns. To be effective, these shadow signals must be oriented to the sun to produce the best shadows. In areas close to the equator, a North—South line gives a shadow anytime except noon. Areas further north or south of the equator require the use of East—West line or some point of the compass in between to give the best result.

(3) Size: The letters should be large as possible for a pilot or crew to spot. Use the diagram below to incorporate the size to ratio for all letter symbols.

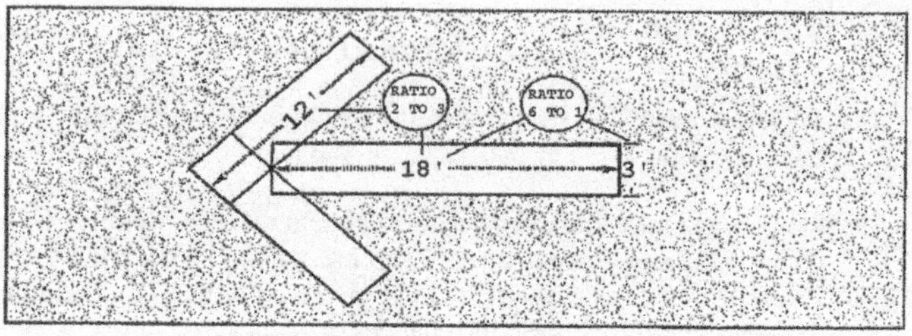

SIZE AND RATIO

(4) Contrast: When constructing letter symbols, contrast the letter from the surrounding vegetation and terrain. Ideally, bring material from another location to build the letter. This could be clothing, air panels, space blanket, etc.

(a) On snow, pile bough or use sea dye from a LPP (Life preserver, personal). Fluorescent sea dye markers have been sighted as far away as 10 miles, although the average detection distance is 3 miles.

4. **AIR TO GROUND COMMUNICATIONS**. Air to ground communications can be accomplished by standard aircraft acknowledgments.

a. Aircraft will indicate that ground signals have been seen and understood by:

(1) Rocking wings from side to side: This can be done during the day or in bright moonlight.

b. Aircraft will indicate that ground signals have been seen but not understood by:

(1) Making a complete, clockwise circle during the day or in bright moonlight.

5. **RECOVERY**. Marines trapped behind enemy lines in future conflicts may not experience quick recovery. Marines may have to move to a place that minimizes risk to the recovery force. No matter what signaling device a Marine uses, he must take responsibility for minimizing the recovery force's safety.

a. Placement Considerations: Improvised signaling devices, in a hostile situation, should not be placed near the following areas due to the possibility of compromise:

(1) Obstacles and barriers.

(2) Roads and trails.

(3) Inhabited areas.

(4) Waterways and bridges.

(5) Natural lines of drift.

(6) Man-made structures.

(7) All civilian and military personnel.

b. Tactical Consideration: The following tactical considerations should be adhered to prior to employing an improvised signaling device.

(1) Use the signals in a manner that will not jeopardize the safety of the recovery force or you.

(2) Locate a position, which affords observation of the signaling device and facilitates concealed avenues of escape (if detected by enemy forces). Position should be located relatively close to extract site in order to minimize "time spent on ground" by the recovery force.

(3) Maintain continuous security through visual scanning and listening while signaling devices is employed. If weapon systems are fire and/or observation should cover available, signaling devices.

(4) If enemy movement is detected in the area, attempt to recover the signaling device, if possible.

(5) Employ improvised signaling devices only during the prescribed times, if briefed in the mission order.

c. Recovery Devices: In mountainous terrain, a helicopter landing may be impossible due to ground slope, snow pack, or vegetation. The survivor must be familiar with recovery devices that may be aboard the aircraft.

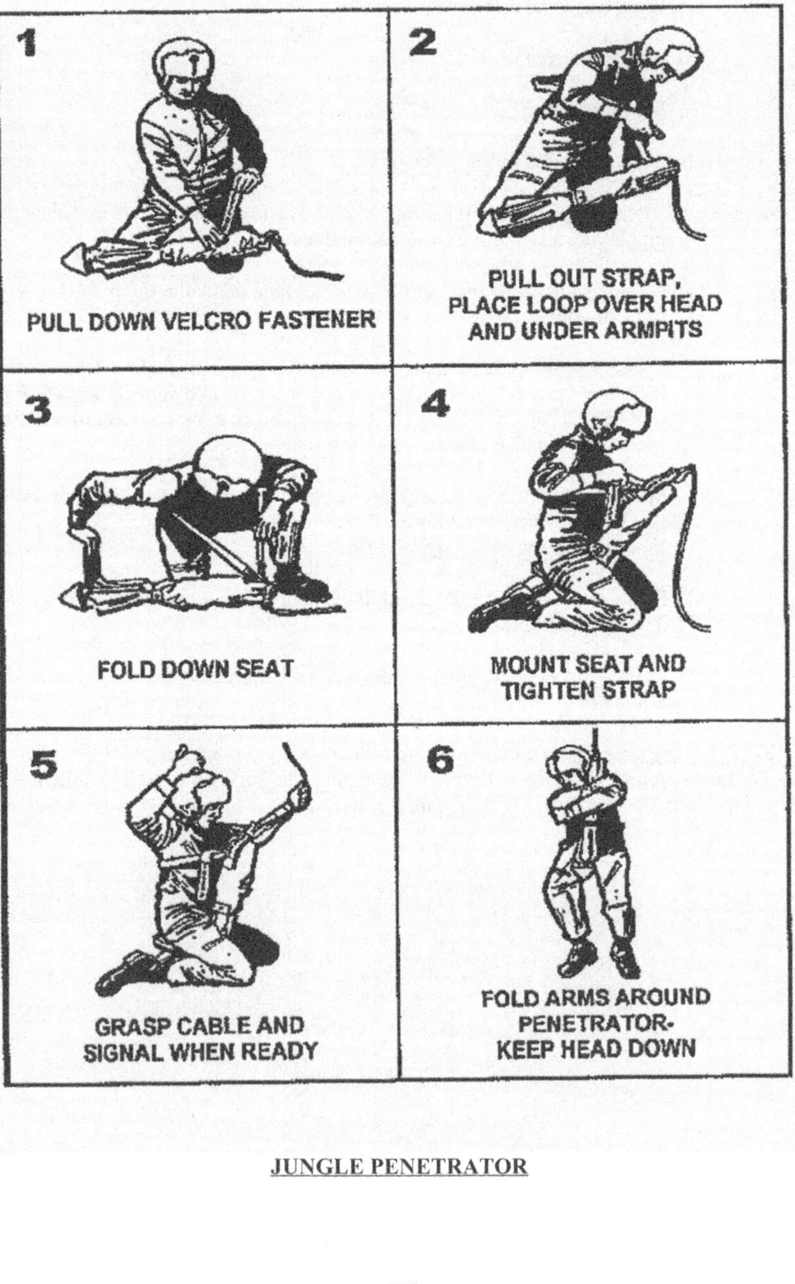

<u>JUNGLE PENETRATOR</u>

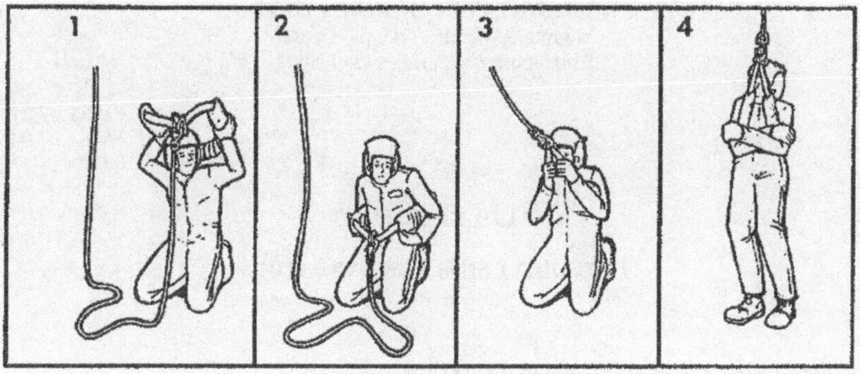

SLING HOIST

d. Recovery by other than aircraft. Recovery by means other than aircraft may occur. Unit Sop's should include signaling and link-up with forces at the following locations:

(1) Border Crossings. The evader who crosses into a neutral country is subject to detention by that country for the duration of the war.

(2) FEBA/FLOT.

(a) Static: Recovery along a static FEBA is always difficult. Under these conditions, enemy and friendly forces can be expected to be densely arrayed and well camouflaged, with good fields of fire. Attempts to penetrate the FEBA should be avoided.

(b) Advancing: Individuals isolated in front of advancing friendly units should immediately take cover and wait for the friendly units to overrun their position.

(c) Retreating: Individuals between opposing forces should immediately take cover and wait for enemy units to pass over their position. After most enemy units have moved on, evaders should try to link up with other isolated friendly elements and return to friendly forces.

(3) Link-up with friendly patrols. Unit authentication numbers and/or locally developed codes may assist the evader to safely make contact in or around the FEBA and when approached by friendly forces.

UNITED STATES MARINE CORPS
Mountain Warfare Training Center
Bridgeport, California 93517-5001

FMST.07.21
FMST.07.04
04/07/02

STUDENT OUTLINE

EXPEDIENT SHELTERS AND FIRES

TERMINAL LEARNING OBJECTIVES

(1) In a summer mountainous environment, construct expedient shelters, in accordance with the references. (FMST.07.21)

(2) In a summer mountainous environment, construct fires, in accordance with the references. (FMST.07.04)

b. **ENABLING LEARNING OBJECTIVES**

(1) Without the aid of references and from a given list, choose the four characteristics of a safe expedient shelter, in accordance with the references. (FMST.07.21a)

(2) Without the aid of references and from a given list, choose the four hazards avoid when using natural shelters, in accordance with the references. (FMST.07.21b)

(3) Without the aid of references and from a given list, choose one of man-made expedient shelters, in accordance with the references. (FMST.07.21c)

(4) Without the aid of references, start a fire using a primitive method, in accordance with the references. (FMST.07.04a)

A4

OUTLINE

1. **BASIC CHARACTERISTICS FOR SHELTER**. (FMST.07.21a) Any type of shelter, whether it is a permanent building, tentage, or an expedient shelter must meet six basic characteristics to be safe and effective. The characteristics are:

   a. Protection from the Elements: The shelter must provide protection from rain, snow, wind, sun, etc.

   b. Heat Retention: It must have some type of insulation to retain heat; thus preventing the waste of fuel.

   c. Ventilation: Ventilation must be constructed, especially if burning fuel for heat. This prevents the accumulation of carbon monoxide. Ventilation is also needed for carbon dioxide given off when breathing.

   d. Drying Facility: A drying facility must be constructed to dry wet clothes.

   e. Free from Natural Hazards: Shelters should not be built in areas of avalanche hazards, under rock fall or "standing dead" trees have the potential to fall on your shelter.

   f. Stable: Shelters must be constructed to withstand the pressures exerted by severe weather.

2. **NATURAL SHELTERS**. Natural shelters are usually the preferred types because they take less time and materials construct. The following may be made into natural shelters with some modification.

   a. Caves or Rock Overhangs: Can be modified by laying walls of rocks, logs or branches across the open sides.

   b. Hollow Logs: Can be cleaned or dug out, then enhanced with ponchos, tarps or parachutes hung across the openings.

   c. Hazards of Natural Shelters: (FMST.04.21b)

      (1) Animals: Natural shelters may already be inhabited (i.e. bears, coyotes, lions, rats, snakes, etc.). Other concerns from animals may be disease from scat or decaying carcasses.

      (2) Lack of Ventilation: Natural shelters may not have adequate ventilation. Fires may be built inside for heating or cooking but may be uncomfortable or even dangerous because of the smoke build up.

      (3) Gas Pockets: Many caves in a mountainous region may have natural gas pockets in them.

(4) Instability: Natural shelters may appear stable, but in reality may be a trap waiting to collapse.

3. **MAN-MADE SHELTERS**. (FMST.04.21c) Many configurations of man-made shelters may be used. Over-looked man-made structures found in urban or rural environments may also provide shelter (i.e. houses, sheds, or barns). Limited by imagination and materials available, the following man-made shelters can be used in any situation.

a. Poncho Shelter.

b. Sapling Shelter.

c. Lean-to.

d. Double lean-to.

e. A-frame Shelter.

f. Fallen Tree Bivouac.

4. **CONSTRUCTION OF MAN-MADE SHELTERS**. To maximize the shelter's effectiveness, Marines should take into consideration the following prior to construction.

a. Considerations.

(1) Group size.

(2) Low silhouette and reduced living area dimensions for improved heat conservation.

(3) Avoid exposed hill tops, valley floors, moist ground, and avalanche paths.

(4) Create a thermal shelter by applying snow, if available, to roof and sides of shelter.

(5) Location of site to fire wood, water, and signaling, if necessary.

(6) How much time and effort needed to build the shelter.

(7) Can the shelter adequately protect you from the elements (sun, wind, rain, and snow). Plan on worst case scenario.

(8) Are the tools available to build it. If not, can you make improvised tools?

(9) Type and amount of materials available to build it.

(10) When in a tactical environment, you must consider the following:

(a) Provide concealment from enemy observation.

(b) Maintain camouflaged escape routes.

(c) Use the acronym BLISS as a guide.

B – Blend in with the surroundings.
L –Low silhouette.
I – Irregular shape.
S – Small.
S – Secluded located.

b. Poncho Shelter: This is one of the easiest shelters to construct. Materials needed for construction are cord and any water-repellent material (i.e. poncho, parachute, and tarp). It should be one of the first types of shelter considered if planning a short stay in any one place.

(1) Find the center of the water-repellent material by folding it in half along its long axis.

(2) Suspend the center points of the two ends using cordage.

(3) Stake the Four Corners down, with sticks or rocks.

c. Sapling Shelter: This type of shelter is constructed in an area where an abundance of saplings are growing. It is an excellent evasion shelter.

(1) Find or clear an area so that you have two parallel rows of saplings at least 4' long and approximately 1 ½' to 2' apart.

(2) Bend the saplings together and tie them to form several hoops, which will form the framework of the shelter.

(3) Cover the hoop with a water-repellent covering.

(4) The shelter then may be insulated with leaves, brush, snow, or boughs.

(5) Close one end permanently. Hang material over the other end to form a door.

d. Lean-to: A lean-to is built in heavily forested areas. It requires a limited amount of cordage to construct. The lean-to is an effective shelter but does not offer a great degree of protection from the elements.

(1) Select a site with two trees (4-12" in diameter) spaced far enough apart that a man can lie down between them. Two sturdy poles can be substituted by inserting them into the ground the proper distance apart.

(2) Cut a pole to support the roof. It should be at least 3-4" in diameter and long enough to extend 4-6" past both trees. Tie the pole horizontally between the two trees, approximately 1 meter off the deck.

(3) Cut several long poles to be used as stringers. They are placed along the horizontal support bar approximately every 1 ½' and lay on the ground. All stringers may be tied to or laid on the horizontal support bar. A short wall or rocks or logs may be constructed on the ground to lift the stringers off the ground, creating additional height and living room dimensions.

(4) Cut several saplings and weave them horizontally between the stringers. Cover the roof with water-repellent and insulating material.

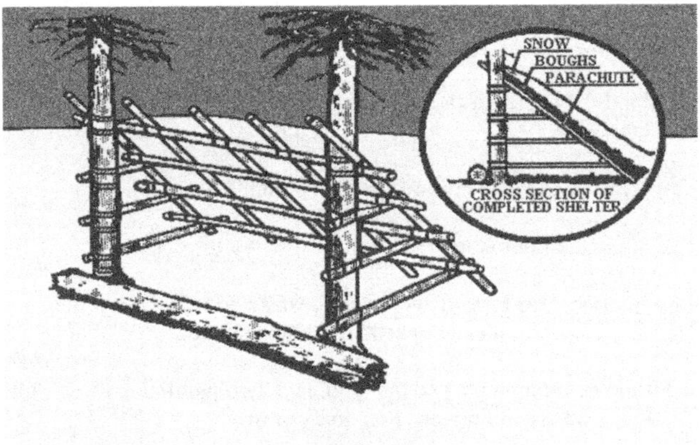

LEAN-TO

e. Double lean-to: The double lean-to shelter is constructed for 2-5 individuals. It is constructed by making two lean-tos and placing them together.

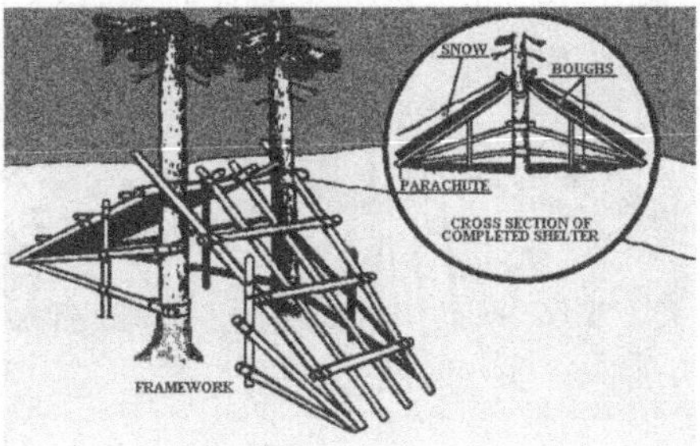

DOUBLE LEAN-TO

f. A-Frame Shelter: An A-Frame shelter is constructed for 1-3 individuals. After the framework is constructed, bough/tentage is interwoven onto the frame and snow, if available, is packed onto the outside for insulation.

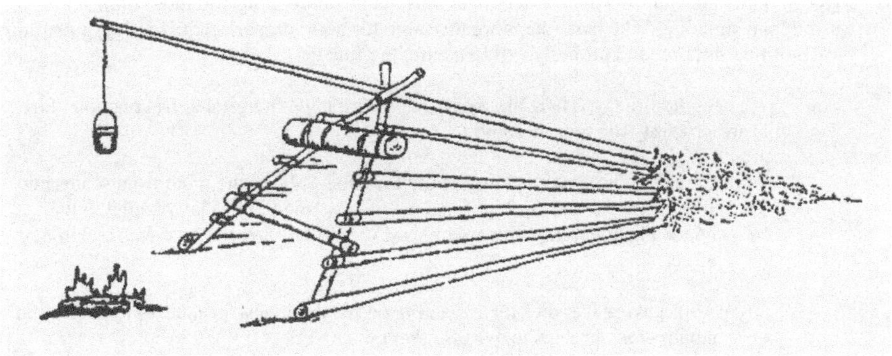

A-FRAME

g. Fallen Tree Bivouac: The fallen tree bivouac is an excellent shelter because most of the work has already been done.

(1) Ensure the tree is stable prior to constructing.

(2) Branches on the underside are cut away making a hollow underneath.

(3) Place additional insulating material to the top and sides of the tree.

(4) A small fire is built outside of the shelter.

FALLEN TREE BIVOUAC

5. **REFLECTOR WALLS.** Heating a shelter requires a slow fire that produces lots of steady heat over a long period of time. A reflector wall should be constructed for all open-ended shelters. A reflector wall is constructed with a flat rock or a stack of green logs propped behind the fire. A surprising amount of heat will bounce back from the fire into the shelter.

6. **FIRES**. Fires fall into two main categories: those built for cooking and those built for warmth and signaling. The basic steps are the same for both: preparing the fire lay, gathering fuel, building the fire, and properly extinguishing the fire.

   a. Preparing the fire lay: There are two types of fire lays: fire pit and Dakota hole. Fire pits are probably the most common.

      (1) Create a windbreak to confine the heat and prevent the wind from scattering sparks. Place rocks or logs used in constructing the fire lay parallel to the wind. The prevailing downwind end should be narrower to create a chimney effect.

      (2) Avoid using wet rocks. Heat acting on the dampness in sandstone, shale, and stones from streams may cause them to explode.

      (3) Dakota Hole: The Dakota Hole is a tactical fire lay. Although no fire is 100% tactical, this fire lay will accomplish certain things:

         (a) Reduces the signature of the fire by placing it below ground.

         (b) Provides more of a concentrated heat source to boil and cook, thus preserving fuel and lessening the amount of burning time.

         (c) By creating a large air draft, the fire will burn with less smoke than the fire pit.

         (d) It is easier to light in high winds.

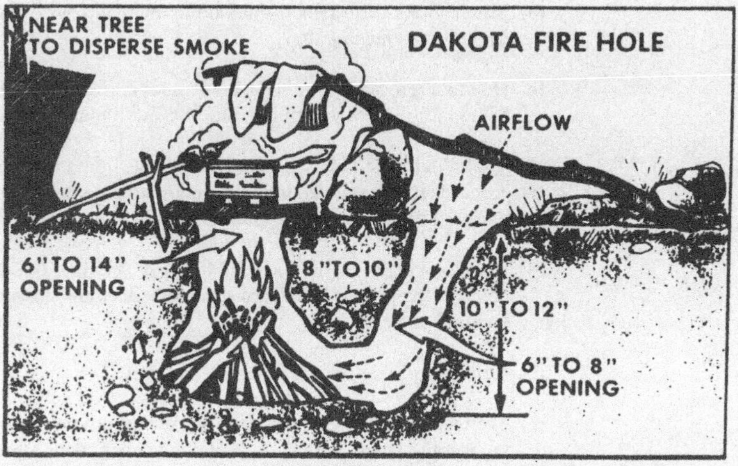

DAKOTA HOLE

b. Gather Fuel: Many Marines take shortcuts when gathering firewood. Taking a few extra minutes can mean the difference between ease and frustration when building a fire.

(1) Tinder: Tinder is the initial fuel. It should be fine and dry. Gather a double handful of tinder for the fire to be built and an extra double handful to be stored in a dry place for the following morning. Dew can moisten tinder enough to make lighting the fire difficult. Some examples are:

(a) Shredded cedar/juniper bark, pine needles.

(b) Dry grass.

(c) Slivers shaved from a dry stick.

(d) Hornet's nest.

(e) Natural fibers from equipment supplemented with pine pitch (i.e., cotton battle dressing).

(f) Cotton balls and petroleum jelly or Char-cloth.

NOTE: Sticks used for tinder should be dry and not larger than the diameter of a toothpick.

(2) Kindling: This is the material that is ignited by the tinder that will burn long enough to ignite the fuel.

(a) Small sticks/twigs pencil-thick up to the thickness of the thumb. Ensure that they are dry.

(b) Due to a typically large resin content, evergreen limbs often make the best kindling. They burn hot and fast, but typically do not last long.

(3) Fuel Wood: Fuel Wood is used to keep the blaze going long enough to fulfill its purpose. Ideally, it should burn slow enough to conserve the woodpile, make plenty of heat, and leave an ample supply of long-lasting coals.

(a) Firewood broken from the dead limbs of standing trees or windfalls held off the ground will have absorbed less moisture and therefore should burn easily.

(b) Refrain from cutting down live, green trees.

(c) Softwoods (evergreens and conifers) will burn hot and fast with lots of smoke and spark, leaving little in the way of coals. Hardwoods (broad leaf trees) will burn slower with less smoke and leave a good bed of coals.

(d) Learn the woods indigenous to the area. Birch, dogwood, and maple are excellent fuels. Osage orange, ironwood, and manzanita, though difficult to break up, make terrific coals. Aspen and cottonwood burn clean but leave little coals.

(e) Stack your wood supply close enough to be handy, but far enough from the flames to be safe. Protect your supply from additional precipitation.

(f) If you happen to go down in an aircraft that has not burned a mixture of gas and oil may be used. Use caution when igniting this mixture.

c. Building the Fire. The type of fire built will be dependent upon its intended use, either cooking or heating and signaling.

(1) Cooking Fires. The following listed fires are best used for cooking:

(a) Teepee Fire: The teepee fire is used to produce a concentrated heat source, primarily for cooking. Once a good supply of coals can be seen, collapse the teepee and push embers into a compact bed.

TEEPEE FIRE

(2) Heating and Signaling Fires.

(a) Pyramid Fire: Pyramid fires are used to produce large amounts of light and heat. They will dry out wet wood or clothing.

PYRAMID FIRE

d. Starting Fires. Starting a fire is done by a source of ignition and falls into two categories, modern igniters and primitive methods.

(1) Modern Methods: Modern igniters use modern devices we normally think of to start a fire. Reliance upon these methods may result in failure during a survival situation. These items may fail when required to serve their purpose.

(a) Matches and Lighters. Ensure you waterproof these items.

(b) Convex Lens. Binocular, camera, telescopic sights, or magnifying lens are used on bright, sunny days to ignite tender.

(c) Flint and Steel: Sometimes known as metal matches or "Mag Block". Scrape your knife or carbon steel against the flint to produce a spark onto the tinder. Some types of flint & steel designs will have a block of magnesium attached to the device, which can be shaved onto the tinder prior to igniting. Other designs may have magnesium mixed into the flint to produce a higher quality of spark.

(2) Primitive Methods. Primitive fire methods are those developed by early man. There are numerous techniques that fall into this category. The only method that will be taught at MCMWTC is the Bow & Drill.

(3) Bow & Drill. The technique of starting a fire with a bow & drill is a true field expedient fire starting method, which requires a piece of cord and knife from your survival kit to construct. The components of the bow & drill are bow, drill, socket, and fireboard, ember patch, and bird's nest.

(a) Bow: The bow is a resilient, green stick about 3/4 of an inch in diameter and 30-36 inches in length. The bowstring can be any type of cord, however, 550 cord works best. Tie the string from one end of the bow to the other, without any slack.

(b) Drill: The drill should be a straight, seasoned hardwood stick about 1/2 to 3/4 of an inch in diameter and 8 to 12 inches in length. The top end is tapered to a blunt point to reduce friction generated in the socket. The bottom end is slightly rounded to fit snugly into the depression on the fireboard.

(c) Socket: The socket is an easily grasped stone or piece of hardwood or bone with a slight depression on one side. Use it to hold the drill in place and to apply downward pressure.

(d) Fire board: The fire board is a seasoned softwood board which should ideally be 3/4 of an inch thick, 2-4 inches wide, and 8-10 inches long. Cut a depression 3/4 of an inch from the edge on one side of the fireboard. Cut a U-shape notch from the edge of the fireboard into the depression. This notch is designed to collect and form an ember, which will be used to ignite the tinder.

(e) Ember Patch: The ember patch is made from any type of suitable material (i.e., leather, aluminum foil, and bark). It is used to catch and transfer the ember from the fireboard to the bird's nest. Ideally, it should be 4 inches by 4 inches in size.

(f) Bird's Nest: The bird's nest is a double handful of tinder, which will be made into the shape of a nest. Tinder must be dry and finely shredded material (i.e., outer bark from juniper/cedar/sage brush or inner bark from cottonwood/aspen or dry grass/moss). Lay your tinder out in two equal rows about 4 inches wide and 8-12 inches long. Loosely roll the first row into a ball and knead the tinder to further break down the fibers. Place this ball perpendicular onto the second row of tinder and wrap. Knead the tinder until all fibers of the ball are interwoven. Insert the drill half way into the ball to form a partial cylinder. This is where the ember will be placed.

(4) Producing a fire using the bow & drill.

(a) Place the ember patch under the U-shaped notch.

(b) Assume the kneeling position, with the left foot on the fireboard near the depression.

(c) Load the bow with the drill. Ensure the drill is between the wood of the bow and bow string. Place the drill into the depression on the fireboard. Place the socket on the tapered end of the drill.

(d) Use the left hand to hold the socket while applying downward pressure.

(e) Use the right hand to grasp the bow. With a smooth sawing motion, move the bow back and forth to twirl the drill.

(f) Once you have established a smooth motion, smoke will appear. Once smoke appears, apply more downward pressure and saw the bow faster.

(g) When a thick layer of smoke has accumulated around the depression, stop all movement. Remove the bow, drill, and socket from the fire board, without moving the fireboard. Carefully remove your left foot off the fireboard.

(h) Gently tap the fireboard to ensure all of the ember has fallen out of the U-shaped notch and is lying on the ember patch. Remove the fireboard.

(i) Slowly fan the black powder to solidify it into a glowing ember. Grasping the ember patch, carefully drop the ember into the cylinder of the bird's nest.

(j) Grasp the bird's nest with the cylinder facing towards you and parallel to the ground. Gently blow air into the cylinder. As smoke from the nest becomes thicker, continue to blow air into the cylinder until fire appears.

(5) Trouble Shooting the Bow & Drill

(a) Drill will not stay in depression- Apply more downward pressure and/or increase width/depth of depression.

(b) Drill will not twirl- lessens the amount of downward pressure and/or tighten bowstring.

(c) Socket smoking- Lessen the amount of downward pressure. Wood too soft when compared to hardness of drill. Add some lubrication: animal fat, oil, or grease.

(d) No smoke- Drill and fireboard are the same wood. Wood may not be seasoned. Check drill to ensure that it is straight. Keep left hand locked against left shin while sawing.

(e) Smoke but no ember- U-shaped notch not cut into center of the depression.

(f) Bowstring runs up and down drill- use a locked right arm when sawing. Check drill to ensure that it is straight. Ensure bowstring runs over the top of the left boot.

(g) Birds nest will not ignite- Tinder not dry. Nest woven too tight. Tinder not kneaded enough. Blowing too hard (ember will fracture).

e. Extinguishing the Fire. The fire must be properly extinguished. This is accomplished by using the drown, stir, and feel method.

(1) Drown the fire by pouring at water in the fire lay.

(2) Stir the ember bed to ensure that the fire is completely out.

(3) Check the bed of your fire by feeling for any hot spots.

(4) If any hot spots are found, start the process all over again.

UNITED STATES MARINE CORPS
Mountain Warfare Training Center
Bridgeport, California 93517-5001

FMST.07.22
04/07/02

STUDENT OUTLINE

SURVIVAL NAVIGATION

LEARNING OBJECTIVES

TERMINAL LEARNING OBJECTIVE. In a summer mountainous environment, navigate in a survival situation, in accordance with the references. (FMST.07.22)

ENABLING LEARNING OBJECTIVES

(1)Without the aid of references and from a given list, choose one of the considerations for travel, in accordance with the references. (FMST.07.22a)

(2)Without the aid of references and from a given list, choose two methods for locating the North Star, in accordance with the references. (FMST.07.22b)

OUTLINE

1. **CONSIDERATIONS FOR STAYING OR TRAVELLING**. (FMST.07.22a)

   a. Stay with the aircraft or vehicle if possible. More than likely somebody knows where it was going. It is also a ready-made shelter.

   b. Leave only when:

      (1) Certain of present location; have known destination and the ability to get there.

      (2) Water, food, shelter, and/or help can be reached.

      (3) Convinced that rescue is not coming.

c. If the decision is to travel, the following must also be considered:

(1) Which direction to travel and why.

(2) What plan is to be followed.

(3) What equipment should be taken

(4) How to mark the trail.

(5) Predicted weather.

d. If the tactical situation permits leave the following information at the departure point:

(1) Departure time.

(2) Destination.

(3) Route of travel/direction.

(4) Personal condition.

(5) Available supplies.

2. DAYTIME SURVIVAL NAVIGATION

a. Sun Movement. It is generally taken for granted that the sun rises in the east and sets in the west. This rule of thumb, however, is quite misleading. In fact, depending on an observer's latitude and the season, the sun could rise and set up to 50 degrees off of true east and west.

b. The following diagram and terms are essential to understanding how the sun and stars can help to determine direction:

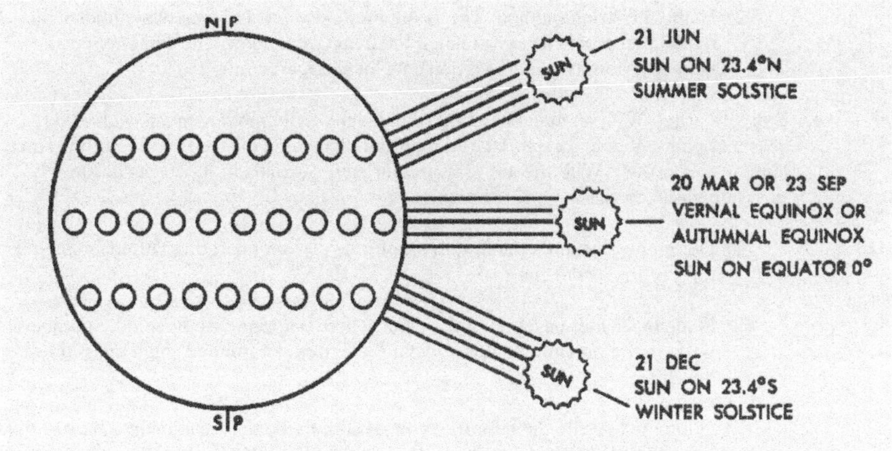

Position of the Sun at Equinox and Solstice

(1) Summer/Winter Solstice: (21 June/21 December) Two times during the year when the sun has no apparent northward or southward motion.

(2) Vernal/Autumnal Equinox: (20 March/23 September) Two times during the year when the sun crosses the celestial equator and the length of day and night are approximately equal.

c. Sun's Movement. As reflected in the diagram above, the sun continuously moves in a cycle from solstice to equinox, throughout each day, however, the sun travels a uniform arc in the sky from sunrise to sunset. Exactly halfway along its daily journey, the sun will be directly south of an observer (or north if the observer is in the Southern Hemisphere). This rule may not apply to observers in the tropics (between 23.5 degrees north and south latitude) or in the Polar Regions (60 degrees latitude). It is at this point that shadows will appear their shortest. The time at which this occurs is referred to as "local apparent noon."

d. Local Apparent Noon. Whenever using any type of shadow casting device to determine direction, "local apparent noon" (or the sun's highest point during the day) must be known. Local apparent noon can be determined by the following methods.

(1) Knowing sunrise and sunset from mission orders, i.e., sunrise 0630 and sunset 1930. Take the total amount of daylight (13 hours), divide by 2 (6 hours 30 minutes), and add to sunrise (0630 plus 6 hours 30 minutes). Based on this example, local apparent noon would be 1300.

(2) Using the string method. The string method is used to find two equidistant marks before and after estimated local apparent noon. The center point between these two marks represents local apparent noon.

e. <u>Sun's Bearing</u>. With an understanding of the sun's daily movement, as well as its seasonal paths, a technique is derived that will determine the true bearing of the sun at sunrise and sunset. With the aid of a circular navigational chart, we can accurately navigate based on the sun's true bearing:

(1) Determine the sun's maximum amplitude at your operating latitude using the top portion of the chart.

(2) Scale the center baseline of the chart where 0 appears as the middle number; write in the maximum amplitude at the extreme north and south ends of the baseline.

(3) Continue to scale the baseline; you should divide the baseline into 6 to 10 tick marks that represent equal divisions of the maximum amplitude.

(4) From today's date along the circumference, draw a straight line down until it intersects the baseline.

(5) The number this line intersects is today's solar amplitude. If the number is left of 0, it is"north" amplitude; if the number is right of 0, it is a "south" amplitude. Use the formula at the bottom of the chart to determine the sun's bearing at sunrise or sunset.

LATITUDE (N OR S)	5	10	15	20	25	30	35	40	45	50	55	60
MAXIMUM AMPLITUDE	24°	24°	24°	25°	26°	27°	29°	31°	34°	38°	44°	53°

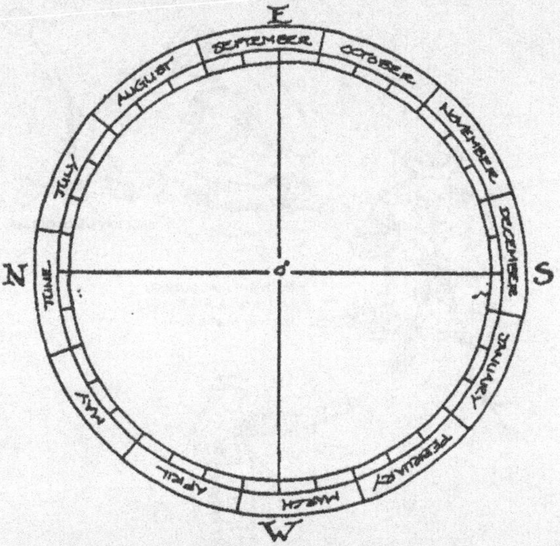

NORTH AMPLITUDES	SOUTH AMPLITUDES
SUNRISE BEARING = 90° - AMPLITUDE	SUNRISE BEARING = 90° + AMPLITUDE
SUNSET BEARING = 270° + AMPLITUDE	SUNSET BEARING = 270° - AMPLITUDE

Circular Navigational Chart

f. Shadow Stick Construction. This technique will achieve a cardinal direction within 10 degrees of accuracy if done within two hours of local apparent noon. Once again, this technique may be impractical near the Polar Regions, as shadows tend to be very long; similarly, in the tropics shadows are generally very small.

(1) Get a straight, 3-6 foot stick free of branches and pointed at the ends and 3-5 small markers: i.e., sticks, rocks, or nails.

(2) Place the stick upright in the ground and mark the shadow tip with a marker.

(3) Wait 10-15 minutes and mark shadow tip again with a marker.

(4) Repeat this until all of the markers are used.

SHADOW STICK METHOD

(5) The markers will form a West—East line.

(6) Put your left foot on the first marker and your right foot on the last marker, you will then be facing north.

3. **POCKET NAVIGATOR** The only material required is a small piece of paper or other flat-surface material upon which to draw the trace of shadow tips and a 1 to 2 inch pin, nail, twig, wooden matchstick, or other such device to serve as a shadow-casting rod.

a. Set this tiny rod upright on your flat piece of material so that the sun will cause it to cast a shadow. Mark the position where the base of the rod sits so it can be returned to the same spot for later readings. Secure the material so that it will not move and mark the position of the material with string, pebbles, or twigs, so that if you have to move the paper you can put it back exactly as it was. Now, mark the tip of the rod shadow.

b. As the sun moves, the shadow-tip moves. Make repeated shadow-tip markings every 15 minutes. As you make the marks of the shadow tip, ensure that you write down the times of the points.

c. At the end of the day, connect the shadow-tip markings. The result will normally be a curved line. The closer to the vernal or autumnal equinoxes (March 21 and September 23) the less pronounced the curvature would be. If it is not convenient or the tactical situation does not permit to take a full day's shadow-tip markings, your observation can be continued on subsequent days by orienting the pocket navigator on the ground so that the shadow-tip is aligned with a previously plotted point.

d. The markings made at the sun's highest point during the day, or solar noon, is the north—south line. The direction of north should be indicated with an arrow on the navigator as soon as it is determined. This north-south line is drawn from the base of the rod to the mark made at solar noon. This line is the shortest line that can be drawn from the base of the pin to the shadow-tip curve.

e. To use your pocket navigator, hold it so that with the shadow-tip is aligned with a plotted point at the specified point. I.e.; if it is now 0900 the shadow-tip must be aligned with that point. This will ensure that your pocket navigator is level. The drawn arrow is now oriented to true north, from which you can orient yourself to any desired direction of travel.

f. The pocket navigator will work all day and will not be out of date for approximately one week.

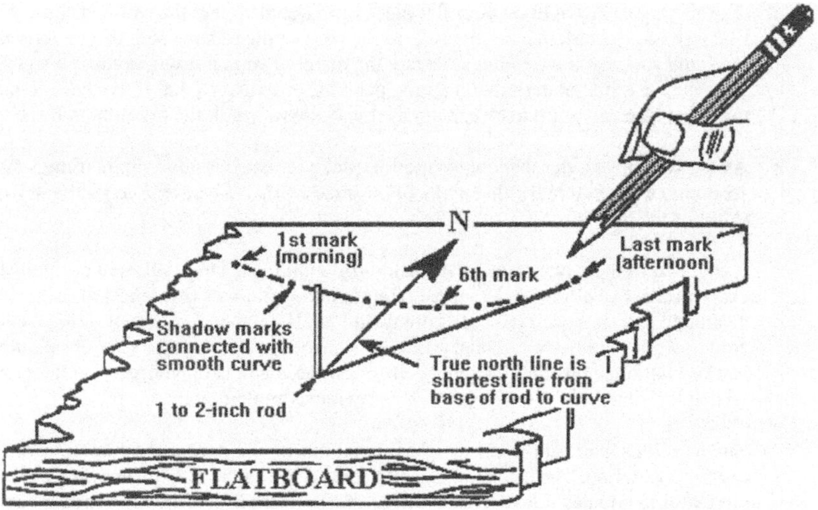

POCKET NAVIGATOR

4. NIGHTTIME SURVIVAL NAVIGATION

a. Mark North. To aid you in navigating at night, it is beneficial to watch where the sun goes down. If you're going to start moving after dark mark the northerly direction.

b. Locating the North Star. There are two methods used in locating the North Star. (FMST.07.22b)

(1) Using the Big Dipper (*Ursa Major*). The best indictors are the two "dippers ". The North Star is the last star in the handle of the Little Dipper, which is not the easiest constellation to find. However, the Big Dipper is one of the most prominent constellations in the Northern Hemisphere. The two lowest stars of the Big Dipper's cup act as pointers to the North Star. If you line up these two stars, they make a straight line that runs directly to the North Star. The distance to the North Star along this line is 5 times that between the two pointer stars.

(2) Using Cassiopeia (*Big M or W*). Draw a line straight out from the center star, approximately half the distance to the Big Dipper. The North Star will be located there.

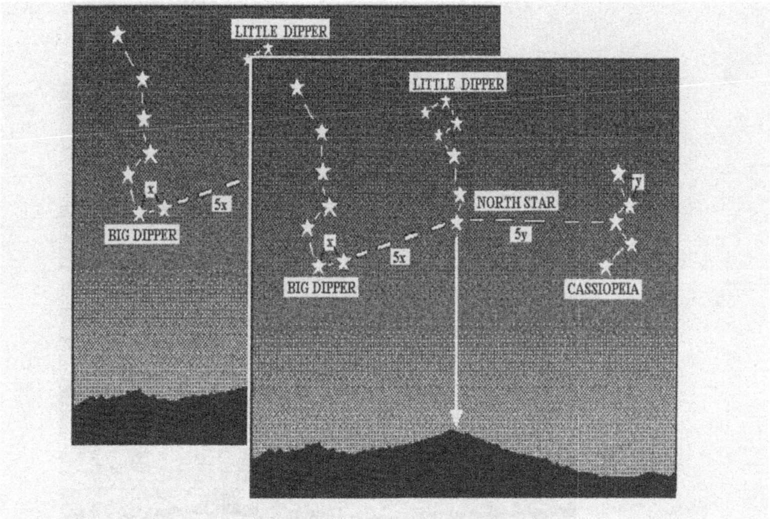

LOCATING THE NORTH STAR

NOTE: Because the Big Dipper and Cassiopeia rotate around the North Star, they will not always appear in the same position in the sky. In the higher latitudes, the North Star is less effective for the purpose of orienting because it appears higher in the sky. At the center of the Arctic Circle, it would be directly overhead and all directions lead south.

c. Southern Cross. In the Southern Hemisphere, Polaris is not visible. There, the Southern Cross is the most distinctive constellation. An imaginary line through the long axis of the Southern Cross, or True Cross, points towards a dark spot devoid of stars approximately three degrees offset from the South Pole. The True Cross should not be confused with the larger cross nearby know as the False Cross, which is less bright, more widely spaced, and has five stars. Two closely spaced, very bright stars that trail behind the crosspiece can confirm the True Cross. These two stars are often easier to pick out than the cross itself. Look for them. Two of the stars in the True Cross are among the brightest stars in the heavens; they are the stars on the southern and eastern arms. The stars on the northern and western arms are not as conspicuous, but are bright.

NOTE: The imaginary point depicted in the picture is the dark spot devoid of stars.

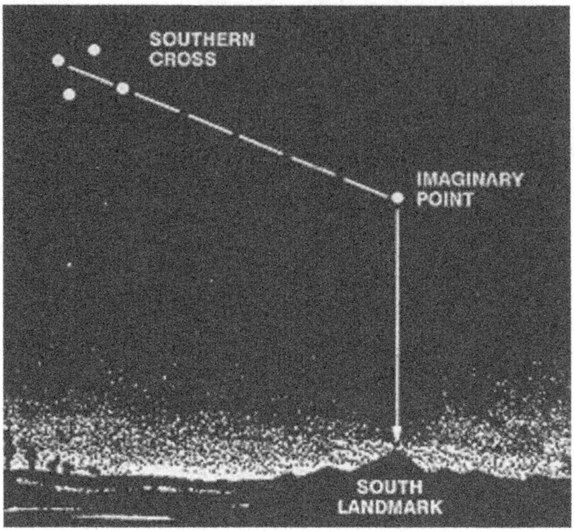

SOUTHERN CROSS

d. Moon Navigator. Like the sun, the moon rises in the east and sets in the west. Use the same method of the shadow stick as you did during the day.

5. **IMPROVISED COMPASSES**. There are three improvised techniques to construct a compass.

a. Synthetic technique. The required items are a piece of synthetic material, (i.e., parachute cloth), and a small piece of iron or steel that is long, thin, and light. Aluminum or yellow metals won't work (only things that rust will do). A pin or needle is perfect, but a straightened paper clip, piece of steel baling wire or barbed wire could also work.

(1) Stroke the needle repeatedly in one direction against the synthetic material. Ensure that you lift the material a few inches up into the air at the end of each stroke, returning to the beginning of the needle before descending for another stroke in the same direction. Do this approximately 30 strokes. This will magnetize the needle.

(2) Float the metal on still water using balled up paper, wood chip, or leaf. Gather some water in a non-magnetic container or a scooped out recess in the

ground, such as a puddle. Do not use a "tin can" which is made of steel. (Aluminum can would be fine.) Place the float on the water, then the metal on it. It will slowly turn to orient itself.

b. Magnet technique. You will achieve the same results by using a magnet. Follow the same steps as you did with the synthetic material. The magnets you are most likely to have available to you are those in a speaker or headphones of a radio.

c. Magnetization through a battery. A power source of 2 volts or more from a battery can be used with a short length of insulated wire to magnetize metal. Coil the wire around a needle. If the wire is non-insulated, wrap the needle with paper or cardboard. Attach the ends to the battery terminals for 5 minutes.

d. Associated problems with improvised compasses. The following are common problems with all improvised compasses.

(1) Soft steel tends to lose its magnetism fairly quickly, so you will have to demagnetize your needle occasionally, though you should not have to do this more than two or three times a day.

(2) Test your compass by disturbing it after it settles. Do this several times. If it returns to the same alignment, you're OK. It will be lined up north and south, though you will have to determine by other means which end is north. Use the sun, stars, or any other natural signs in the area.

(3) Remember that this will give magnetic north. In extreme northern latitudes, the declination angle can be extreme.

6. SURVIVAL NAVIGATION TECHNIQUES

a. Navigator.

(1) Employ a navigation method.

(2) Find the cardinal direction.

(3) Pick a steering mark in the desired direction of travel.

b. Maintain a Log: The possibility may arise when you will not have a map of the area. A log will decrease the chance of walking in circles.

(1) Construction.

(a) Use any material available to you i.e., paper, clothing, MRE box, etc.

(b) Draw a field sketch annotating North, prominent terrain features, and Friendly/enemy position.

(2) Maintenance.

(a) Annotate distance traveled, elevation gained and lost, and cardinal directions.

(b) Maintain and update field sketch as movement progresses.

(c) Ensure readability of your field sketch. (I.e., don't clutter the sketch so much that it can't be read.)

c. <u>During Movement Constantly Refer To</u>.

(1) Log.

(2) Steering marks.

d. <u>Actions If You Become Lost</u>.

(1) Immediate action

(a) Orient your sketch. This will probably make your mistake obvious.

(2) Corrective action

(a) Backtrack using steering marks until you have determined the location of your error.

(b) Re-orient your sketch.

(c) Select direction of travel and continue to march.

UNITED STATES MARINE CORPS
Mountain Warfare Training Center
Bridgeport, California 93517-5001

FMST.07.08
04/07/02

STUDENT OUTLINE

WATER PROCUREMENT

LEARNING OBJECTIVES

TERMINAL LEARNING OBJECTIVE Given a unit in a cold weather environment and the necessary equipment and supplies, perform cold weather preventive medicine, in accordance with the references. (FMST.07.08)

ENABLING LEARNING OBJECTIVES

(1) Without the aid of references from a given list, choose one of the techniques for finding incidental water, in accordance with the references. (FMST.07.08i)

(2) Without the aid of references from a given list, choose one of the four hazardous fluids to avoid substituting for potable water, in accordance with the references. (FMST.07.08j)

(3) Without the aid of references, construct a water-producing device, in accordance with the references. (FMST.07.08k)

OUTLINE

1. WATER INTAKE

a. Thirst is not a strong enough sensation to determine how much water you need.

b. The best plan is to drink, utilizing the OVER DRINK method. Drink plenty of water anytime it is available and particularly when eating.

A6

c. Dehydration is a major threat. A loss of only 5 % of your body fluids causes thirst, irritability, nausea, and weakness; a 10% loss causes dizziness, headache, inability to walk, and a tingling sensation in limbs; a 15% loss causes dim vision, painful urination, swollen tongue, deafness, and a feeling of numbness in the skin; also a loss of more than 15% body fluids could result in death.

d. Your water requirements will be increased if:

(1) You have a fever.

(2) You are experiencing fear or anxiety.

(3) You evaporate more body fluid than necessary. (i.e., not using the proper shelter to your advantage)

(4) You have improper clothing.

(5) You ration water.

(6) You overwork.

2. **INCIDENTAL WATER**. (FMST.07.08.i)

a. During movement, you may have to ration water until you reach a reliable water source. Incidental water may sometimes provide opportunities to acquire water. Although not a reliable or replenished source, it may serve to stretch your water supply or keep you going in an emergency. The following are sources for incidental water:

(1) Dew: In areas with moderate to heavy dew, tying rags or tuffs of fine grass around your ankles can collect dew. While walking through dewy grass before sunrise, the rags or grass will saturate and can be rung out into a container. The rags or grass can be replaced and the process is repeated.

(2) Rainfall: Rainwater collected directly in clean container or in plants that contain no harmful toxins is generally safe to drink without disinfecting. The survivor should always be prepared to collect rainfall at a moment's notice. An inverted poncho works well to collect rainfall.

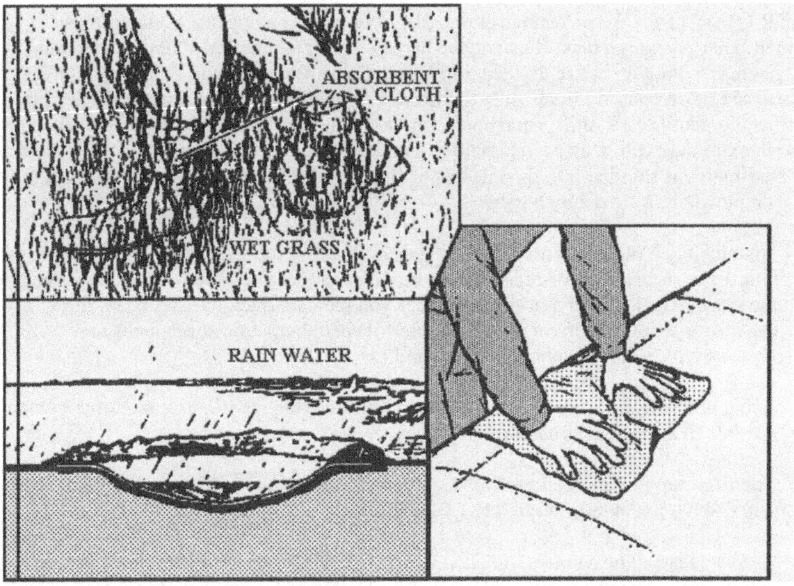

3. **HAZARDOUS FLUIDS** (FMST.07.08j)

a. Survivors have occasionally attempted to augment their water supply with other fluids, such as alcoholic beverages, urine, blood, or seawater. While it is true that each of these fluids has a high water content, the impurities they contain may require the body to expend more fluid to purify them. Some hazardous fluids are:

(1) Seawater: Seawater in more than minimal quantities are actually toxic. The concentration of sodium and magnesium salts is so high that fluid must be drawn from the body to eliminate the salts and eventually the kidneys cease to function.

(2) Alcohol: Alcohol dehydrates the body and clouds judgment. Super-cooled liquid, if ingested, can cause immediate frostbite of the throat, and potential death.

(3) Blood: Blood, besides being salty, is a food. Drinking it will require the body to expend additional fluid to digest it.

(4) Urine: Drinking urine is not only foolish, but also dangerous. Urine is nothing more than the body's waste. Drinking it only places this waste back into the body, which requires more fluid to process it again.

4. **WATER QUALITY**. Water contains minerals, toxins, and pathogens. Some of these, consumed in large enough quantities may be harmful to human health. Pathogens are our primary concern. Pathogens are divided into Virus, Cysts, Bacteria, and Parasites. Certain pathogens are more resistant to chemicals and small enough to move through microscopic holes in equipment (i.e., T-shirt, parachute). Certain pathogens also have the ability to survive in extremely cold water temperatures. Pathogens generally do not live in snow and ice. Water quality is divided into three levels of safety with disinfecting as the most desired level, then purified, followed by potable.

a. Disinfecting. Water disinfecting removes or destroys harmful microorganisms. Giardia cysts are an ever-present danger in clear appearing mountain water throughout the world. By drinking non-potable water you may contract diseases or swallow organisms that could harm you. Examples of such diseases or organisms are Dysentery, Cholera, Typhoid, Flukes, and Leeches.

b. Remember impure water, no matter how overpowering the thirst, is one of the worst hazards in a survival situation.

c. The first step in disinfecting is to select a treatment method. The two methods we will discuss are as follows: (FMST.07.08j)

(1) Heat. The Manual of Naval Preventive Medicine (P-5010) states that you must bring the water to a rolling boil before it is considered safe for human consumption. This is the most preferred method.

(a) Bringing water to the boiling point will kill 99.9% of all Giardia cysts. The Giardia cyst dies at 60°C and Cryptosporidium dies at 65C. Water will boil at 14,000' at 86°C and at 10,000' at 90C. With this in mind you should note that altitude does not make a difference unless you are extremely high.

(2) Chemicals. There are numerous types of chemicals that can disinfect water. Below are a few of the most common. In a survival situation, you will use whatever you have available.

(a) Iodine Tablets.

(b) Chlorine Bleach.

(c) Iodine Solution.

(d) Betadine Solution.

(e) Military water purification tablets. These tablets are standard issue for all DOD agencies. These tablets have a shelf life of four years from the date of manufacture, unless opened. Once the seal is broken, they have a shelf life of one year, not to exceed the initial expiration date of four years.

49703 Month / Year / Batch Number

(3) Water Disinfecting Techniques and Halogen Doses.

Iodination techniques Added to 1 liter or quart of water	Amount for 4 PPM	Amount for 8 PPM
Iodine tablets Tetraglcine hydroperiodide EDWGT Potable Aqua Globaline	½ tablet	1 tablet
2% iodine solution (tincture)	0.2 ml 5 gtts	0.4 ml 10 gtts
10% povidone-iodine solution*	0.35 ml 8 gtts	0.70 ml 16 gtts
Chlorination techniques	Amount for 5 PPM	Amount for 10 PPM
Household bleach 5% Sodium hypochlorie	0.1 Ml 2 gtts	0.2 ml 4 gtts
AquaClear Sodium dichloroisocyanurate		1 tablet
AquaCure, AquaPure, Chlor-floc Chlorine plus flocculating agent		8 PPM 1 tablet

*Providone-iodine solutions release free iodine in levels adequate for disinfecting, but scant data is available.

Measure with dropper (1 drop=0.05 ml) or tuberculin syringe

PPM-part per million gtts-drops ml-milliliter

Concentration of Halogen **Contact time in minutes at various water temperatures**

	5 C / 40 F	15 C / 60 F	30 C / 85 F
2 PPM	240	180	60
4 PPM	180	50	45
8 PPM	60	30	15

NOTE: These contact times have been extended from the usual recommendations to account for recent data that prolonged contact time is needed in very cold water to kill *Giardia* cysts.

NOTE: Chemicals may not destroy Cryptosporidium.

d. Purification. Water purification is the removal of organic and inorganic chemicals and particulate matter, including radioactive particles. While purification can eliminate offensive color, taste, and odor, it may not remove or kill microorganisms.

(1) Filtration. Filtration purifying is a process by which commercial manufacturers build water filters. The water filter is a three-tier system. The first layer, or grass layer, removes the larger impurities. The second layer, or sand layer, removes the smaller impurities. The final layer, or charcoal layer (**not the ash but charcoal from a fire**), bonds and holds the toxins. All layers are placed on some type of straining device and the charcoal layer should be at least 5-6 inches thick. Layers should be changed frequently and straining material should be boiled. Remember that this is not a disinfecting method, cysts can possibly move through this system.

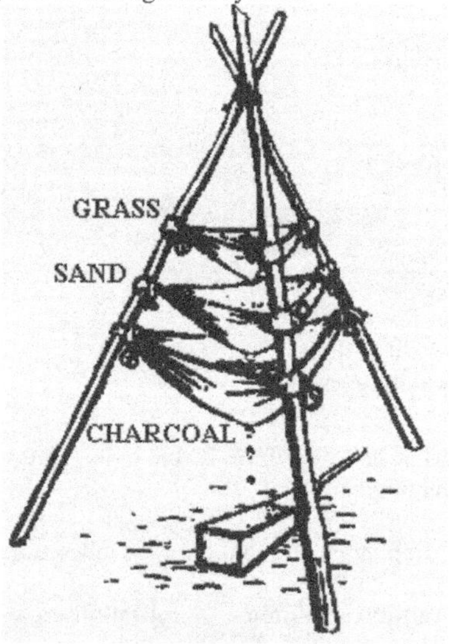

WATER FILTER

(2) Commercial Water Filters. Commercial water filters are generally available in most retail stores and may be with you. Understanding what the filter can do is the first step in safeguarding against future illnesses.

(a) A filter that has a .3 micron opening or larger will not stop Cryptosporidium.

(b) A filter system that does not release a chemical (i.e., iodine) may not kill all pathogens.

(c) A filter that has been overused may be clogged. Usage may result in excessive pumping pressure that can move harmful pathogens through the opening.

e. Potable. Potable indicates only that a water source, on average over a period of time, contains a "minimal microbial hazard," so the statistical likelihood of illness is acceptable.

(1) Sedimentation. Sedimentation is the separation of suspended particles large enough to settle rapidly by gravity. The time required depends on the size of the particle. Generally, 1 hour is adequate if the water is allowed to sit without agitation. After sediment has formed on the bottom of the container, the clear water is decanted or filtered from the top. Microorganisms, especially cysts, eventually settle, but this takes longer and the organisms are easily disturbed during pouring or filtering. Sedimentation should not be considered a means of disinfecting and should be used only as a last resort or in an extreme tactical situation.

5. **SOLAR STILLS**. (FMST.07.08k)

a. Solar stills are designed to supplement water reserves. Contrary to belief, they will not provide enough water to meet the daily requirement for water.

b. AboveGround Solar Still. This device allows the survivor to make water from vegetation. To make the aboveground solar still, locate a sunny slope on which to place the still, a clear plastic bag, green leafy vegetation, and a small rock.

(1) Construction.

(a) Fill the bag with air by turning the opening into the breeze or by "scooping" air into the bag.

(b) Fill the bag half to three-quarters full of green leafy vegetation. Be sure to remove all hard sticks or sharp spines that might puncture the bag.

CAUTION
Do not use poisonous vegetation. It will provide poisonous liquid.

(c) Place a small rock or similar item in the bag.

(d) Close the bag and tie the mouth securely as close to the end of the bag as possible to keep the maximum amount of air space. If you have a small piece of tubing, small straw, or hollow reed, inserts one end in the mouth of the bag before tying it securely. Tie off or plug the tubing so that air will not escape. This tubing will allow you to drain out condensed water without untying the bag.

(e) Place the bag, mouth downhill, on a slope in full sunlight. Position the mouth of the bag slightly higher than the low point in the bag.

(f) Settle the bag in place so that the rock works itself into the low point in the bag.

(g) To get the condensed water from the still, loosen the tie and tip the bag so that the collected water will drain out. Retie the mouth and reposition the still to allow further condensation.

(h) Change vegetation in the bag after extracting most of the water from it.

(i) Using 1-gallon zip-loc. bag instead of trash bags is a more efficient means of construction.

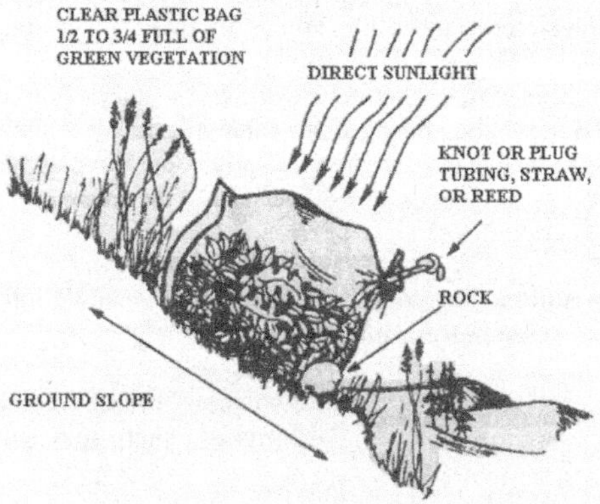

ABOVE GROUND SOLAR STILL

c. BelowGround Solar Still. Materials consist of a digging stick, clear plastic sheet, container, rock, and a drinking tube.

(1) Construction.

(a) Select a site where you believe the soil will contain moisture (such as a dry streambed or a low spot where rainwater has collected). The soil should be easy to dig, and will be exposed to sunlight.

(b) Dig a bowl-shaped hole about 1 meter across and 24 inches deep.

(c) Dig a sump in the center of the hole. The sump depth and perimeter will depend on the size of the container you have to place in it. The bottom of the sump should allow the container to stand upright.

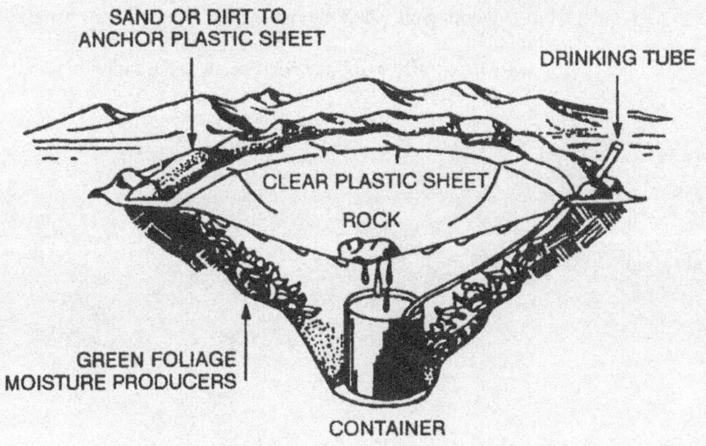

BELOW GROUND SOLAR STILL

(d) Anchor the tubing to the container's bottom by forming a loose overhand knot in the tubing. Extend the unanchored end of the tubing up, over, and beyond the lip of the hole.

(e) Place the plastic sheet over the hole, covering its edges with soil to hold in place. Place a rock in the center of the plastic sheet.

(f) Lower the plastic sheet into the hole until it is about 18 inches below ground level. Make sure the cone's apex is directly over the container. Ensure the plastic does not touch the sides of the hole because the earth will absorb the moisture.

(g) Put more soil on the edges of the plastic to hold it securely and prevent the loss of moisture.

(h) Plug the tube when not in use so that moisture will not evaporate.

(i) Plants can be placed in the hole as a moisture source. If so, dig out additional soil from the sides.

(j) If polluted water is the only moisture source, dig a small trough outside the hole about 10 inches from the still's lip. Dig the trough about 10 inches deep and 3 inches wide. Pour the polluted water in the trough. Ensure you do not spill any polluted water around the rim of the hole where the plastic touches the soil. The trough holds the polluted water and the soil filters it as the still draws it. This process works well when the only water source is salt water.

(k) Three stills will be needed to meet the individual daily water intake needs.

UNITED STATES MARINE CORPS
Mountain Warfare Training Center
Bridgeport, California 93517-5001

WMC
04/07/02

STUDENT HANDOUT

FORAGING PLANTS AND INSECTS FOR SURVIVAL USES

PURPOSE The purpose of this period of instruction is to familiarize the Wilderness Medicine Student with the potential resources for survival sustenance.

OUTLINE

1. GENERAL CONSIDERATIONS. There are few places without some type of edible vegetation. Plants contain vitamins, minerals, protein, carbohydrates, and dietary fiber. Some plants also contain fats. The following are general considerations:

a. Do not assume that because birds or animals have eaten a plant, it is edible by humans.

b. Poor plant recognition skills will seriously limit your ability to survive.

c. Plant dormancy and snowfall make foraging plants difficult during the winter months.

d. Plants generally poison by:

(1) Ingestion: When a person eats a part of a poisonous plant.

(2) Contact: When a person makes contact with a poisonous plant that cause any type of skin irritation or dermatitis.

(3) Absorption: When a person absorbs the poison through the skin, which can interrupt a bodily function.

(4) Inhalation: Poisoning can occur through the inhalation of smoke that contains poisonous plant residue.

A7

e. Plant properties can change throughout the growing season. Plants can be edible during certain periods while poisonous in others.

2. PLANT RECOGNITION. Plant recognition can take years of study to learn. The following is used as a general guideline.

a. Types of plants that should be avoided are:

(1) All mushrooms.

(2) Plants with a milky sap.

(3) White and yellow berries should be avoided, as they are almost always poisonous.

(4) Plants with shiny leaves should be considered poisonous.

(5) Plants that are irritants to the skin should not be eaten.

(6) All beans and peas should be avoided.

b. Edible Plants: There are several edible plants found at the Marine Corps Mountain Warfare Training Center. All are easy to recognize and many of the same plants are located throughout the United States and abroad.

(1) Wild Onions - Complete plant.

(2) Dandelion - Roots and leaves.

(3) Mallow - Leaves, stem and flowers.

(4) Watercress -Complete plant.

(5) Cattail - Root, stalk and stem.

(6) Bull Thistle - The flower.

(7) Juniper Tree - Berries and cambium layer.

(8) All Conifers - Cambium layer, needles, and nuts within the cones.

(9) Currants - Berries.

c. Preparing an edible plant. All edible plants should be boiled twice with a change of water to remove any bitterness, acids, or bacteria that the plant may contain.

3. PLANT TESTING PROCEDURE. Always adopt the following procedure when trying unknown plants as food.

a. Wait 8 hours without eating.

b. Select a plant that grows in sufficient quantity in the local area. Separate the part of the plant you wish to test root, stem, leaf, or flower. Certain parts of plants are poisonous while the, other parts may be edible.

c. Rub a portion of the plant you have selected on your inner forearm. Wait 15 minutes and look for any swelling, rash, or irritation.

d. Boil the plant or plant part in changes of water. The toxic properties of many plants are water-soluble or can be destroyed by heat. Cooking and discarding two changes of water can lessen the amount of poisonous material or remove it completely. These boiling periods should last at least 5 minutes each.

e. Place 1 teaspoon of the prepared plant food in the mouth and chew for 5 minutes, but do not swallow. If unpleasant effects occur (burning, bitter, or nauseating taste), remove the plant from the mouth at once and discard it as a food source. If no unpleasant effects occur, swallow the plant material and wait 8 hours.

f. If after 8 hours no unpleasant effects have occurred (nausea, cramps, and diarrhea), eat two tablespoonfuls and wait 8 hours.

g. In no unpleasant effects have occurred at the end of this 8-hour period, the plant may be considered edible.

h. Completely document and sketch the plant in a logbook to refer to for future use. This will aid in future procurement of this plant. If plant properties have changed, you will have to repeat the plant testing procedure.

NOTE: Have charcoal ready to eat. If any side effects occur, attempt to induce vomiting and immediately eat charcoal. Charcoal will absorb toxins and reduce the chance of Death.

4. MEDICINAL PLANTS One of the many things that hinder a survivor's ability to survive is medical problems. Injuries incurred will reduce survival expectancy and the ability to evade.

a. Coyote Willow: Coyote Willow is a thick forming shrub with clustered stems and very narrow leaves. It habits wet soils, especially riverbanks, sandbars, and silt flats.

(1) Drinking tea made from the inner bark can reduce fever.

(2) The dried and powdered inner bark can be used to stop bleeding.

(3) Administering a half-cup of willow charcoal dissolved in water can treat diarrhea.

(4) The Coyote Willow roots can be mashed and applied to tooth aches.

b. Yarrow: Yarrow is a plant (tall, usually not branched, with many white, slat topped group flowers). Do not confuse Yarrow with Water Hemlock.

(1) Rubbing a handful of crushed Yarrow flowers and leaves on any exposed can make insect repellent skin.

(2) Placing a Yarrow leaf poultice on the wound can stop bleeding.

(3) Relief from many rashes, including poison oak and ivy, can be achieved by applying a Yarrow leaf compress to the effected area.

(4) The Yarrow root can be chewed to relieve the pain of a toothache or break a fever.

(5) A potent anesthetic can be made by scrubbing fresh Yarrow roots in water to clean them. Once the roots are clean, crush them into a spongy mass and apply gently to the wounded area.

c. Tea To make a tea, place the plants in a container and pour in two cups of boiling water. Let the mixture steep for 20 minutes and drink.

d. Poultice: Place plant (Coyote Willow, Yarrow, etc.) in gauze or other similar material and fold it so the gauze will hold the plant in place. Put gauze in a larger cloth, about 6"x8", and roll the sides inward. Fold the cloth over without losing the plant. While boiling water, dip the bottom portion of the cloth containing the plant into the hot water by holding the edges. Keep the plant submerged in the boiling water until it becomes saturated. Bring the cloth straight up and with a twisting motion, wring the excess water back into the pot. Apply the poultice to the affected area as soon as it has cooled enough to place on the wound. To be effective, the poultice should be as hot as you can tolerate it.

e. Compress: A compress is made just like a poultice; except it is cold when applied to the wound. NOTE: Poultice or Compress should be applied for 1 to 24 hours, as needed. When applying a poultice, you may experience a throbbing pain as it draws out the infection and neutralizes toxins. When the pain subsides, the poultice has accomplished its task and should be removed. Apply a fresh poultice as needed until the desired level of healing has been reached.

5. POISONOUS PLANTS. Rarely will a survivor have an ideal means of killing large game, however there are certain plants that can aid the survivor. The two plants that we will talk about are the Water Hemlock and Monkshood.

a. Preparing Poisonous Plants: Once a poisonous plant has been located, dig up the root of the plant. The roots of the Water Hemlock and Monkshood generally grow 8 to 10 inches deep. Extreme caution must be used when handling the root. Do not handle the root without a barrier between your hands and the root. A barrier can be gloves, socks, T-shirt, or even moss. Split the root lengthwise to expose the inside of the root where the toxin is located. With the root split, rub the tip of your spear/arrow inside of the root opening. In a slow controlled manner, work from the bottom of the tip to the top. Once the tip is thoroughly coated, allow the toxin to dry and apply another coat to the tip. Continue to apply coat after coat of toxin, until the root is completely drained of its' toxin. You are now finished and the tip is ready for use.

NOTE: Do not eat the meat in the vicinity of the poisoned wound.

CAUTION: Do not apply poison to the actual point and edges of tip. This avoids accidental puncture to you. Wash and boil all gloves and clothing used as a barrier before wearing

6. INSECTS. Insects are the most abundant life form on earth and are an excellent survival food. They are easy to catch and provide 65-80% protein; compared to 20% for beef. They aren't too appetizing, but personal bias has no place in a survival situation. The focus must remain on maintaining your health.

a. Insects to avoid.

(1) All adults that sting or bite.

(2) Hairy or brightly colored insects.

(3) All Caterpillars.

(4) Insects that have a pungent odor.

(5) All spiders.

(6) Disease carriers like ticks, flies, or mosquitoes.

b. Edible Insects.

(1) Insect Larvae.

(2) Grasshoppers.

(3) Beetles.

(4) Grubs.

(5) Ants.

(6) Termites.

(7) Worms.

c. Foraging for Insects. One must be careful not to expend more energy harvesting food than can be replaced. For example, catching insects such as grasshoppers can become exasperating and tiring.

(1) At night grasshoppers climb tall plants and clings to the stalks near the top. They can be picked from the plants in the early morning while they are chilled and dormant.

(2) Dig for worms in damp humus soil, under rocks/logs or look for them on the ground after it has rained.

(3) Carpenter ants are found in dead trees and stumps, which can be gathered by hand.

(4) Most other insects can be found in rotten logs, under rocks, and in open grassy areas.

d. Preparing Edible Insects.

(1) Insects with a hard outer shell have parasites. Remove the wings and barbed legs before cooking.

(2) Drop worms into potable water for at least a half-hour. They will naturally purge themselves. You can either cook or eat them raw.

(3) Most other insects can be eaten raw. Cooking insects will improve their taste. If the thought of eating insects is unbearable, grind them into a paste and mix with other foods.

UNITED STATES MARINE CORPS
Mountain Warfare Training Center
Bridgeport, California 93517-5001

WMC.A8.0
05/02/01

STUDENT HANDOUT

SURVIVAL TRAPS AND SNARES

PURPOSE: The purpose of this class is to familiarize the student with some of the techniques used to take game. Game can be useful for both food and clothing.

OUTLINE

1. GENERAL TIPS AND TECHNIQUES FOR TRAPPING

a. General Tips. Knowing a few general hints and tips will make the trapping of animals much easier and considerably more effective. The eight general tips for trapping are:

(1) Know your game: Knowing the habits of the animal you want to trap will help lure it into your trap. Such things as when and where they move, feed, and water will help you determine where the trap can be most effectively placed.

A8

(2) Keep things simple: You don't have time in a survival situation to construct elaborate traps and they do not necessarily do a better job.

(3) Set traps in the right place: Animals will travel and stop in certain locations. That is where to best employ traps.

(4) Cover up your scent: Animals will avoid an area, which smells strange to them. Smoke from your fire is the best cover to use.

(5) Use the right type of trap: Some traps work better than others do (i.e. a deadfall works better on squirrels than a spring pole).

(6) Use the right size trap: Adjust the size of your traps to the size of the animal.

(7) Check traps: Check your traps twice daily: morning and evening. Checking your traps less than twice a day can allow your game to escape, rot, or be taken by other predators.

(8) Bait your traps: Bait of any type will add to your chances of success.

b. General Techniques. A general technique is the method in which the trap is intended to kill or hold the animal.

(1) Strangulation: This method strangles the animal, such as a snare.

(2) Mangling: This method crushes the animal, such as a deadfall.

(3) Entanglement: This method entangles the animal, such as a net.

(4) Live: This method holds the animal, such as a box trap.

c. Types of Triggers. There are three basic triggers used for all traps and path guards. Depending on the situation, variations of these triggers can be used.

(1) Puite Figure 4.

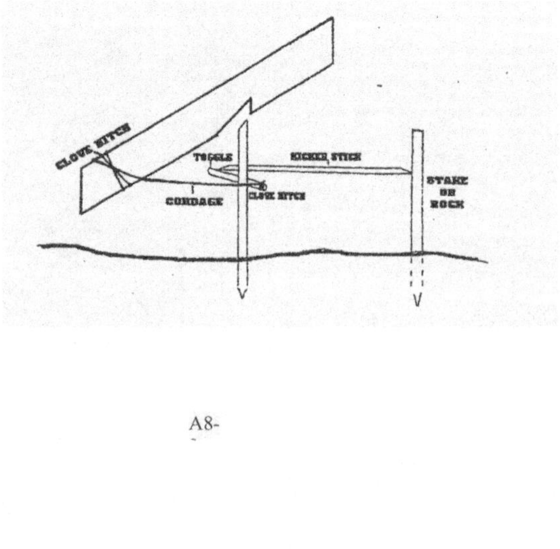

(2) <u>Toggle</u>.

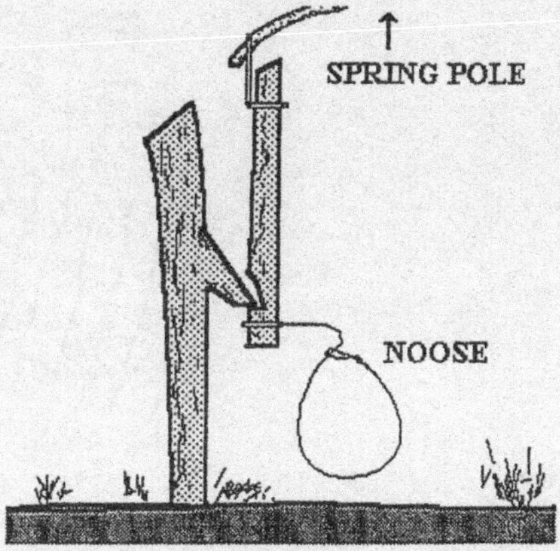

<u>TOGGLE</u>

(c) <u>Universal</u>.

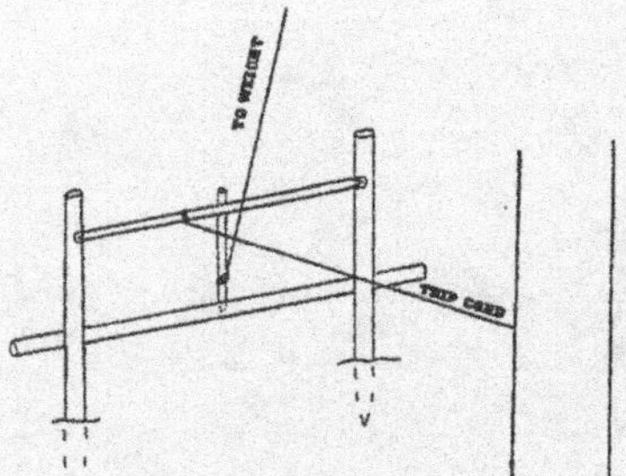

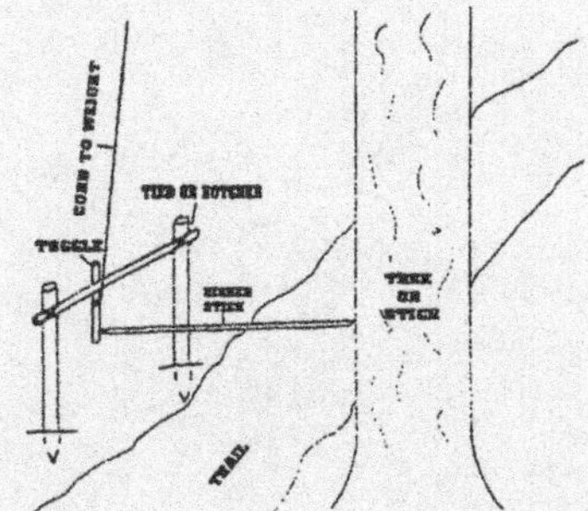

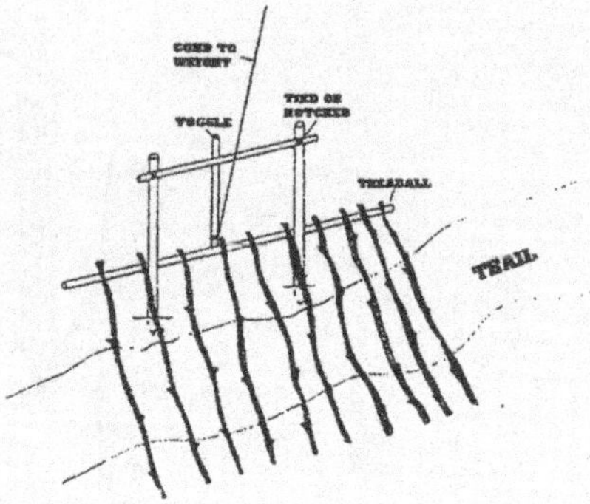

2. TRAPS. Traps are c
action is generally caused by either a weight or spring loaded device.

a. Puite Deadfall: The Puite deadfall requires a knife and piece of cord to construct. It is designed to mangle small rodents. The trigger for the deadfall is the Puite figure 4.

b. Spring Pole: The spring pole requires a small sapling and cordage to construct. The trigger for the spring pole is the toggle. It is designed to lift the animal off the ground; not allowing predatory animals to take your game. Remember, the trigger can not be so tight that the intended game can not set it off.

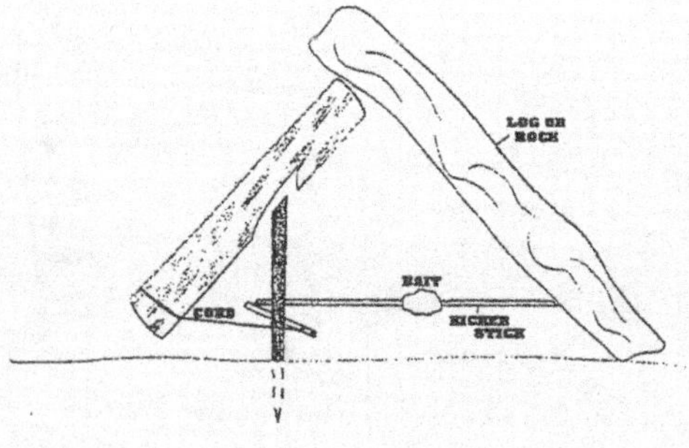

SPRING POLE

c. Box Trap: The box trap requires limited cordage to construct. It is designed to hold live small rodents and birds. The box trap trigger is the Puite figure 4.

3. PATH GUARDS Path guards is designed to protect and provide security for your shelter area against the enemy and predatory animals. They are classified into noise and casualty producing path guards.

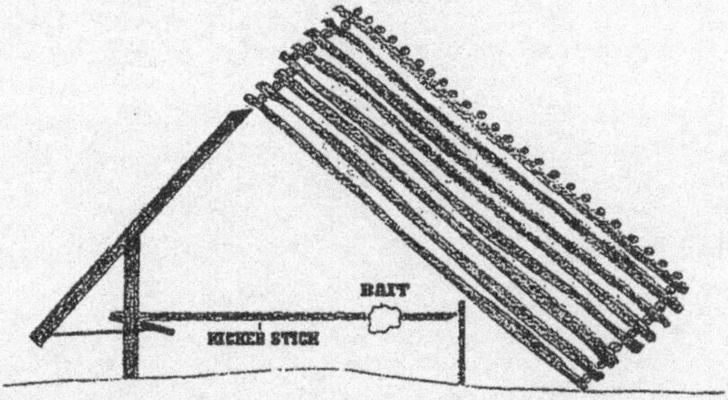

BOX TRAP

a. Noise producing path wards. Noise producing pathguards serve as an alarm for your shelter area. When triggered, it should produce some type of loud noise. Although construction can vary, depending on materials available, one example is as follows:

(1) Secure a young sapling to a universal trigger.

(2) At the end of the sapling, tie several pieces of metal to the sapling. Use what ever is available for metal.

(3) Camouflage the metal on the ground.

(4) When triggered, the sapling should swing back and forth, causing the metal rattle.

b. Casualty producing path wards. Casualty producing path guards, when triggered, should cause death or injury to the enemy or predatory animal. Tips should be poisoned as discussed in Survival Plant Uses class. Triggers for this type of path guard should be the universal triggers.

(1) Log Jerk

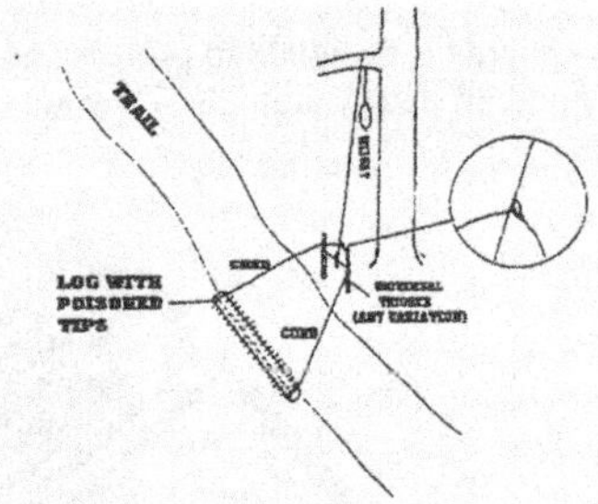

(2) Fish Hook Nightmare

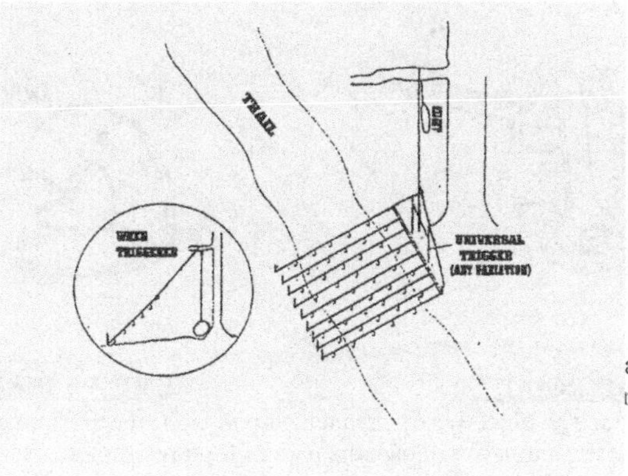

4. SNARING. A snare … a loop at one end, which tightens, … nd less time consuming to constr…

a. Requirements for snaring there are five requirements too effectively employ snares. They are as follows:

(1) Wire: Although snares can be used with rope or cord, they are less effective than wire. Wire should have memory and resist kinking. Aircraft cable type in size 1/32 to 3/16 inch should be used. This type of wire prevents animal chew out and resists breakage. Remember, you want the smallest diameter cable capable of holding the animal.

(2) Locking device: A locking device is imperative for a snare to work properly. Locking devices secure the snare around the animal's neck. There are several methods available for a locking device.

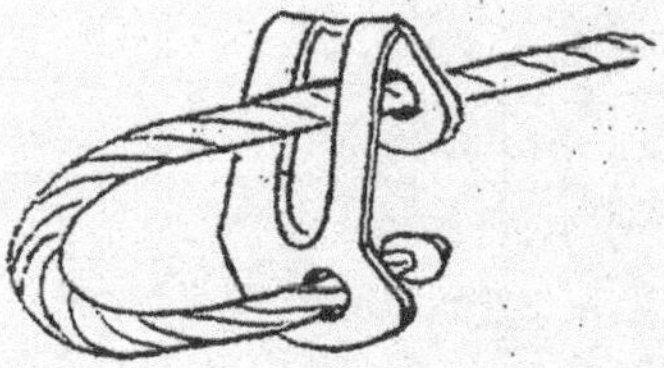

COMMERCIAL

BENT WASHER

(3) Channelization: The objective of channelization is to force the animal to move through the snare. Remember, animals will take the path of least resistance.

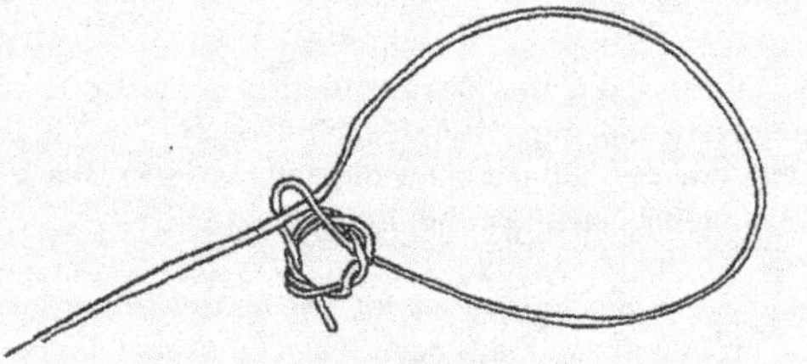

KNOT

(4) Loop size: Probably the single most important requirement in snaring. A correctly employed snare will have the snare holding the animal around the neck. Loop size is placed on the snare according to the intended animal. Too large will result in a body or leg catch, resulting in possible chew out or breakage. Too small will enable the animal to force the snare to the side, resulting in a miss. Additionally, the loop must be placed with specific ground clearance. The following chart will give a general guideline.

ANIMAL	NOOSE SIZE	GROUND CLEARANCE
SQUIRREL	2 1/2 TO 3 INCHES	1/2 TO 1 1/2 INCHES
RABBIT	4 TO 5 1/2 INCHES	1 1/2 TO 3 INCHES
RACCOON	6 INCHES	3 TO 4 INCHES
FOX	7 TO 10 INCHES	8 TO 10 INCHES
COYOTES	12 TO 14 INCHES	12 INCHES
BOBCAT	9 INCHES	8 INCHES

Note: (1) Noose size is the diameter of the snare loop.

(2) Ground clearance is measured from the bottom of the loop to the ground.

(5) Bait. In a survival situation, you will not be able to employ numerous snares. Baiting all snares increases your chances of success.

b. Snares although there are numerous ideas to employ snares, here are a few ideas.

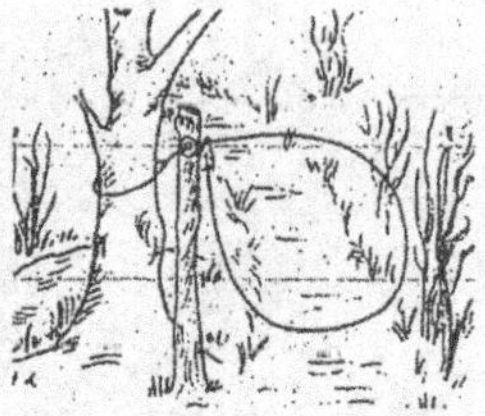

TRAIL SET

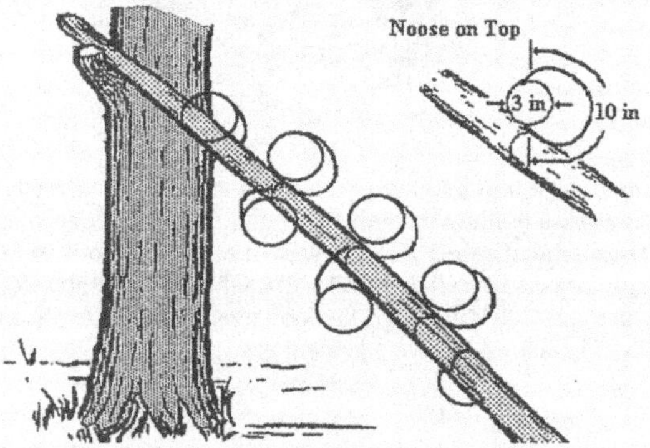

SQUIRREL POLE

DEN SET

CUBBIE SET

5. FISH. In a mountainous region, fish are normally an abundant resource. Not only are they a food source, the "left-overs" provide an excellent bait for traps and snares.

a. Fishing locations: Fishing in mountain streams is generally best done with a hand line. When fishing these streams, always look for these places to fish.

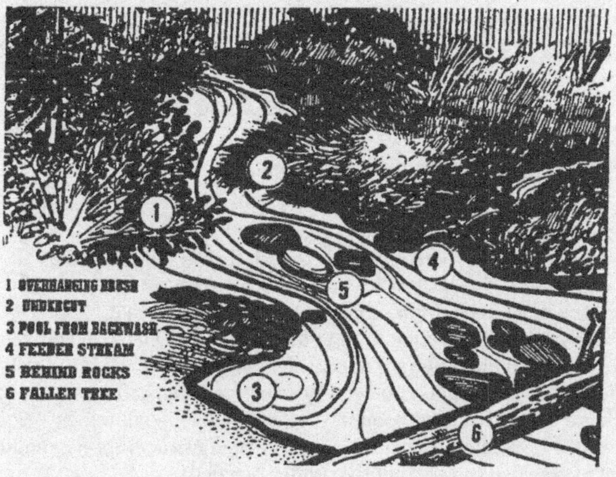

FISHING LOCATIONS

b. Expedient hooks: Although hooks should be carried in a survival kit, the survivor should be able to construct additional hooks if the situation arises. Expedient hooks are made to become lodged in the throat of the fish. Below are a few examples.

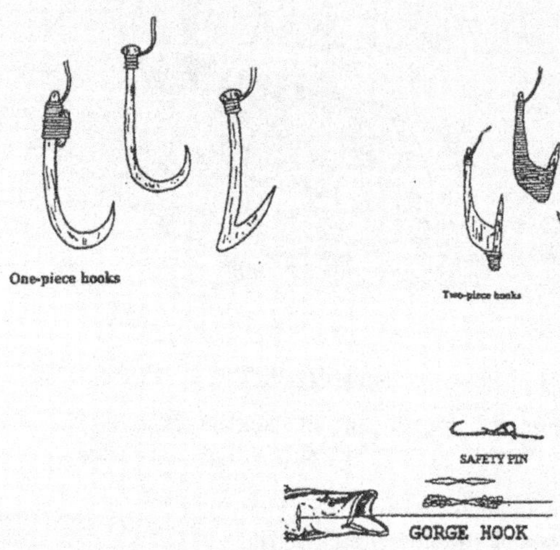

EXPEDIENT HOOKS

c. Fish traps: A fish trap can be effective if you have a shallow stream and time to construct it. A basic fish trap in nothing more than a barricade of rocks or sticks across a stream with another barricade using a funnel-type entrance which fish can be driven into but have a difficult time finding their way out. Once fish are trapped between these two barricades they may be speared, clubbed, or grabbed. This can be very effective when certain types of fish are moving in large groups to spawn. This type of trap is very effective in catching fish. A door can be constructed at the mouth of the trap so that excess fish can be kept live until needed.

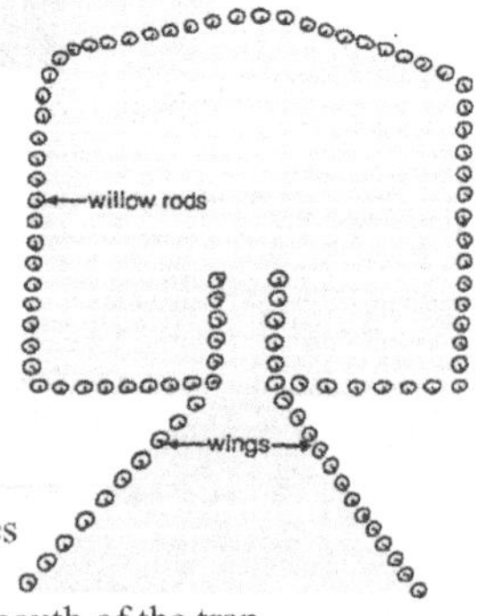

FISH TRAP

d. Set Lines: Set lines are an effective method of fishing which conserves energy. Put them out over night with several boated hooks attached. Place them with the hooks either on the bottom or suspended off the bottom, until you have determined where the fish are feeding.

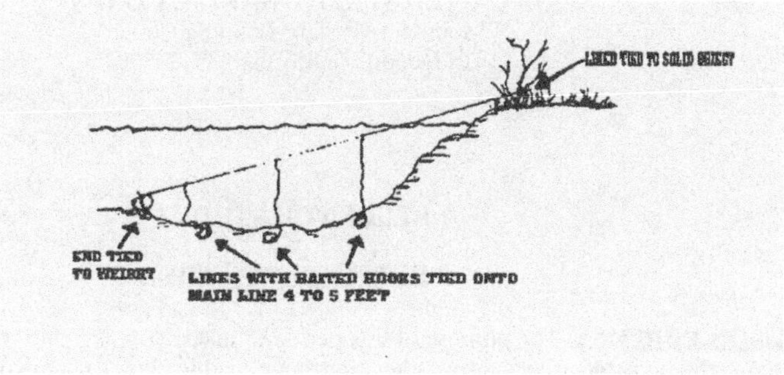

SET LINES

e. Netting: A gill net is most effective in still water, e.g., a lake (near the inlet and outlet are good locations) or back water in a large stream (for survival doesn't hesitate to block the stream). Nets can be constructed using the inner cords of parachute shroud lines. Place floats on top and weights on the bottom to keep the net vertical in the water. When the ice is on a lake, the fish are inclined to stay deeper. The smaller the mesh, the smaller the fish you can catch, but a small mest will still entangle a large fish. A mesh of 3 inches is a good standard. A gill net is very time consuming and requires allot of material to construct.

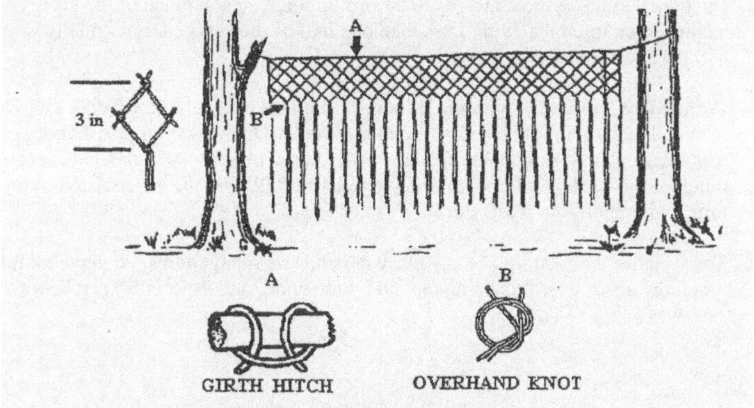

GILL NET

UNITED STATES MARINE CORPS
Mountain Warfare Training Center
Bridgeport, California 93517-5001

WMC
04/07/02

STUDENT HANDOUT

MOUNTAIN WEATHER

LESSON PURPOSE: The purpose of this period of instruction is to introduce you to weather patterns, the invisible aspects of weather, and some visible clues to help you forecast the weather. This lesson relates to all operations in the mountains at MWTC.

OUTLINE

1. GENERAL

a. The earth is surrounded by the atmosphere, which is divided, into several layers. The world's weather systems are in the troposphere, the lower of these layers. This layer reaches as high as 40,000 feet.

b. Dust and clouds in the atmosphere absorb or bounce back much of the energy that the sun beams down upon the earth. Less than one half of the sun's energy actually warms the earth's surface and lower atmosphere.

c. Warmed air, combined with the spinning (rotation) of the earth, produces winds that spread heat and moisture more evenly around the world. This is very important because the sun heats the Equator much more than the poles and without winds to help restore the balance, much of the earth would be impossible to live on. Where the air-cools, you can get clouds, rain, snow, hail, fog, frost, etc.

d. The weather that you find in any place depends on many things, i.e. how hot the air is, how moist the air is, how it is being moved by the wind, and especially, is it being lifted or not?

A10

2. PRESSURE

a. All of these factors are related to air pressure, which is the weight of the atmosphere at any given place. The lower the pressure, the more likely are rain and strong winds.

b. In order to understand this we can say that the air in our atmosphere acts very much like a liquid.

c. Areas with a high level of this liquid would exert more pressure on the Earth and be called a "high pressure area".

d. Areas with a lower level would be called a "low pressure area".

e. In order to equalize the areas of high pressure it would have to push out to the areas of low pressure.

f. The characteristics of these two pressure areas are as follows:

(1) High-pressure area. Flows out to equalize pressure.

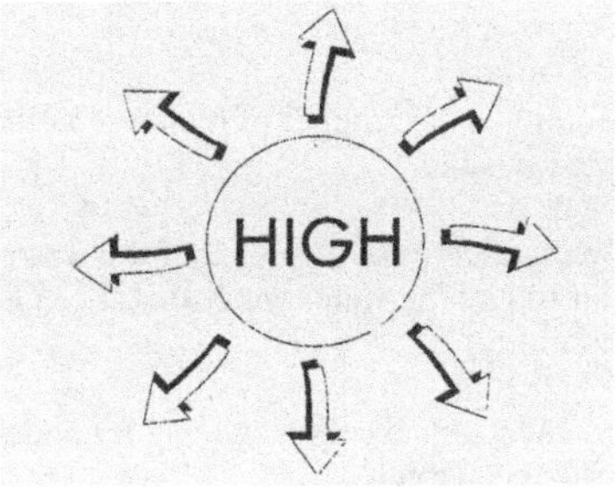

(2) Low pressure area. Flows in to equalize pressure.

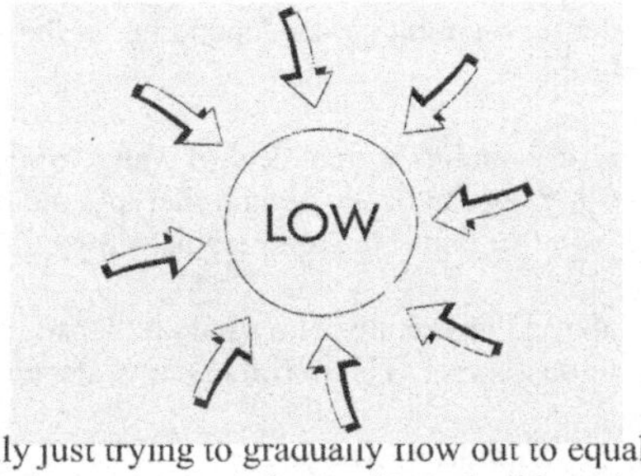

g. The air from the high-pressure area is basically just trying to gradually flow out to equalize its pressure with the surrounding air, while the low pressure is beginning to build vertically. Once the low has achieved equal pressure, it can't stop and continues to build vertically; causing turbulence, which results in bad weather.

NOTE: When looking on the weather map, you will notice that these differences in pressure resemble contour lines. These contour lines are called isobars and are translated to mean equal pressure area.

h. Isobars: Pressure is measured in millibars or another more common measurement is inch mercury.

i. Fitting enough, areas of high pressure are called ridges and areas of low pressure are called troughs.

NOTE: The average air pressure at sea level is:

29.92 inches mercury.
1,013 millibars.

j. As we go up in altitude, the pressure (or weight) of the atmosphere decreases.

EXAMPLE: At 18,000 feet in elevation it would be 500 millibars vice 1,013 millibars at sea level.

3. HUMIDITY Humidity is the amount of moisture in the air. All air holds water vapor, although it is quite invisible.

a. Air can hold only so much water vapors, but the warmer the air, the more moisture it can hold. When the air has all the water vapor that it can hold, the air is said to be saturated (100% relative humidity).

b. If the air is then cooled, any excess water vapor condenses; that is, it's molecules join to build the water droplets we can see.

c. The temperature at which this happens is called the "condensation point". The condensation point varies depending on the amount of water vapor and the temperature of the air.

d. If the air contains a great deal of water vapor, condensation will form at a temperature of 20°C (68°F). But if the air is rather dry and does not hold much moisture, condensation may not form until the temperature drops to 0°C (32°F) or even below freezing.

e. Adiabatic Lapse Rate: The adiabatic lapse rate is the rate that air will cool on ascent and warm on descent. The rate also varies depending on the moisture content of the air.

(1) Saturated Air = 2.2°F per 1,000 feet.

(2) Dry Air = 5.5°F per 1,000 feet.

4. WINDS. As we stated earlier, the uneven heating of the air by the sun and rotation of the earth causes winds. Much of the world's weather depends on a system of winds that blow in a set direction. This pattern depends on the different amounts of sun (heat) that the different regions get and also on the rotation of the earth.

a. Above hot surfaces, air expends (air molecules spread out), and move to colder areas where it cools and becomes denser, and sinks to the earth's surface. This forms a circulation of air from the poles along the surface or the earth to the Equator, where it rises and moves towards the poles again.

b. Once the rotation of the earth is added to this, the pattern of the circulation becomes confusing.

c. Because of the heating and cooling, along with the rotation of the earth, we have these surface winds. All winds are named from the direction they originated from:

(1) Polar Easterlies: These are winds from the polar region moving from the east. This is air that has cooled and settled at the poles.

(2) Prevailing Westerlies: These winds originate from approximately 30 degrees North Latitude from the west. This is an area where prematurely cooled air, due to the earth's rotation, has settled back to the surface.

(3) Northeast Tradewinds: These are winds that originate from approximately 30 degrees North from the Northeast. Also prematurely cooled air.

d. Jet Stream: A jet stream can be defined as a long, meandering current of high speed winds near the tropopause (transition zone between the troposphere and the stratosphere) blowing from generally a westerly direction and often exceeding 250 miles per hour. The jet stream results from:

(1) Circulation of air around the poles and Equator.

(2) The direction of airflow above the mid latitudes.

(3) The actual path of the jet stream comes from the west, dipping down and picking up air masses from the tropical regions and going north and bringing down air masses from the Polar Regions.

NOTE: The average number of long waves in the jet stream is between three and five depending on the season. Temperature differences between polar and tropical regions influence this. The long waves influence day to week changes in the weather; there are also short waves that influence hourly changes in the weather.

5. AIR MASSES. As we know, all -of these patterns move air. This air comes in parcels known as "air masses". These air masses can vary in size from as small as a town too as large as a country. These air masses are named for where they originate:

a. Maritime: Over water.

b. Continental: Over land.

c. Polar: Above 60 degrees North.

d. Tropical: Below 60 degrees North.

e. Combining these give us the names and description of the four types of air masses:

(1) Continental Polar: Cold, dry air mass.

(2) Maritime Polar: Cold wet air mass.

(3) Continental Tropical: Dry, warm air mass.

(4) Maritime Tropical: Wet, warm air mass.

ABBREVIATED MEANINGS
CP = Continental Polar.
MP = Maritime Polar.
CT = Continental
Tropical. MT =
Maritime Tropical.

f. The thing to understand about air masses; they will not mix with another air mass of a different temperature and moisture content. When two different air masses collide, we have a front, which will be covered in more detail later in this period of instruction.

g. Here are some other types of winds that are peculiar to mountain environments but don't necessarily affect the weather:

(1) Anabolic wind: These are winds that blow up mountain valleys to replace warm rising air and are usually light winds.

(2) Catabolic wind: These are winds that blow down mountain valley slopes caused by the cooling of air and are occasionally strong winds.

6. LIFTING/COOLING. As we know, air can only hold so much moisture depending on its temperature. If we cool this air beyond its saturation point, it must release this moisture in one form or another, i.e. rain, snow, fog, dews, etc. There are three ways that air can be lifted and cooled beyond its saturation point.

a. Orographic uplift: This happens when an air mass is pushed up and over a mass of higher ground such as a mountain. Due to the adiabatic lapse rate, the air is cooled with altitude and if it reaches its saturation point we will receive precipitation.

OROGRAPHIC UPLIFT

b. Convection effects: This is normally a summer effect due to the sun's heat radiating off of the surface and causing the air currents to push straight up and lift air to a point of saturation.

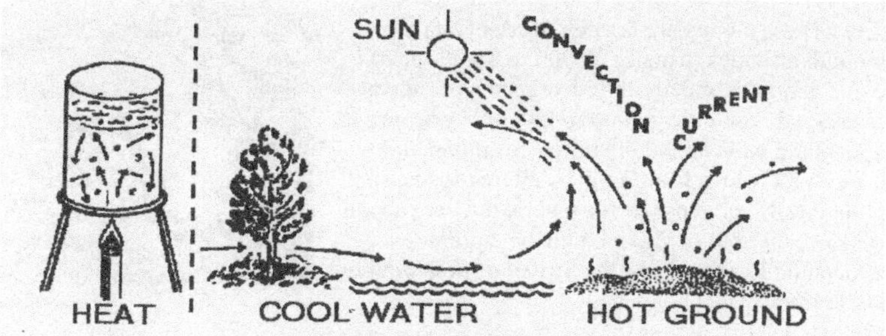

CONVECTION EFFECTS

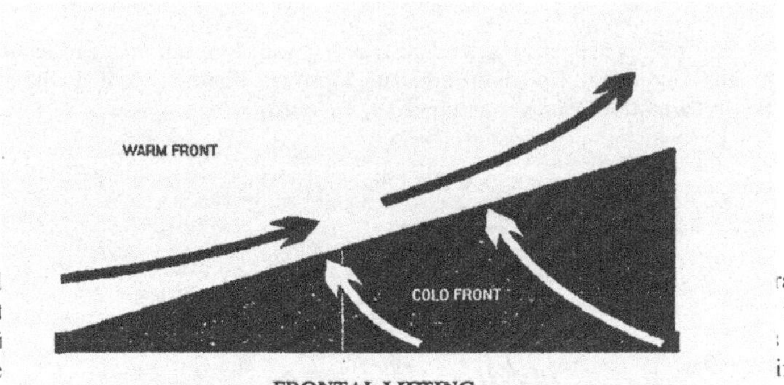

FRONTAL LIFTING

c. Frontal ... rature content ... is forced aloft; fi ... where we receive ... ifting is not uncomı

7. CLOUDS. Anytime air is lifted or cooled beyond its saturation point (100% relative humidity), clouds are formed. Clouds are one of our signposts to what is happening. Clouds can be described in many different ways; they can also be classified by height or appearance, or even by the amount of area covered, vertically or horizontally.

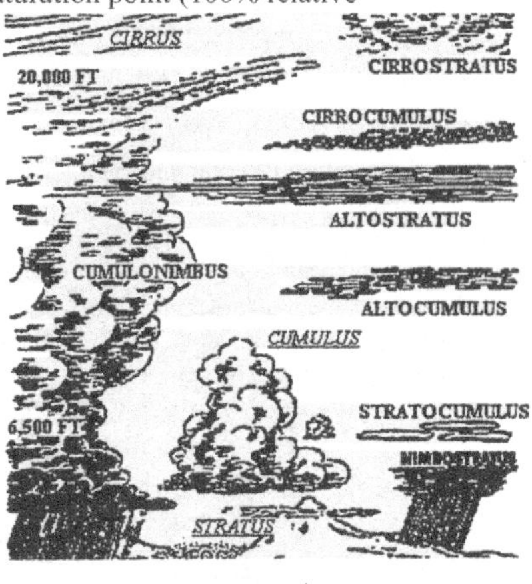

a. Cirrus: These clouds are formed of ice crystals at very high altitudes (usually 20,000 to 35,000 feet) in the mid-latitudes and are thin, feathery type clouds. These clouds can give you up to 24 hours warning of approaching bad weather, hundreds of miles in advance of a 'warm front. Frail, scattered types, such as "mare-tails" or dense cirrus layers, tufts are a sign of fair weather but predictive may be a prelude to approaching lower clouds, the arrival of precipitation and the front.

b. Cumulus: These clouds are formed due to rising air currents and are prevalent in unstable air that favors vertical development. These currents of air create cumiliform clouds that give them a piled or bunched up appearance, looking similar to cotton balls. Within the cumulus family there are three different types to help us to forecast the weather:

(1) Cotton puffs of cumulus are Fair Weather Clouds but should be observed for possible growth into towering cumulus and cumulonimbus.

(2) Towering cumulus are characterized by vertical development. Their vertical lifting is caused by some type of lifting action, such as connective currents found on hot summer afternoons or when wind is forced to rise up the slope of a mountain or possibly the lifting action that

may be present in a frontal system. The towering cumulus has a puffy and "cauliflower-shaped" appearance.

(3) Cumulonimbus clouds are characterized in the same manner as the towering cumulus, form the familiar "thunderhead" and produce thunderstorm activity. These clouds are characterized by violent updrafts, which carry the tops of the clouds to extreme elevations. Tornadoes, hail and severe rainstorms are all products of this type of cloud. At the top of the cloud, a flat anvil shaped form appears as the thunderstorm begins to dissipate.

c. Stratus: Stratus clouds are formed when a layer of moist air is cooled below its saturation point. Stratiform, clouds lie mostly in horizontal layers or sheets, resisting vertical development. The word "stratus" is derived from the Latin word "layer". The stratus cloud is quite uniform and resembles fog. It has a fairly uniform base and a dull, gray appearance. Stratus clouds make the sky appear heavy and will occasionally produce fine drizzle or Very light snow with fog. However, because there is little or no vertical movement in the stratus clouds, they usually do not produce precipitation in the form of heavy rain or snow.

d. As previously stated, clouds are formed when air is lifted to a point where it cools to its saturation point. We also know that frontal lifting affects our fronts, which produce the largest portion of our precipitation.

8. FRONTS. As we know, fronts often happen when two air masses of different moisture and temperature content interact. One of the ways we can identify that this is happening is by the progression of the clouds.

a. Warm Front: A warm front occurs when warm air moves into and over a slower (or stationary) cold air mass. Since warm air is less dense, it will rise naturally so that it will push the cooler air down and rise above it. The cloud you will see at this stage is cirrus. From the point where it actually starts rising, you will see stratus. As it continues to rise, this warm air-cools by the cold air and, this, receiving moisture at the same time. As it builds in moisture, it darkens becoming "nimbus-stratus", which means rain of thunderclouds. At that point some type of moisture will generally fall.

b. Cold Front. A cold front occurs when a cold air mass (colder than the ground that it is traveling over) overtakes a warm air mass that is stationary or moving slowly. This cold air, being denser, will go underneath the warm air, pushing it higher. Of course, no one can see this "air", but they can see clouds and the clouds themselves can tell us what is happening. The cloud progression to look for is cirrus to cirrocumulus to cumulus and, finally, to cumulonimbus.

c. Occluded Front. Cold fronts move faster than warm ones so that eventually a cold front overtakes a warm one and the warm air becomes progressively lifted from the surface. The zone of division between cold air ahead and cold air behind is called a "cold occlusion". If the air behind the front is warmer than ahead, it is a warm occlusion. Most land areas experience more occlusions than other types of fronts. In the progression of clouds leading to fronts, orographic uplift can play part in deceiving you of the actual type of front, i.e. progression of clouds leading to a warm front with orographic cumulus clouds added to these. The progression of clouds in an occlusion is a combination of both progressions from a warm and cold front.

9. USING PRESSURE AS AN INDICATOR. A very important factor of telling us what might happen is the pressure. As we know, low pressure or dropping pressure normally indicates deteriorating weather whereas high pressure usually gives us more good weather or clearing of bad weather. There are a couple of ways to monitor our pressure and are as follows:

a. Barometer: A barometer could be described as a pan of mercury with a tube leading out of the pan. Pressure from the atmosphere causes the mercury to rise in the tube.

(1) The tube is marked in millibars and the station that's reading these millibars will know how much it should rise for that location. Once again, if it rises more than normal, it would be considered a high-pressure reading.

b. Altimeter: Another means that is used to measure pressure is an altimeter, which is commonly used by mountaineers. It works like this:

(1) As you rise in elevation the pressure becomes less, thus allowing the needle in the altimeter to rise. If the needle rises without you rising with it, there is less pressure in the atmosphere than before and thus, a low-pressure area.

c. Contrail Lines: A basic way of identifying a low-pressure area is to note the contrail lines from jet aircraft. If they don't dissipate within two hours, that indicates a low-pressure area in your area. This usually occurs about 24 hours prior to an oncoming front.

d. Lenticulars: These are optical, lens-shaped cumulus clouds that have been sculpted by the winds. This indicates moisture in the air and high winds aloft. When preceding a cold front, winds and clouds will begin to lower.

10. USING SIGNS FROM NATURE. These signs will give you a general prediction of the incoming weather conditions. Try to utilize as many signs together as possible, which will improve your prediction. All of these signs have been tested with relative accuracy, but shouldn't be depended on 100%. But in any case you will be right more times than wrong in predicting the weather. From this we can gather as much information as needed and compile it along with our own experience of the area we are working in to help us form a prediction of incoming weather. The signs are as follows:

a. A spider's habits are very good indicators of what weather conditions will be within the next few hours. When the day is to be fair and relatively windless, they will spin long filaments over which they scout persistently. When precipitation is imminent, they shorten and tighten their snares and drowse dully in their centers.

b. Insects are especially annoying two to four hours before a storm.

c. If bees are swarming, fair weather will continue for at least the next half day.

d. Large game such as deer, elk, etc., will be feeding unusually heavy four to six hours before a storm.

e. When the smoke from a campfire, after lifting a short distance with the heated air, beats downward, a storm is approaching. Steadily rising smoke indicates fair weather.

f. A red sun or sky in the morning indicates nearing rain due to the excessive amount of moisture in the air.

g. A red evening sky indicates that the air contains so little moisture that rain within the next 24 hours is highly improbable.

h. A gray, overcast evening sky indicates that moisture carrying dust particles in the atmosphere have become overloaded with water; this condition favors rain.

i. A gray morning sky indicates dry air above the haze caused by the collecting of moisture on the dust in the lower atmosphere; you can reasonably a fair day.

j. When the setting sun shows a green tint at the top as it sinks behind clear horizon, fair weather is probable for most of the next 24 hours.

k. A rainbow in the late afternoon indicates fair weather ahead. However, a rainbow in the morning is a sign of prolonged bad weather.

l. A corona is the circle that appears around the sun or the moon. When this circle grows larger and larger, it indicates that the drops of water in the atmosphere are evaporating and that the weather will probably be clear. When this circle shrinks by the hour, it indicates that the water drops in the atmosphere are becoming larger, forming into clouds, rain is almost sure to fall.

m. In fair weather, air currents flow down streams and hillsides in the early morning and start drifting back up towards sunset. Any reversal of these directions warns of a nearing storm.

n. When the breeze is such that the leaves show their undersides, a storm is likely on the way.

o. It is so quiet before a storm, that distant noises can be heard more clearly. This is due to the inactivity of wildlife a couple of hours before a storm.

p. When in the mountains, the sight of morning mist rising from ravines is a good sign of clear weather the rest of the day.

q. A heavy dew or frost in the morning is a sign of fair weather for the rest of the day. This is due to the moisture in the atmosphere settling on the ground vice in the form of precipitation such as rain, snow, etc.

UNITED STATES MARINE CORPS
Mountain Warfare Training Center
Bridgeport, California 93517-5001

FMST.07.42
03/14/02

STUDENT HANDOUT

RESCUE PHILOSOPHY FOR SWIFTWATER RESCUE TEAMS

TERMIANL LEARNING OBJECTIVE In a swift water rescue scenario, organize a swift water rescue, in accordance with the reference. (FMST.07.42)

ENABLING LEARNING OBJECTIVES.

(1) With out the aid of reference, from a given list select the correct components of rescue preparation, in accordance with the reference. (FMST.07.42a)

(2) With out the aid of reference, from a given list select the correct role description of the rescue team members, in accordance with the reference. (FMST.07.42b)

(3) With out the aid of reference, from a given list select the correct methods of rescue, in accordance with the reference. (FMST.07.42c)

OUTLINE.

1. **COMPONENTS OF RESCUE PREPARATION**. (FMST.07.42a) These components comprise the backbone of both technical skills and the proper mental attitude required of a rescue professional. Remember that this course only provides you with the first component.

a. Training. Provides the students with the basic technical skills required to perform a safe rescue, these can be gained by attending an approved SWIFTWATER RESCUE COURSE.

b. Practice. Use of the skills learned in training.

c. Experience. Gained through advanced training, practical exercises and real-world situations.

B1

d. Judgment. The combination and application of skills for a given situation. Remember that every rescue situation is different. Don't get into a SOP mindset.

2. **PRIORITIES OF SAFETY**. Most of you have heard the saying that "It is better to be judged by twelve then carried by six". With that in mind you must realize that, at times, there is no way to affect a safe rescue. If you place yourself or your team in a hostile environment, from which you cannot escape, you are adding to the problem, not assisting in the solution. For this reason we set priorities of safety.

a. Self-aid. Be able to rescue yourself.

b. Buddy-aid. Ensure all safety measures are in force and that team members can rescue each other in an emergency.

c. Victim-aid. Only after self and buddy-aid are ensured should you attempt to rescue the victim. This will enable you to concentrate and focus your entire effort on the rescue.

3. **ROLES OF THE TEAM MEMBERS**. (FMST.07.24b) Often when dealing with rescues, you will find yourself working with outside agencies and organizations. In order to successfully coordinate the entire effort the Incident Command System (ICS) was developed. In swiftwater rescue situations there are four specific roles that must be filled.

a. Incident Commander. Is in overall command of the rescue efforts. Does not necessarily have the technical knowledge to conduct the rescue, but must be able to coordinate the rescue effort.

b. Rigger. Personnel responsible for equipment and rope systems, i.e. establishing the tension diagonal or the tyrolean high line.

c. Gofer. This group of individuals fills all other supporting roles, i.e. Communications, logistics, medevac, accounting, equipment accountability etc.

d. Rescuer. The individuals who are actually conducting the rescue.

4. **COMPONENTS OF SEARCH AND RESCUE**. In order to successfully complete a rescue there is four objectives that must be met. In order to remember them we use the acronym LAST.

a. Locate. This is the search in the search and rescue business.

b. Access. The determination of the assets, equipment, technical skills, and personnel required safely reaching and extracting the victim.

c. Stabilization. Providing the on scene medical attention required in order to extract the victim without causing further injury or aggravating an existing condition.

d. Transport. Removing the victim from the hostile environment to a safe one where further treatment can be provided.

5. **METHODS OF RESCUE**. (FMST.07.42c) In any rescue, we must access the situation to determine the best techniques to apply. Bearing in mind the self-aid, buddy-aid, victim-aid concept as well as time, weather, victim's condition, etc. In all cases we will attempt the lowest risk methods first, while preparing for a higher risk option. In all cases remember the KISS principle.

a. Reach. Use a ladder, fire hose, stick, paddle, rifle or any other sturdy object to extend your reach in order to reach the victim.

b. Throw. Get some type of floatation device to the victim. Or throw a line to the victim in order to pendulum or pull him/her to the shore.

c. Row. Putting a boat in the water to reach the victim.

d. Go and Tow. These are the contact rescues that, while look good on film, put you at the extreme risk of becoming a second victim.

e. Helo. Most civilian helicopter crews are not trained in SAR evolutions. Military aircraft and crews are a better bet. Several uses for helos are in the support role. Locating the victim, inserting and extracting teams and moving equipment are some of these roles. If using the helo in a direct role it is better to use a qualified SAR swimmer/diver to hook-up or pick-up victims.

UNITED STATES MARINE CORPS
Mountain Warfare Training Center
Bridgeport, California 93517-5001

FMST.07.25
03/14/02

STUDENT HANDOUT

PERSONAL AND TEAM EQUIPMENT

TERMINAL LEARNING OBJECTIVE As a member of a swiftwater rescue team, maintain swiftwater rescue equipment, in accordance with the reference. (FMST.07.25)

ENABLING LEARNING OBJECTIVES.

(1) Without the aid of references, from a given identify the correct items of personal equipment, in accordance with the reference. (FMST.07.25a)

(2) Without the aid of references, from a given identify the correct items of team equipment, in accordance with the reference. (FMST.07.25b)

OUTLINE.

1. **PERSONAL EQUIPMENT**. (FMST.07.25a) The following items of personal equipment should be in each man's possession for a swiftwater rescue.

    a. Dry Suit. Warmest, most versatile, full body protection available. Extra thermal layers can be added for varying water temperatures. Provides wind protection out of the water and provides protection against contaminated water.

    b. Wet Suit. Excellent protection from flotsam. Variety of thick nesses for thermal protection. Insulating quality is degraded in moving water due to the flushing effect.

    c. Footwear. Primary is full sole scuba booties. You can use tennis shoes or jungle boots with heavy wool socks.

    d. Gloves. Neoprene is recommended.

    e. Hoods. Wetsuit hood or pile hood for insulation. Remember 30-50% of your body heat losses can occur through the head and neck region of your body if left unprotected.

B2

f. Helmet. Protects you from flotsam and impact injuries. Should be padded and ventilated to allow water to flow through.

g. Knife. Fixed or locking blade attached to the vest, handle down.

h. Whistle. Required for communication. Best without "pea".

i. Eye Protection. Protects subject from debris.

j. Personal Floatation Device. This is the single most important piece of equipment. Jackets should provide at least 15 lbs. of floatation. Proper fit is a priority. Rescue vests with a "Blow-Out" belt should only be used by trained professionals.

2. **TEAM EQUIPMENT**. (FMST.07.25b) The following is a list of some of the team equipment for a typical four-man team. Of course the situation will dictate which equipment will be required for the mission on hand, but all hands should be familiar with this equipment. This is only a list. The uses and application of this gear, along with practical application exercises later in this course will provide you with the hands-on training and working knowledge to properly and safely use this equipment.

QTY	ITEM
1	300 ft, 1 Static Rope
4	150 ft, 11 mm Static Rope
25	85 Locking Stubai Carabiners
4	2" Pulleys
2	Rescue Figure 8
6	3 ft, 7mm Prussic
20	15 ft, 1" Tubular Nylon Runners
2	Ascenders
3	Throw Bags
2 pr	Fins
1	Rescue PFD
1	Self Bailing Raft (13 or 14 foot)
1	Oar Frame w/ Oars, or Seven Paddles
3	Radios (waterproofed)
1	Medical Kit
1	Litter

NOTE: This is not a complete list of rescue equipment. However, it is a good basic load.

UNITED STATES MARINE CORPS
Mountain Warfare Training Center
Bridgeport, California 93517-5001

FMST.07.41
03/14/02

STUDENT HANDOUT

SWIFTWATER TERMINOLOGY AND DYNAMICS

TERMINAL LEARNING OBJECTIVES In a swift water rescue scenario, evaluate swift water river conditions in accordance with the reference. (FMST.07.41)

ENABLING LEARNING OBJECTIVES.

1) Without the aid of reference, from a given list correctly identify the swiftwater terminology, in accordance with the reference. (FMST.07.41a)

2) Without the aid of reference, from a given list correctly identify the classifications of rivers, in accordance with the reference. (FMST.07.41b)

OUTLINE.

1. **SWIFTWATER TERMINOLOGY**. (FMST.07.41a)

a. Hydraulic or Hydraulic effect: Movement of water caused by pressure.

b. Eddy: Horizontal reversal. Current is blocked by an obstacle creating a vacuum which, when filled, causes water to flow upstream.

c. Eddy Fence: Reversal line between downstream current and the eddy. This fence can be very strong and the water flow may vary in height by several inches.

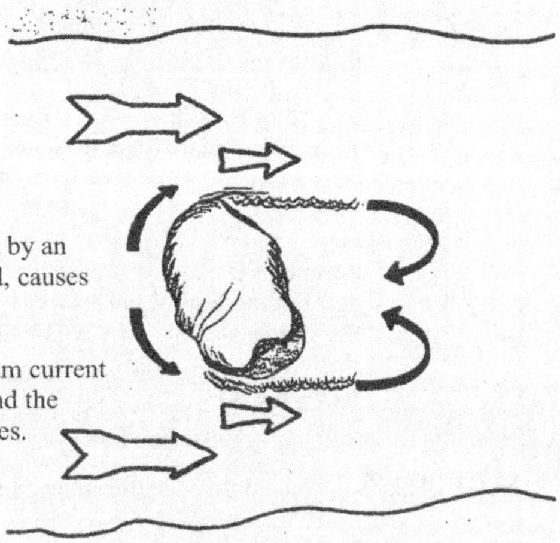

EDDY/EDDY FENCE

B3

d. Hole, Stopper, Keeper: Vertical reversal as water flows over an obstacle the main current follows the base of the river causing a surface vacuum. Downstream water flows back upstream to fill this void. This sets up a recirculating effect.

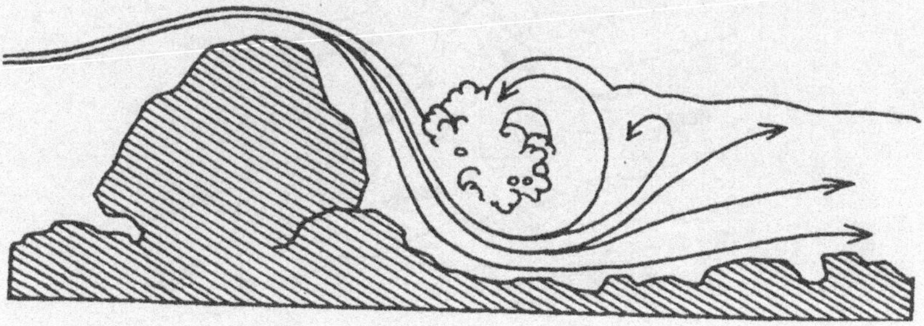

HOLE/STOPPER/KEEPER

e. Standing Wave. Haystack: Series of waves in the main channel caused by obstacles, ledges, or increases in the flow rate, gradient, or volume.

STANDING WAVES/HAYSTACKS

f. Downstream V. Tonal: Represent largest volume of water or main channel.

g. Upstream V: Caused by obstacles submerged just below the surface.

h. Strainer: Debris, logs, small trees on flooded banks, which allow water to flow through but will not allow a body to pass.

STRAINER / GUARD RAIL

STRAINER / TREE

i. Cushion, Pillow: Build-up of water on the upstream side of obstacles or the outside of curves.

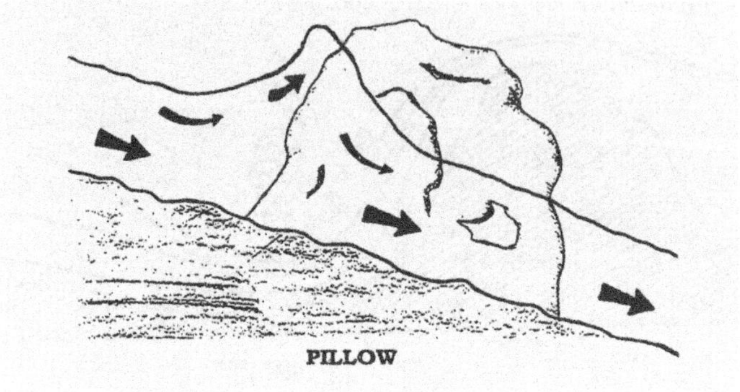

j. Confluence: Merging of streams.

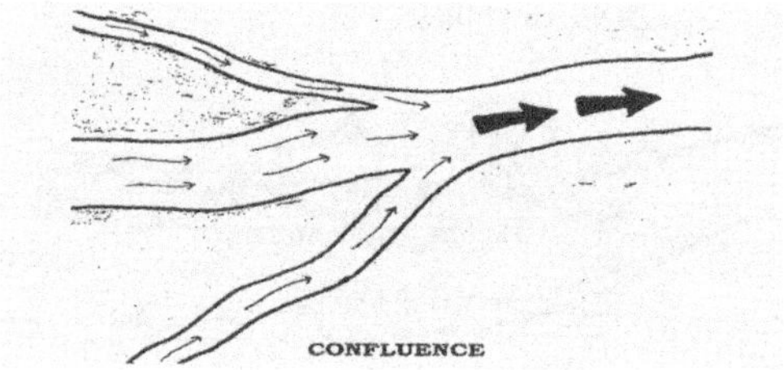

k. River Right. On the right side of the river looking downstream.

1. River Left. On the left side of the river looking downstream.

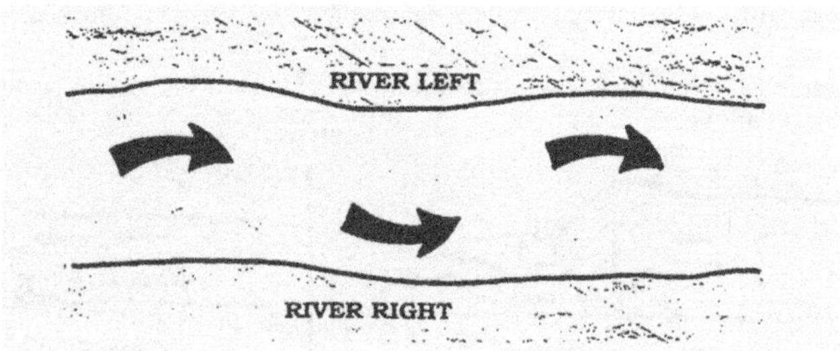

2. **CLASSIFICATIONS OF RIVERS**: (FMST.07.41b)

a. Class 1. Moving water, few obstacles, small ripples or waves.

b. Class 2. Moving water, clear channels, wide waves to 3', no scouting required.

c. Class 3. Rapids, narrow channels that require complex maneuvering, waves capable of flipping rafts or swamping canoes, scouting required.

d. Class 4. Long hazardous rapids, precise boat handling in rough water, scouting required, rescues hazardous.

e. Class 5. Severe, extended, extremely violent rapids, professional boat handlers and scouting required. Rescues extremely hazardous, situations are life threatening.

f. Class 6. Only Expert/Professional River Runners. Rescue may be impossible.

g. CLASSES U DON'T EVEN THINK ABOUT IT! Waters are not navigable.

3. **RIVER STRENGTH**.

a. Volume of Water. The volume of water in a given stretch of the river between two points is based on a simple formula and is measured in Cubic Feet Per Second (CFS). Simply take the river depth x river width x river velocity = volume of water in CFS.

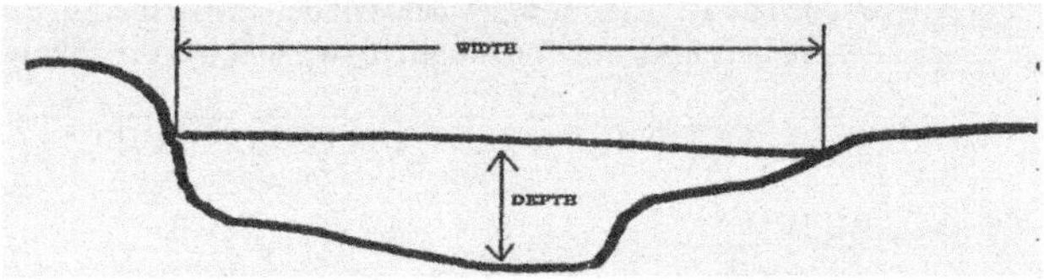

For example: 6 feet deep x 100 feet wide x 2 feet per second = 1,200 CFS

b. River Velocity. In order to determine the velocity or current speed measure a 100-foot section of the river and drop a float at the upstream end. Then, time its movement to the lower end and use the chart below.

TIME IN SECONDS	VELOCITY FT PER SEC	VELOCITY MPH
5	20	13
10	10	7
15	6.7	5
17	5.9	4
20	5	3.5
25	4	3
29	3.5	2.5
37	2.7	2
50	2	1.5
80	1.3	1
110	0.7	0.5

NOTE: Should the shape of the river change by becoming wider or narrower, once you have the flow in CFS, you can determine the new velocity by using the following formula:

$$\text{Velocity} = \frac{\text{Volume in CFS}}{\text{Width x Depth}}$$

c. Force Exerted. The velocity of the water will determine the force exerted X's the amount of exposed area under pressure. For simplicity we will use the following chart:

CURRENT VELOCITY	FORCE EXERTED ON OBJECT		
MPH	LEGS RAFT	BODY	WRAPPED
3	17	34	168
6	67	134	672
9	151	302	1,512
12	269	538	2,688

4. **LOW HEAD DAMS**. Low head dams are used throughout the world for irrigation, flood control, at hydroelectric plants and for many other uses. Basically they are pouring over obstacles that create hydraulics, keeper holes, and recirculating currents at their bases. Unlike natural obstacles, that rarely span the entire river and that allow you to be "flushed" out of the bottom or sides, low head dams create a hydraulic that extends from shore to shore.

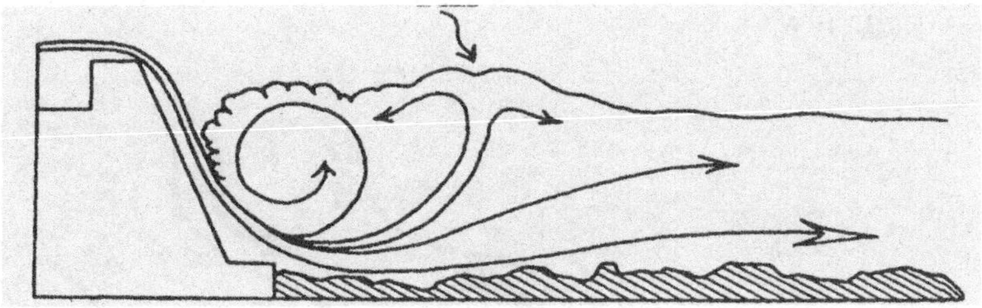

a. The size of the hydraulic created is based on several denominators:

(1) Height of drop.

(2) Angle of drop.

(3) Volume of water CFS.

(4) Velocity of water MPH.

(5) Bottom composition of runout.

b. The degree of difficulty for the rescuer is also compounded by:

(1) Water depth below dam.

(2) Debris build-up.

(3) Dam construction/conformation.

(4) Lack of access.

(5) Lack of natural anchors.

(6) Distance between riverbanks.

c. The basic hydrology is the same as any other pour over obstacle. As the river flows over the dam, the water follows the face of the dam to the river bottom. When sufficient speed diminishes the water rebounds to the surface and continues downstream. This flow has caused a vacuum or pocket directly below the dam that must be filled. Therefore, when the water rebounds to the surface, some of it flows back upstream to form a vertical reversal or pour-over eddy. The line of demarcation for the upstream/downstream flow is called the boil line. The shape and angle of the dam face and the volume and velocity of the water flowing over the dam create the distance and width of the boil line.

5. **REVIEW**.

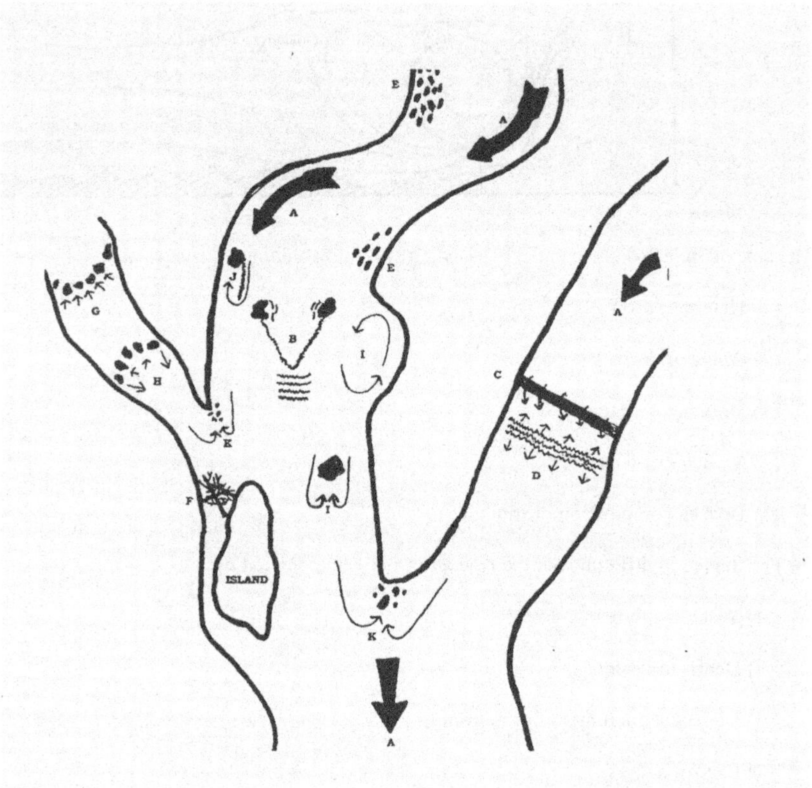

A. Main Current
B. Downstream V
C. Low head Dam
D. Boil Line
E. Rock Garden
F. Debris Pile/Strainer
G. Frowning Hole
H. Smiling Hole
I. Eddy
J. Eddy Fence
K. Confluence

UNITED STATES MARINE CORPS
Mountain Warfare Training Center
Bridgeport, California 93517-5001

FMST.07.24
04/14/02

STUDENT HANDOUT

COMMUNICATIONS

TERMINAL LEARNING OBJECTIVES Given a casualty in a swift water environment, assignment as a member of a rescue team, and a whistle, employ swift water rescue communication signals to support mission requirements per the reference. (FMST.07.24)

ENABLING LEARNING OBJECTIVES.

(1) In a summer mountainous environment, prior to swift water rescue operations, when the instructor verbally states a hand and arm signal, the student will demonstrate that signal, in accordance with the references. (FMST.07.24a)

(2) In a summer mountainous environment, standing at the edge of a swift water stream, when the instructor blasts a whistle a varied amount of times, the student will focus their attention in the proper direction, in accordance with the references. (FMST.07.24b)

OUTLINE.

1. **RELIABLE COMMUNICATION**. In many cases time is critical in a SWR scenario. Therefore immediate, understandable communication is a must. Radios, bull horns, and voice work well in some situations, but wet equipment, dead batteries, and ambient noise levels can render these methods ineffective. Therefore, rescue professionals have developed a series of simple hand and arm, and whistle signals. Any rescue team can develop their own signals. The only requirement is that all hands understand the signals and that they are briefed to any other agency you are working with.

B4

2. **HAND AND ARM SIGNALS**. (FMST.07.24a)

ACTION	MEANING
1 hand extended over head	Require Assistance
Both hands extended forming an "O" with one hand touching The head	Okay
Extend both hands above head, then point left/right	Move in direction indicated
Extend both hands above head, wave, then point left/right	Eddy out in direction indicated
Cross both arms over chest	Require medical assistance or Corpsman

3. **WHISTLE BLASTS**. (FMST.07.24b)

ACTION	MEANING
One Blast	Stop/Attention
Two Blasts	Upstream
Three Blasts	Downstream
Three Blasts Repeated	Emergency

UNITED STATES MARINE CORPS
Mountain Warfare Training Center
Bridgeport, California 93517-5001

FMST.0
7.27
04/14/02

STUDENT HANDOUT

THROW BAGS AND FOOT ENTANGLEMENTS

TERMINAL LEARNING OBJECTIVES In a swiftwater rescue scenario employ a throw bag, in accordance with the reference. (FMST.07.27)

ENABLING LEARNING OBJECTIVES.

1) Without the aid of reference, provided with a throw bag and swimming victim, rescue the victim, in accordance with the reference. (FMST.07.27a)

OUTLINE.

1. **THE THROW BAG**: The throw bag is very simply a cloth bag with a small float insert, packed with 50 to 70 feet of 1/4" to 3/8" floating polypropylene or spectra/poly line. The concept is to hold the exposed end of the line and throw the packed bag to a swimmer. As the bag flies, the rope deploys enabling you to reach a victim 50 to 70 feet from shore. It is used either from a standing position, or a moving position on open shores.

   a. Standing Procedures. (FMST.07.27a)

1. Assume a stable belay position and ensure your footing is solid. Avoid loose, sloped or wet rock.

2. Attract victim by blowing your whistle (1 blast). Show him/her the throw bag and yell, "grab the rope". NOTE: If the victim grabs the bag rather than the rope, the rope will continue to deploy to its full length.

3. Grasping the exposed end of the rope, throw the bag underhand in a smooth motion directly at the victim. You may need to lead them in order to score a direct hit.

4. Tell the victim to place the rope over his/her upstream shoulder. This will establish a good ferry angle.

B5

5. Do not wrap the rope around yourself or tie it off at anytime. You may have to release the victim to avoid an obstacle or strainer. Use a dynamic belay and pull the victim to shore or pendulum the victim to other rescuers downstream.
6. If available have other rescuers downstream to recover the victim or repeat the procedure if you miss!

b. Moving Procedures. The moving procedures are the same as the standing procedures except you are on a smooth, open shoreline, so you move slightly downstream of the victim as you pull them gently ashore.

2. <u>**FOOT ENTRAPMENTS AND VERTICAL PINNINGS**</u>. These scenarios are caused by a variety of actions. Swimmers put their feet down or fishermen, in waders, get their feet stuck between rocks and are pushed over by the current. Swimmers are stuck in a strainer. Rafters get wrapped on a rock with victims pinned between the raft and rock or kayakers drop down a vertical shoot and get lodged in the rocks at the bottom. Whatever way it occurs you must respond quickly in order to save the victim. Initially take the following precautions:

    1. Position upstream spotters, to provide warning of boats or debris.

    2. Plan alternate methods if initial try doesn't work.

    3. Have back-up rescuers downstream in order to retrieve the victim.

  a. Once these actions are complete conduct the rescue using one of the five following methods.

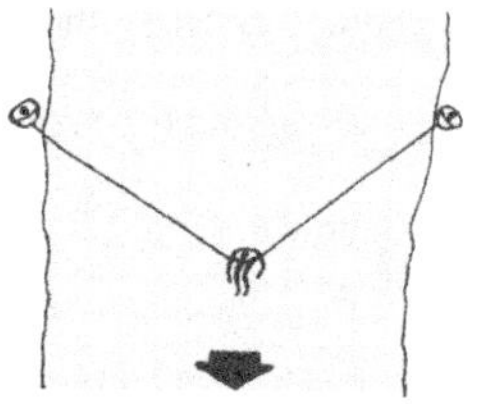

    1. <u>Bechel-Ray One Line Method</u>. The simplest and safest method. Get a rescuer on both sides of the river. Establish a rope between the two rescuers and downstream of the victim. Walk upstream until the rope hooks the victim. If you can position the line near the victims ankles you may be able to pop him/her loose and the downstream safety can pick him/her up. If not you can haul the victim upstream until he/she is clear of the obstacle, then by one rescuer playing out line and the other retrieving it you can bring the victim ashore. Remember to bring the victim to the shore with the medevac capability.

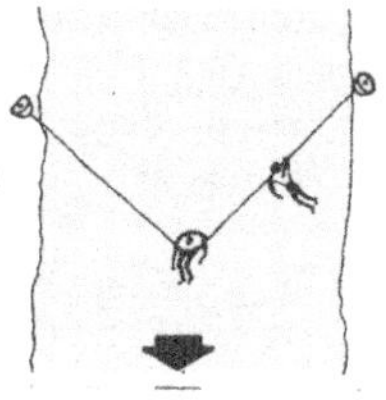

    2. <u>The Strong Swimmer-variation</u>. If the previous method doesn't work, leave the line in place and anchor it to stabilize the victim above the surface. Then have a "strong swimmer" go down one side of the rope, as would a tension diagonal. This rescuer is not tied in. Upon arrival the rescuer can readjust the line on the victim, attempt to pull them free, give them a PFD and helmet, or if a debris pile or rock is near, get on it and assist in further efforts.

3. Two Line Loop. There are two variations on this method both of which are applied using the initial rope used in the Bechel-Ray attempt.

(1) Deploy a second rope across the river and hook it onto the first line. Allow it to go one-half way down the rope to the victim. Pull at a 90° angle to rope #1. This will establish a mechanical advantage and pull the victim loose.

(2) Once the victim is free walk the line downstream trapping the victim in a loop. Then by releasing slack on line one, you can bring him/her ashore.

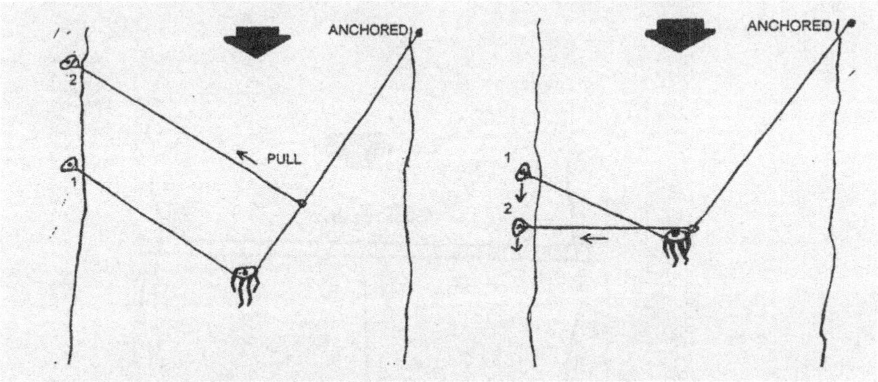

4. European Tethered Rescuer Lower. Using a rescue vest with a "blow-out" belt establish an upstream belay and lower the tethered rescuer to the victim. The rescuer's body will block the water forming an eddy around the victim enabling him to be pulled free. Ensure that the rescuer has a working rescue vest, as this will enable him to cut away if the situation dictates.

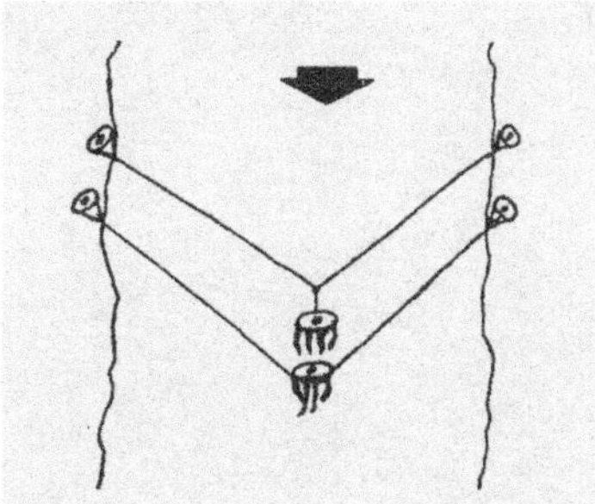

5. Mechanical Advantage Systems. As a last resort you can use a mechanical advantage system. This method is usually used only for body recoveries, as the amount of force exerted on the victim is tremendous and can cause severe injuries.

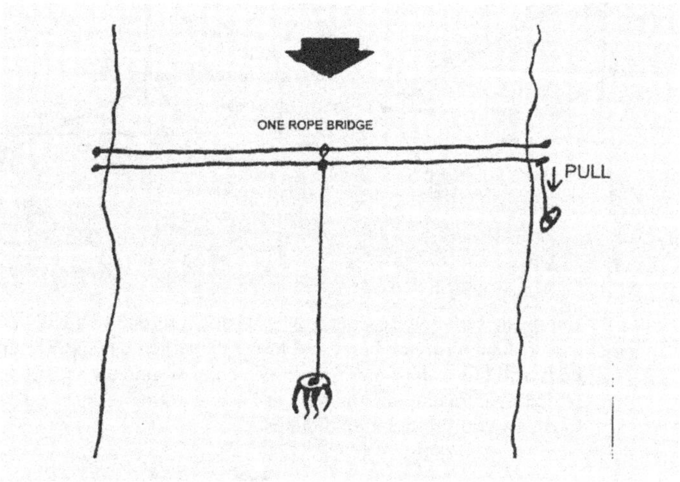
ONE ROPE BRIDGE
PULL

UNITED STATES MARINE CORPS
Mountain Warfare Training Center
Bridgeport, California 93517-5001

FMST.07.28
04/14/02

STUDENT OUTLINE

DEFENSIVE SWIMMING AND FERRY ANGLE

TERMINAL LERNING OBJECTIVES Given a swift water river and throw bags, perform aggressive swift water swimming to meet mission requirements per the reference. (FMST.07.28)

ENABLING LEARNING OBJECTIVES.

(1) Without the aid of reference, demonstrate the defensive swimming position, in accordance with the reference. (FMST.07.28a)

(2) Without the aid of reference, demonstrate use of the ferry angle to cross the river, in accordance with the reference. (FMST.07.28b)

OUTLINE

1. **THE DEFENSIVE SWIMMING POSITION (DSP).** (FMST.07.28a) The DSP is assumed by lying flat on your back, chin on chest (to read the river & select a route), with your feet up, heels just below the level of your buttocks. Holding your feet in this position will enable you to fend off of shallow rocks and will reduce the likelihood of a foot entrapment or pinning. Your arms and hands should be kept close to the body but may be used like oars on a boat to maneuver. Your head will remain upstream except to cross a strainer or to over take a victim in a swimming rescue.

2. **FERRY ANGLE.** (FMST.07.28b) A ferry angle is used to maneuver diagonally across a river. By placing your body at a 45° angle to the current, the river will push you across the river. This is because the force of the current will travel down the exposed, upstream, side of the body moving it laterally more rapidly then the current pushing straight. This is like traversing a slope with edged skis vice sideslipping down on flat skis. In order to cross faster or to reach an eddy you may roll to your belly and execute the rescue crawl stroke or breaststroke. Remember to maintain your ferry angle with your head upstream this will enable you to travel diagonally across the river.

3. **STRAINER CROSSING**. The first rule is to avoid the strainer or shallow submerged obstacle. Read the river and maneuver to the main channel by using downstream "V's" or tongues. If the obstacle cannot be avoided, roll on your belly, head downstream and use the rescue crawl stroke. You must get your body moving faster than the current. Upon arrival at the obstacle use both hands to "vault" yourself over. A well-timed dolphin kick will assist in keeping your legs clear of entanglements. Once clear of the obstacle, resume the defensive swimming position. Bear in mind that it is usually safer to remain in the main current or to eddy hop downstream than it is to attempt to get out on a flooded riverbank, the outside of a bend or on slippery rocks. Also, use the dolphin kick rather than placing your feet on the bottom and kicking off. This will reduce the likelihood of foot entrapments.

UNITED STATES MARINE CORPS
Mountain Warfare Training Center
Bridgeport, California 93517-5001

FMST.07.25
03/14/02

STUDENT HANDOUT

FEET-WET RESCUES AND C-SPINE ROLLS

TERMINAL LEARNING OBJECTIVES In a swift water rescue scenario, conduct a swimming rescue, in accordance with the reference. (FMST.07.25)

ENABLING LEARNING OBJECTIVES.

(1) Without the aid of reference, while in a river and provided with a victim, perform a contact rescue, in accordance with the reference. (FMST.07.25a)

(2) Without the aid of reference, while in a river and provided with a simulated unconscious victim, perform a C-spine roll, in accordance with the references. (FMST.07.25b)

OUTLINE.

1. **FEET WET RESCUE GENERAL CONSIDERATIONS**. (FMST.07.25a)

   a. Always wear the proper personal equipment, i.e. PFD, helmet, drysuit, etc.

   b. Don't make physical contact if it can be avoided. Make him/her "chase" you ashore. Use a life ring, can, or boogie board, and tow it along with the victim. Toss him a throw bag and swim/tow him/her until you reach the shore, then pendulum or haul the victim in. Remember, to a panicked victim YOU are a floatation device.

2. **CONTACT RESCUE**.

   a. Approach the victim from upstream. Use the rescue crawl stroke to catch-up to him/her. Talk to the victim and attempt to get him/her to swim to shore or "chase" you to the shore.

B7

b. Explain to the victim the defensive swimming position and attempt to get them in this position. Assume this position yourself, in contact rescues this is called the "Reverse and Ready" position. Tell the victim what you are going to do prior to grabbing them.

c. Gain contact by grabbing the shoulders of the life jacket or utilize a cross-chest carry. If the victim is unconscious and face down use a C-Spine roll down and maintain stability as much as possible.

d. Assume a good ferry angle with the victim downstream of you and use his/her body to protect you from obstacles.

e. If the victim panics and attempts to climb on top of you, break away or lock the victim up and go for an alligator roll. Reestablish control and calm the victim as soon as possible.

f. Once contact is made remember that you may be held accountable under the "abandonment" laws if you quit.

3. **TOWED VICTIM SWIM**. When towing a victim it is essential that you establish a ferry angle. If the victim is on a boogie board or float, explain to him/her what you are doing and have them help by kicking their feet. If you are towing by the PFD grasp the shoulder towards the far bank and remain in the DSP (defensive swimming position) in case your victim panics.

4. **LIVE BAIT RESCUES**. The final go and tow method is the "live bait" rescue. Using a rescue vest with a "blow out" belt and a mule team on the shore to belay you, swim to the victim as explained in contact rescues. Once you establish contact you are, in effect, a throw bag with hands. The belayer will then either pendulum you ashore or hauls you in. In an emergency you can cut away and conduct a normal contact rescue.

5. **C-SPINE ROLLS AND STABILITY**.

a. Basic Rules: (FMST.07.25b)

(1) Suspect C-Spine involvement in any head injury.

(2) Stabilize and immobilize, use C-Collar, Sam Splints, or a field expedient method.

(3) Keep the victim in the water until a spine board or equivalent is available, or there are enough rescuers to move the victim safely.

(4) Observe for other signs, i.e. cranial bleeding, bleeding from ears, eyes, nose (Do not attempt to stop this bleeding) or unequal/unresponsive pupil response.

b. Methods:

(1) First, approach the victim from behind. Reach under the arms and grasp the victim's head with both hands, then roll the victim. This is the least desirable method as you cannot swim properly or evaluate the ABC's of the victim.

(2) Second, approach the victim from the side, extend his/her arms above and alongside his/her head, reach across and roll the victim onto your arm maintaining the victims arms in place to ensure stabilization. Evaluate the ABC's, provide rescue breathing and use your free arm to swim.

(3) Finally, approach a victim wearing a PFD from the head. If face down, cross your arms thumbs inboard and use your forearms to immobilize the neck. Then roll the victim over. If face up, simply approach as if gaining contact on a contact rescue and maintain control of the C-Spine.

UNITED STATES MARINE CORPS
Mountain Warfare Training Center
Bridgeport, California 93517-5001

WMC

04/14/02

STUDENT HANDOUT

BOATS AND BOAT HANDLING

LESSON PURPOSE. The purpose of this period of instruction is to introduce the student to the boats available at MWTC for river rescue evolutions.

OUTLINE

1. **GENERAL**. There are a wide variety of boats and hybrid river craft available to rescue crews. Whitewater rafts, rigid raiders, zodiacs, airboats, and regular pleasure boats can all be used in the proper situations. All have been used both safely and have caused deaths and injuries when used improperly. Wherever you are, learn the equipment available, both its assets and its limitations.

2. **WHITEWATER RAFTS**. The whitewater raft of today is an inflatable, self-bailing, multi-chambered vessel with several tie rings and foot cups. Additionally, the materials used in construction resist rips, tears, and punctures much better than their predecessors. The size most often found on local rivers is 12 to 14 feet long due to the size of the rivers. In the MWTC SAR locker, we have a 13-foot self-bailer. This will be our primary rescue boat.

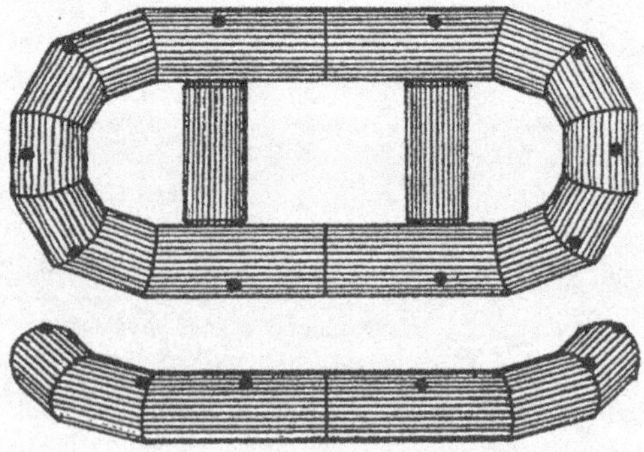

B8

3. **RAFTING COMMANDS**. In order to control the direction of a peddler powered boat the coxswain or guide must use a series of simple, concise, and understandable commands.

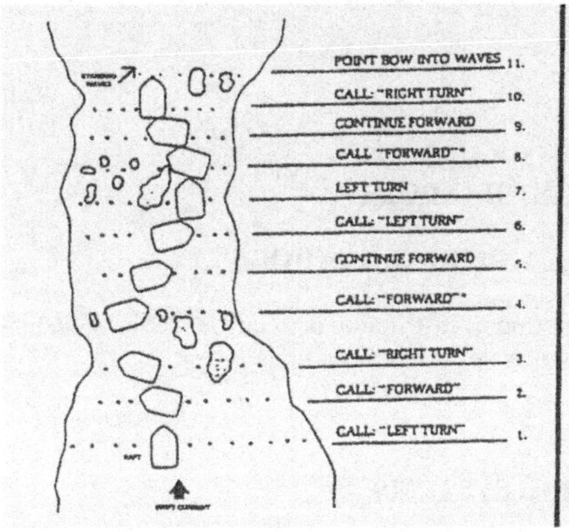

NORMAL TACTICS APPROACH

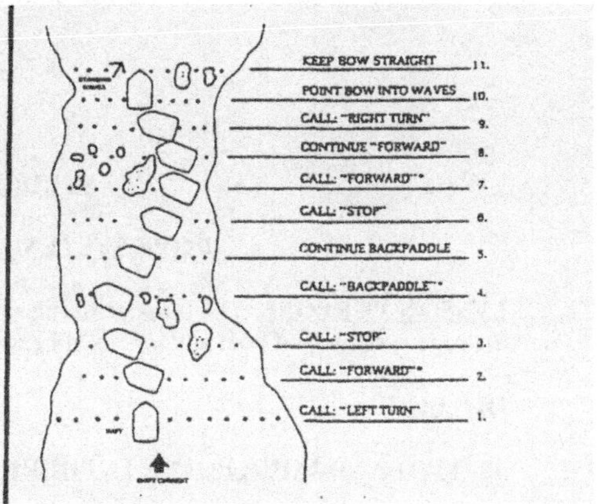

CONSERVATIVE/FAST CURRENT

a. Forward - All hands paddle forward together.

b. Back Paddle - All hands back paddle together.

c. Stop - Stop paddling - Remove blades from water.

d. Left Turn - Left side back paddles, right side forward.

e. Right Turn - Right side back paddles, left side forward.

f. High Side - Move quickly to the down streamside of the boat.

4. **PADDLE STROKES**. In order to control the raft from the coxswain position, either with a crew or alone on a tether system, you will need to apply the following strokes:

a. Sweep - Stroke in a wide arc around the fantail of the boat.

b. Draw - Reach out 90° away from the boat and "draw" it straight back to you.

c. Pry - Put paddle blade beside raft and "pry" the water away from the boat.

d. "J" - Used by lone guide to keep boat straight. Regular straight stroke finished with an outboard stroke or tail.

UNITED STATES MARINE CORPS
Mountain Warfare Training Center
Bridgeport, California 93517-5001

WMC
04/14/02

STUDENT HANDOUT

ROPE SYSTEMS AND BOAT ANCHORS

LESSON PURPOSE: The purpose of this period of instruction is to introduce the student to the rope systems and boat anchors used in swiftwater rescue situations.

OUTLINE

1. **TWO AND FOUR POINT TETHERING SYSTEMS**.

a. With all rope systems the first order of business is to get a rope across the river. Best case is to throw a rope bag across to someone already on the other side. Next best is to use another boat or raft to ferry the rope across. Last we use a swimmer to drag the rope across. Remember to use a smaller line on the swimmer and not to tie it to him/her. When swimming, start high and pendulum across using a good ferry angle.

b. Clip in on the side "D" rings forward for a two point or on the two forward and two aft "D" rings for a four-point tether system. For a two point you now need to simply load in a rescuer (if used) and move the boat into the river by taking slack on the far side and giving slack from the near side. For a four point man the two nearside lines and load a far side aft lineman and a rescuer (if used) into the boat and ferry the boat across. The far side aft belayer will get out and the boat is on a four point system ready for action.

c. This system is good for Class II Rivers with clear banks, few mid stream obstacles, and little floating debris. Use more belayers as required by river speed and boatload.

d. In all rope systems the whistle and hand and arm signals used to communicate are those taught in the communications class earlier in this course.

e. To break down the system simply reverse the procedures. The only difference is that the last man rides in the boat instead of swimming.

B9

2. **THREE POINT AND BOATMEN'S SELF-EQUALIZING ANCHORS.**

a. These anchors have in common the fact that they are self equalizing. This means that, regardless of the direction of pull, all anchor points ("D" rings, thwarts, etc) will remain under equal tension.

b. Using ropes the knots required are the double figure 8 on a bight, a directional figure 8, and a figure 8 on a bight.

c. Using webbing (one inch tubular nylon) a water tape knot is required.

d. If you have an abundance of carabineers, simply clip into the anchor points and back into the small loop on the double figure 8 on a bight, or into the directional figure 8. If not run the rope through the small loop. Remember that this is not preferred as it causes rope on rope stress and rubbing that can melt the rope.

e. When using webbing the same concept applies. Ensure you put a twist in the webbing at the contact point to ensure that a single blow-out will not collapse the entire system.

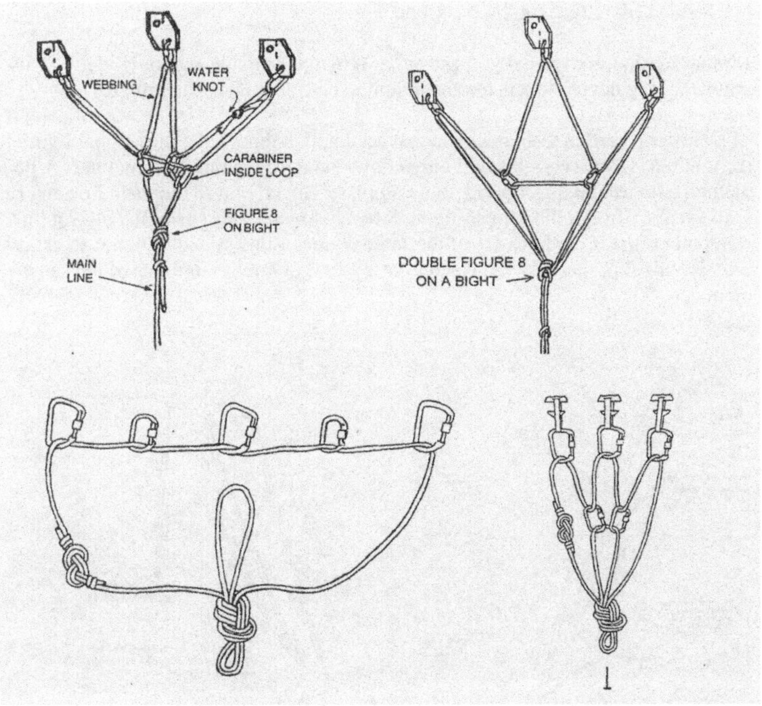

3. **PIG-RIG FOR MECHANICAL ADVANTAGE.** At times you will find yourself with ropes that are long enough to span the river, but not long enough to establish a mechanical advantage system. In these situations, you can use a piggy-back line or "pig-rig" by following the procedures below:

a. Tie off the mainline on either ends, or tie off the far side and set up a belay device on the near side using a munter hitch or rescue 8.

b. Establish an anchor for the pig-rig below the near side anchor.

c. Tie a prussic on the mainline and clip in a carabiner. Tie a figure 8 on a bight into the end of the pig-rig line and clip it into the prussic's carabineer.

d. Run the pig-rig line back to the anchor carabineer and then back through the prussic carabiner.

e. Pull on the running end towards the anchor. This will give you a 4:1 mechanical advantage on the main line.

f. If more tension is required you can either retie the nearside anchor (if tied), or use your belaying device to tension the mainline before moving the prussic.

g. If no prussic cord is available, you can accomplish this with a directional figure 8 on the mainline. However, this will only allow you to tighten the system once. A better method is to use the pig-rig line to tie a prussic directly on to the main line and run your system off of a directional figure 8 tied below the prussic on the pig-rig line. Remember that if the lines are of the same diameter this system will not work as the prussic will slip, therefore, you should use a kragur knot with ropes of the same diameter.

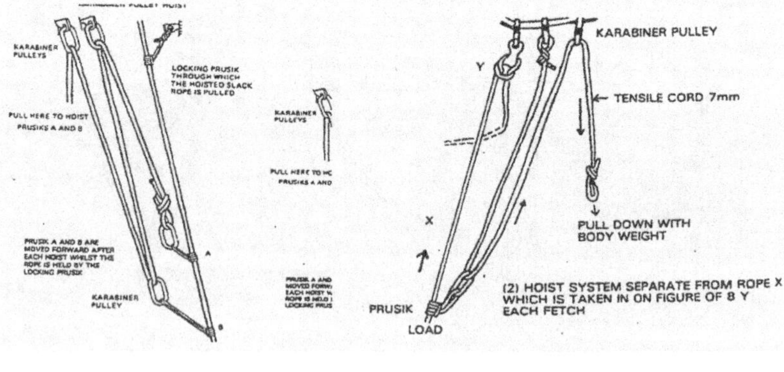

4. **TYROLEAN HIGHLINE AND TELFER LOWERING SYSTEMS** Both of these systems are used to maintain a boat as a stable rescue platform. The lateral movement (side to side/shore to shore) in both systems is controlled by mule teams on the shore. The main difference is that in the tyrolean highline the movement up and down-stream is also controlled from the shore, while in the telfer lowering system, the rescuer in the boat controls that aspect of movement. Both systems have good and bad points and both should be mastered. The situation will dictate which system you will use. Follow the steps below to establish these systems:

a. First you must establish TWO lines across the river. You can use any method taught to accomplish this.

b. Establish one line as the anchor line and tighten it as taught in the "one rope bridge" class or use a pig-rig, if required. If you have uneven mule teams set it at a slight down-stream angle towards the weak team's shore (no more then 10 to 15 degrees) to assist the weak team.

c. Rig the boat with a self-equalizing anchor system.

d. Rig the lowering systems as shown in the diagrams. Either tyrolean or telfer.

e. Load a rescuer or rescue team into the boat and execute. Remember the rescuer can act as coxswain and assist in maneuvering the boat by establishing a proper ferry angle.

f. Communications will be as taught earlier. Ensure you have a shore commander who maintains eye contact with the rescuers and mule teams on both the near and far shores.

g. When using the telfer system, a counter hitch, stitch plate or rescue 8 can be used as a belay device. This enables you to tie-off when you reach the victim so that you can affect the rescue without worrying about controlling the boat.

h. When using either of these systems it is a good idea to set-up a mechanical advantage system between the control (central) pulleys aid the raft/rescue platform.

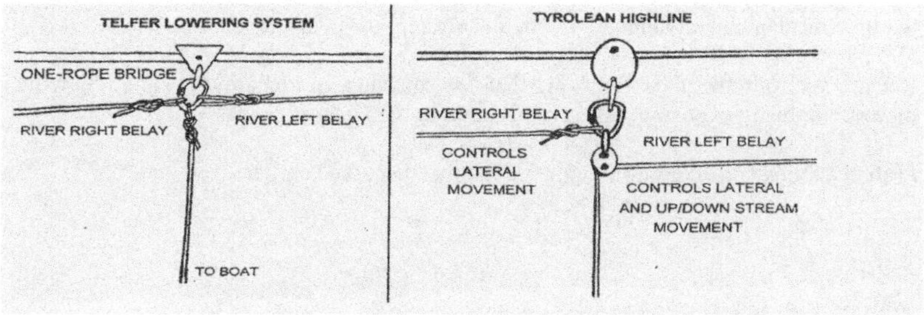

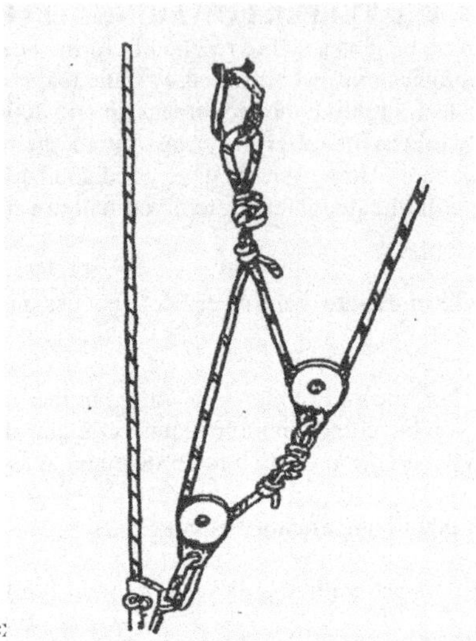

5. **BOAT WRAPS**. As logy and Dynamics" class, the river can put an e: /rapped around an obstacle in the river. You first job is to ensure that NO ONE is trapped between the boat and the obstacle. If you do not have full accountability of personnel, don't hesitate! CUT THE BOAT! The easiest method to do this is to cut the floor out of the older "bucket" boats, or to cut the laces holding in the floor of a "self bailing" raft. Depending on how tightly wrapped the boat is you have several options available:

   a. First ensure the crew is accounted for and treated for any injuries.

   b. Establish a tag-line and attempt to bounce and pull the raft free.

   c. Set a Z-drag inside the boat to dump water.

   d. Use an oar or tree to lever the boat off the obstacle.

e. Set up a mechanical advantage system and attempt to "peal" the boat free.

f. Release air from the tubes. Be aware that this can cause more harm than good by reducing tube strength and exposing more surface area to the water pressure.

g. If all else fails, consider cutting out the floor as discussed earlier.

UNITED STATES MARINE CORPS
Mountain Warfare Training Center
Bridgeport, California 93517-5001

WMC
04/14/02

STUDENT HANDOUT

VEHICLE RESCUES

LESSON PURPOSE. The purpose of this period of instruction is to train the students in the procedures used to rescue personnel from vehicles trapped in swift water.

1. **VEHICLE BEHAVIOR**. There are three types of hazards in a river. Floating objects such as trees and branches, suspended objects or tumbling objects such as sawyers, rocks, and other debris, and stationary objects such as live trees on flooded banks and boulders. A vehicle will become all of these "loads". For the first minute, or so, they will float, as they become partially filled with water, they will be suspended. This is normally when they tumble or roll, and finally they will become stationary as they lodge against another stationary object or fill completely. Once they become stationary, depending on the water depth, they may form an eddy, a pour through or pour over, creating a hydraulic or a standing wave if fully covered. Also, the straighter a vehicle is (head-on) to the water flow the more stable it is. Finally, bear in mind that changes in water flow or displacement (weight) of the vehicle (taking a victim off it) may cause the vehicle to once again become a suspended load and tumble.

2. **RESCUE PROCEDURES**

a. The initial action is to secure the vehicle. Ensure that you have adequate anchors. Trees, guardrails, and bridge abutments are good. Avoid using another vehicle if possible.

b. Approach the vehicle on the downstream side by using the eddy created by the vehicle. Use a boat, either tethered or under power, a strong swimmer or boogie board. Another method is to use the line securing the vehicle as a tension diagonal to reach the vehicle. Then establish a second tension diagonal or endless rope to move the victims ashore. Ensure your upstream spotters are in place to warn of other suspended debris coming your way.

B10

c. If the vehicle is sealed tight, crack one window slightly to allow water to get in the vehicle slowly and equalize the pressure. Warning: If you pop or shatter a window you may cause a sympathetic detonation of the remaining windows.

d. Equip the victims with PFD's and remove to the boat or onto the rope system. Consider the following points:

(1) The victims weight might be the only thing stopping it from rolling. You might need to transfer a rescuer in a one for one exchange.

(2) On hard pack or asphalt the vehicle will shift more easily.

(3) If the vehicle is sideways keep the victims on the upstream side. The high side rule is reversed.

(4) Vehicles roll slowly.

(5) Helo's are a last resort.

UNITED STATES MARINE CORPS
Mountain Warfare Training Center
Bridgeport, California 93517-5001

FMST.07.3
0
04/14/02

STUDENT HANDOUT

STREAM CROSSING

TERMINAL LEARNING OBJECTIVE: Given a mountain swift water environment, and a staff, conduct a mountain stream crossing to meet mission requirements per the reference. (FMST.07.30)

ENABLING LEARNING OBJECTIVES:

(1) Without the aid of references, choose from a given list the considerations in site selection for a stream crossing, in accordance with the references. (FMST.07.30a)

(2) Without the aid of the reference, choose from a given list the safety precautions taken while conducting a stream crossing, in accordance with the reference. (FMST.07.30b)

(3) In a summer mountainous environment, execute the individual preparations taken prior to crossing a stream, in accordance with the reference. (FMST.07.30c)

(4) In a summer mountainous environment and given a staff, cross a mountain stream using the staff method, in accordance with the reference. (FMST.07.30d)

(5) In a summer mountainous environment, cross a mountain stream using team methods, in accordance with the reference. (FMST.07.30e)

1. **SITE SELECTION** (FMST.07.30a)

a. Mountain streams and rivers are military obstacles and therefore are danger areas for units crossing them. In order to reduce the time in the vicinity of the danger area, a recon team should precede the main body and select the best crossing site. The site selection for a stream crossing should include these eight considerations:

B11

(1) Look for logjams, rocks or fallen trees that will provide a dry crossing if possible.

(2) If a dry crossing is not possible, select a crossing point at a wide and shallow point where the current is slower.

(3) Avoid sharp bends. They can be deep with a strong current on the outside of the bend.

(4) Look for a firm, smooth bottom. This is because large rocks and boulders provide poor footing and cause a great deal of turbulence in the water.

(5) It may be easier to cross several small channels of water rather than one large one.

(6) Do not cross just above rapids, falls, or logjams, taking a fall or slipping could have serious consequences.

(7) Cross in the early morning. The water level will be lower since there has been less daylight for the snow to melt. Also, on sunny days, you will have more time to dry clothing and equipment.

(8) There should be a suitable spot downstream for safety swimmers.

2. **SAFETY PRECAUTIONS** (FMST.07.30b). The following two safety precautions must be taken while conducting stream crossings.

   a. There must be a safety line at a 45-degree angle downstream across the stream. This is for anyone who slips and is swept downstream to grab so that he is capable to stopping himself.

   b. There must be safety swimmers downstream. These are strong swimmers who are positioned downstream to help anyone who is swept downstream. The safety swimmers will use throw bags.

NOTE: The safety line, as well as any other lines that must be taken across the stream, will be taken across by using the lead swimmer method taught in ONE-ROPE BRIDGE.

3. **INDIVIDUAL CROSSING PREPARATIONS.** (FMST.07.30c) Prior to beginning to cross a stream, there are certain preparations that each individual should take. The five preparations are as follows:

   a. Wear pack with shoulder straps fastened snugly. Waterproof the pack for buoyancy if possible.

   b. Weapons will be slung diagonally over the shoulder with the weapon itself being between the pack and the individual's back.

c. Button all pockets and remove blousing garters. This prevents the water from creating added drag against the individual, which it would do if it could flow into open pockets.

d. Wear boots to protect the feet, but remove socks and insoles to keep them dry:

e. Wear the minimum amount of clothing. This reduces the amount of clothing that must be dried after the crossing.

f. Wear helmet in slow moving, shallow water.

NOTE: If you are in a tactical situation, the actual situation will dictate which of the above precautions will be taken.

4. **INDIVIDUAL CROSSING METHODS**. There are three individual methods that may be used.

a. Staff Method. (FMST.07.30d) A strong staff or pole about 6 feet long is used as a crossing aid. It should be strong enough to support the Marine's weight and trimmed clean of any branches. Placing both hands on the pole, the Marine should place the staff just upstream of his intended path. He should use the staff as the third leg of a tripod and should move only one leg or the staff at a time. He should face upstream using the staff to retain his balance. The staff is also used as a probe to discover bottom irregularities that could trip the Marine. The Marine should drag his feet instead of picking them up.

STAFF CROSSING METHOD

b. Swimming. This is an obvious method, if your Marines are good swimmers. This is not always the case, so usually this method is not a preferred one.

(1) In fast, shallow water, the Marine should angle across on his back with his feet downstream and his head up. He should use his hands to tread water and his feet to fend off obstructions.

(2) In fast, deep water, the Marine should angle across the stream on his stomach with his head upstream, to establish a proper ferry angle.

c. Belayed Method. In chest deep or water with a strong current, a rope can assist greatly. A rope will be secured from bank to bank, with the far anchor slightly downstream from the near anchor. The rope will be anchored off so that it lays at a minimum of 45 degrees. The Marine attaches himself to the rope by using a sling rope as safety line, tying a bowline around his waist and a figure-of-eight loop with a steel locking carabiner inserted. He then attaches the carabiner to the crossing line and crosses using the current to assist him.

5. **TEAM CROSSING METHODS** (FMST.07.30e)

a. Line Abreast. Small units (squad to platoon) can cross in moderate currents up to chest deep, by lining arms in a line abreast or chain method. The largest man of the chain is placed on the upstream side of the group. The group will enter the stream parallel to the flow of the stream. The middleman of the chain will control the group's movement and give the command when to step.

LINE ABREAST

b. Line Astern Method. Three or more men can line up facing the current. The upstream man, who should be the largest man in the group, breaks the current while the downstream men hold him steady. The upstream man may use a staff, similar to the individual staff method, to steady himself. All men side step at the same time with one man calling the cadence.

LINE ASTERN

c. Huddle Method. Between three and eight men can face inboard as in a football huddle. They will wrap their arms around each other's shoulders and cross the stream in this formation. The upstream man should change position as they cross because the entire formation will rotate. This prevents one man from becoming exhausted in the upstream position.

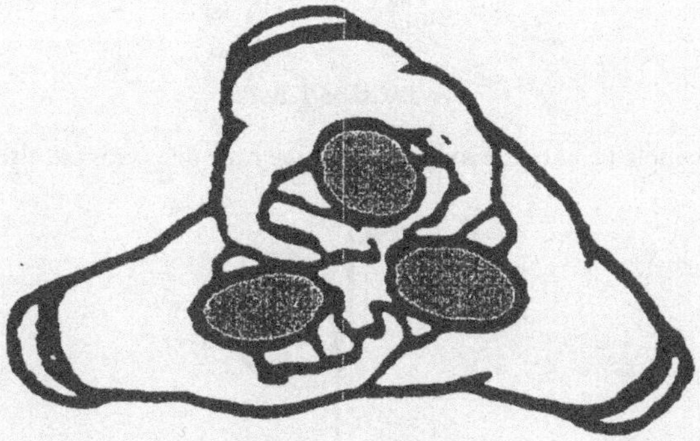

HUDDLE METHOD

UNITED STATES MARINE CORPS
Mountain Warfare Training Center
Bridgeport, California 93517-5001

WMC
04/09/02

STUDENT OUTLINE

ROPE MANAGEMENT

LESSON PURPOSE: To familiarize the student with some of the terms and procedures for working with a rope.

BODY

1. **TERMS USED IN ROPE WORK**

   a. Bight. A simple bend in the rope in which the rope does not cross itself.

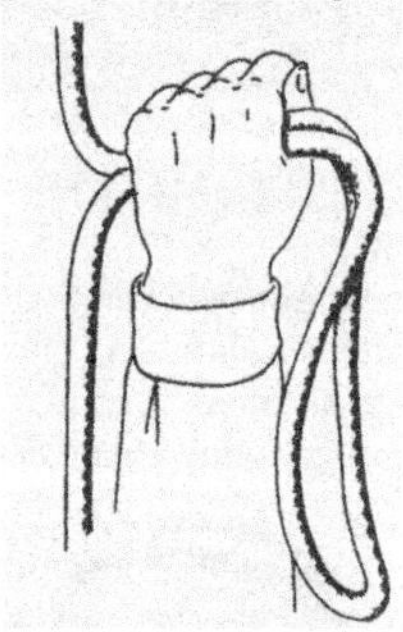

BIGHT OF ROPE

   b. Loop. A simple bend in the rope in which the rope does cross itself.

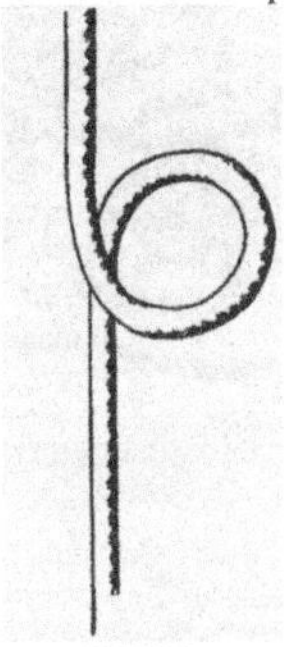

LOOP

c. Half Hitch. A loop which runs around an object in such a manner as to bind on itself.

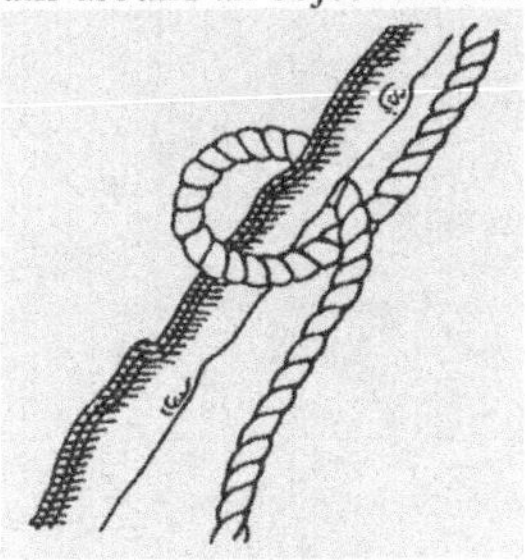

HALF HITCH

d. Standing End. The part of the rope which is anchored and cannot be used, also called the static end.

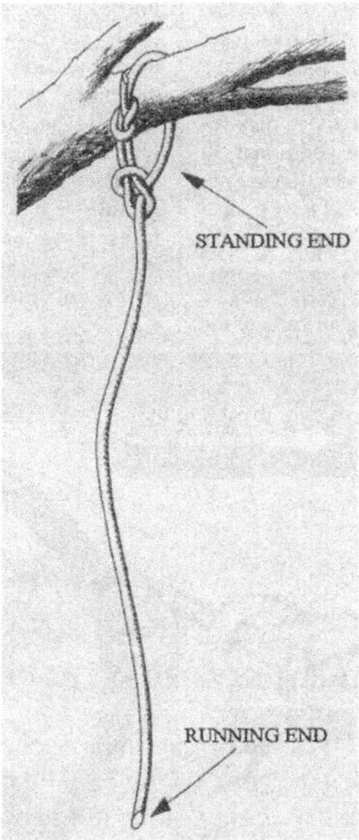

e. Running End. The free end of the rope which can be used.

f. Lay. The same as the twist of the rope. (Applies only to hawser laid ropes, such as manila.)

g. Pigtail. The short length left at the end of a rope after tying a knot or coiling a rope. It may or my not be tied off with a secondary knot, depending on the circumstance.

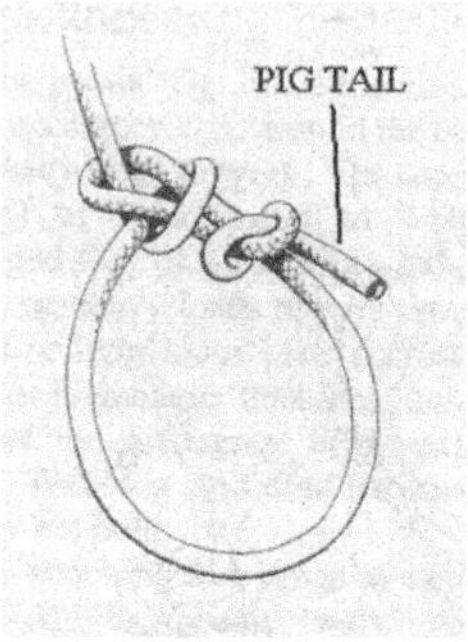

PIG TAIL

h. Stacking (or Flaking). Taking off one wrap at a time from a coil, and letting it fall naturally to the ground.

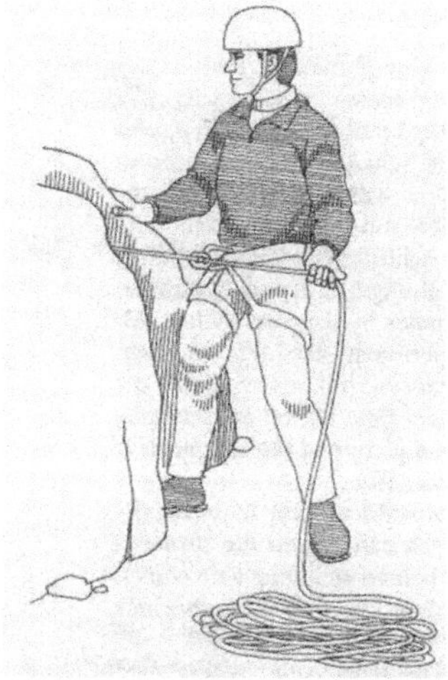

FLAKED OUT ROPEON THE GROUND

i. Dressing the knot. This involves the orientation of all of the knot parts so that they are properly aligned, straightened, or bundled, and so the parts of the knot look like the accompanying pictures. Neglecting this can result in an additional 50% reduction in knot strength.

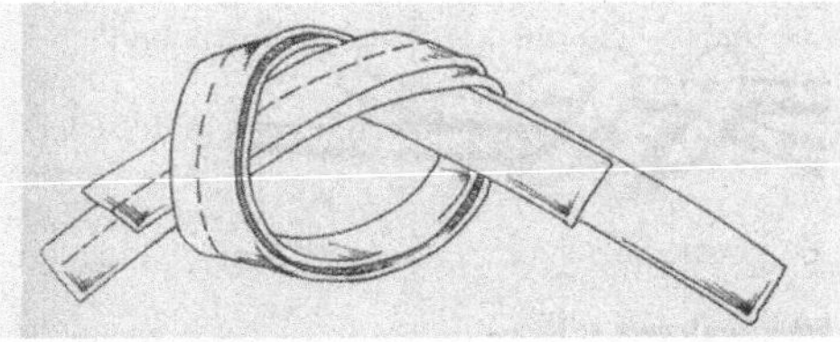

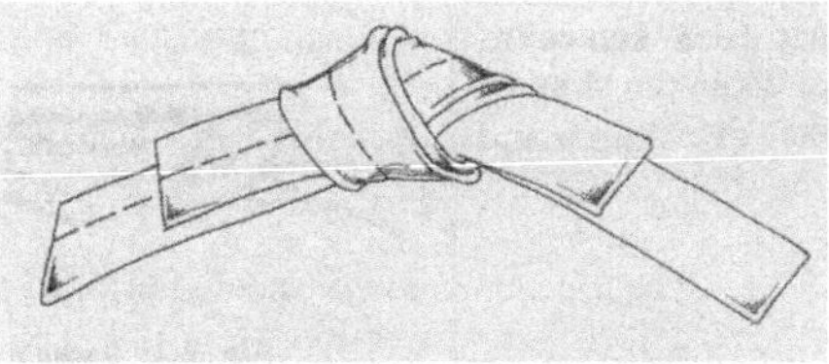

NON-DRESSED DOWN WATER TAPE KNOT **DRESSED DOWN WATER TAPE KNOT**

j. Setting the knot. This involves tightening all parts of the knot so that all of the rope parts bind upon other parts of the knot so as to render it operational. A loosely tied knot can easily deform under strain and change character.

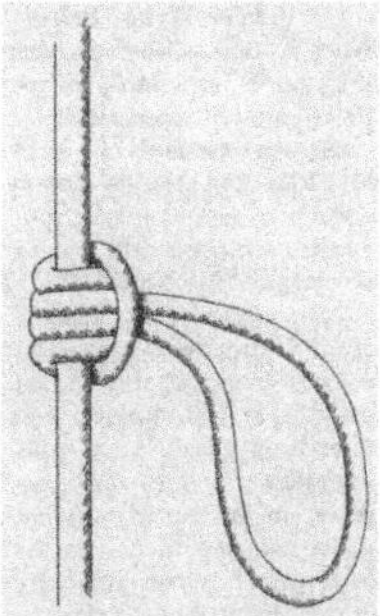

A FULLY SET PRUSIK KNOT

2. CONSIDERATIONS FOR THE CARE OF ROPE

a. The rope should not be stepped on or dragged on the ground unnecessarily. Small particles of dirt will get into and through the sheath causing unnecessary wear to the rope within.

b. The rope should never come in contact with sharp edges of any type. Nylon rope is easily cut, particularly when under tension. If a rope must be used around an edge which could cut it, then that edge must be padded or buffed using fire hose if available, or several small sticks.

c. Keep the rope as dry as possible. If it should become wet, hang it in large loops, above the ground, and allow it to dry. A rope should never be dried out by an open flame, or be hung to dry on metal pegs, as this will cause rust to get in the rope thus rendering it unserviceable.

d. Never leave a rope knotted or tightly stretched longer than necessary.

e. When using rope installations, never allow one rope to rub continually against another.

NOTE: With manila ropes this will cause the rope to fray, whereas nylon ropes can melt under the friction that this causes.

f. The rope should be inspected prior to each use for frayed or cut spots, mildew, rot or defects in construction.

g. Mark all climbing ropes at their midpoints to facilitate establishing the midpoint for a procedure requiring you to use the middle of the rope. The rope should be marked with a bright colored adhesive tape.

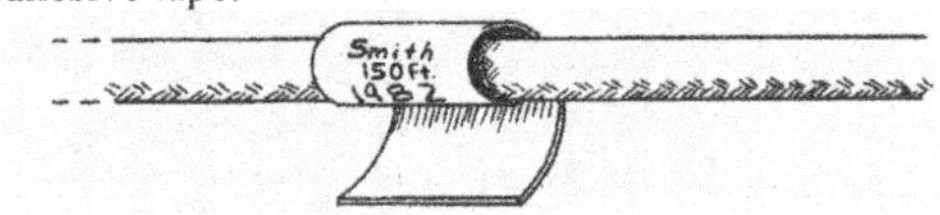

MARKING THE MIDDLE OF THE ROPE

h. The rope should not be marked with paints or allowed to come in contact with oils or petroleum products for these products will weaken it.

i. A climbing rope should NEVER be used for any other purpose except for mountaineering, i.e., towing vehicles.

j. The ends of a new rope or ends caused by a cut should be cut with the rope cutter contained in the MACK and marked with a serial number.

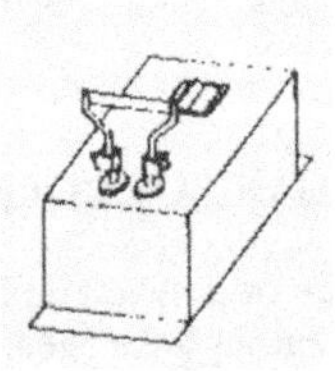

ROPE CUTTER

k. The rope should never be subjected to high heat or flame as this can significantly weaken it.

l. To clean rope use mild soap and rinse thoroughly with water. A rope washer can be used to clean or rinse the rope.

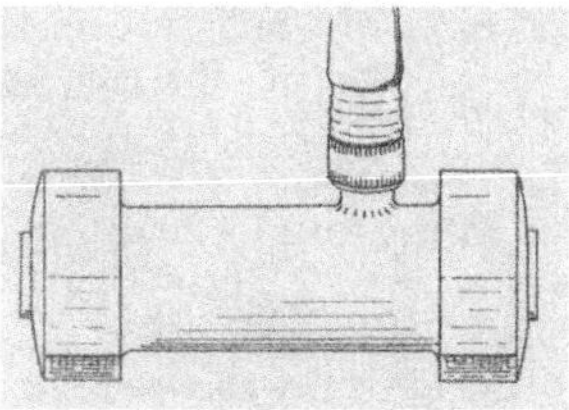

ROPE WASHER

m. When not in use, ropes should be coiled and hung on wooden pegs rather than on nails or any other metal object. They should be stored in a cool place out of the direct rays of the sun.

n. When in areas of loose rock, the rope must be inspected frequently for cuts and abrasions.

o. Always maintain an accurate Rope Log whenever using a rope.

p. Ropes 300-600 foot in length should be Mountain Coiled.

3. **INSPECTION OF ROPE**

a. All ropes have to be inspected before, during, and after all operations. Kernmantle rope is harder to inspect than a laid rope. I.e. green line. The Assault Climber must know what to look and feel for when inspecting a rope. Any of the below listed deficiencies can warrant the retirement of a rope.

(1) Excessive Fraying. Indicates broken sheath bundles or PIC breakage.

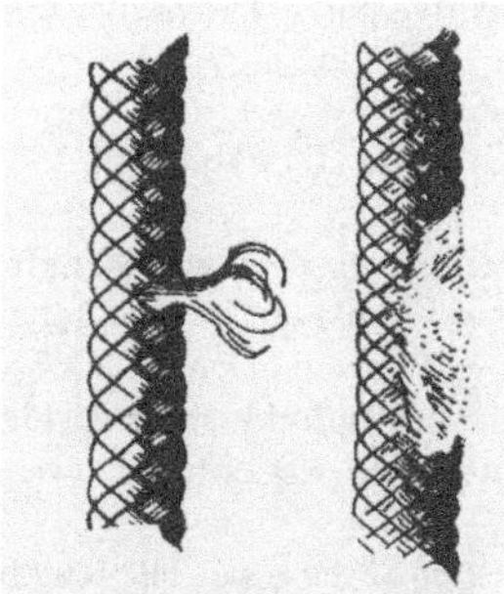

DAMAGED AND EXPOSED ROPE

(2) Exposed Core Fibers. Indicates severe sheath damage. (When you can see the inner core fibers)

(3) Glossy Marks. Signify heat fusion damage, also called a booger.

(4) Uniformity of Diameter / Size. May indicate core damage, noted by an obvious depression (hour glass) or exposure of white core fibers protruding from the sheath (puff).

(5) Discoloration. A drastic change from the ropes original color may indicate chemical change or damage.

(6) Stiffness or Soft Spots. Could signify core damage.

NOTE: Dynamic ropes measuring between 10mm and 12mm are marked at each end of its pigtails with a number "1" indicating that the rope is UIAA approved for single rope lead climbing. Dynamic ropes measuring between 8mm and 9mm are marked at each end of its pigtails with a "1/2" number indicating that two of these ropes are required to conduct a lead climb.

4. **ROPE LOG.** The purpose of the Rope Log is to maintain an accurate record for the use of each rope contained within the Marine Assault Climbers Kit (MACK). Due to the turnover of personnel, and the fact that no one person may have the same rope twice, the rope log is used to ensure the safe use, serviceability, and account of each rope.

a. Serial number. By assigning each rope a serial number, responsible units can determine information about a rope. As soon as the ropes are cut to the desired length for their intended purpose, each rope will be assigned a serial number by the responsible unit. That rope should then be labeled with that serial number in some permanent manner. The best method for this is as soon as the ends of the new rope have been whipped and fused, mark both ends of the rope with the serial number and then dip the ends of the rope, in effect, laminating the serial number to the rope ends.

(1) A rope serial number has five parts. The proper format for a rope serial number is as follows:

1. Type of rope: (S) - Static or (D) - Dynamic.

2. Last two digits of the year the rope was manufactured. Each rope has a shelf life of two years, after that it must not be used for any mountaineering purpose.

3. Four-digit number for that individual rope which is assigned by the responsible unit, and should be assigned sequentially as new ropes are issued out.

4. The length and diameter of the rope. The length may be recorded in feet, and the diameter in millimeters.

5. Responsible unit code. Example: S-96-0001-150/11mm-F2/8

Example meaning: This rope is a 150 foot, 11mm, static rope, manufactured in 1996. It is the first rope issued by Fox Co 2/8.

b. Recording Information in a Rope Log. Once a new rope has been serialized, the rope log for that rope should be started up. At a minimum, it should contain the following information:

(1) The rope serial number.

(2) Manufacturer. Depending on the manufacturer, ropes of the same type and diameter may vary in tensile strength, stretch factor, and durability.

(3) Date of manufacture. Five years from this shelf life date, ropes are considered to have reached their normal expiration date, and should be destroyed.

(4) Date in service. This is to be recorded for tracking purposes to establish how long a rope is in service. Two years is considered maximum rope life in service.

(5) Each time a rope is used, the using unit is responsible to record how that rope was used and how much use the rope received. Additionally, before checking out a rope and prior to turning it back in, the rope must be inspected by qualified personnel, and initial in the "Inspected By" block of the rope log.

NOTE: ANY TIME A DYNAMIC ROPE IS SUBJECTED TO A FALL FACTOR 2, THAT ROPE SHOULD NOT BE USED AGAIN FOR MOUNTAINEERING.

		DYNAMIC			
RIAL #					
ANUFACTURER					
ATE OF MANUFACTURE					
ATE IN SERVICE					
	INDICATE THE DATE AND NUMBER OF EACH SPECIFIC USE				
CLIMBS	**FALLS (FACTOR)**	**TOP ROPES**	**RAPPELS**	**REMARKS**	**INSPECTED BY**

ROPE LOG

5. **COILING A ROPE**. There are two types of rope coils frequently used at MWTC. The Mountain Coil and the Butterfly Coil.

a. Mountain Coil. This coil is useful for carrying the rope over a pack or over a climber's shoulder and neck. It can be used for short time storage. The mountain coil can be tied in the following manner:

(1) Sit down with your leg bent at a 90-degree angle, heel on the deck. Starting at one end, the rope is looped around the leg in a clockwise fashion, going over the knee and under the boot sole until the entire rope is coiled.

(2) If coiling a 150-foot rope, use only one leg and offset the other, when coiling a 300-foot rope, use two legs and keep them together.

(3) With the starting end of the rope, form a 12-inch bight on the top of the coils.

(4) Uncoil the last loop and along the top of the coils, wrap 4-6 times towards the closed end of the bight.

(5) The end of the rope being wrapped is then placed through the closed end of the bight.

(6) The running end of the bight is then pulled snuggly to secure the coil.

(7) To prevent the coil from unraveling, the two pigtails are tied together with a square knot.

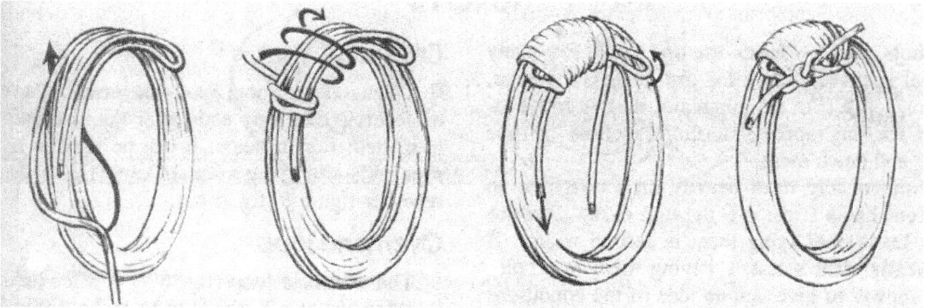

MOUNTAIN COIL

b. Butterfly Coil. This method is used for carrying a rope when the individual needs to have maximum use of his upper body, (i.e. while climbing), without the encumbrance of a large rope coil hanging across his chest.

(1) Coiling the Butterfly Coil

(a) Step 1: Find the middle of the rope, then form a three foot bight laying both ropes in the upraised palm at the two foot point.

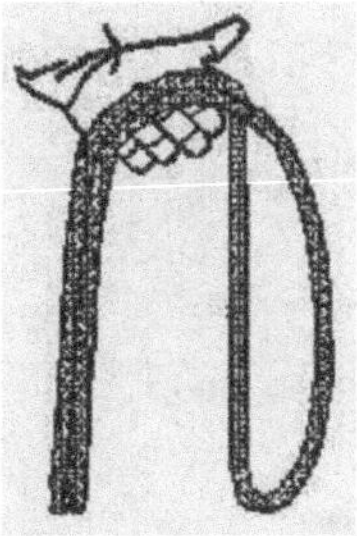

STARTING THE BUTTERLFY COIL

(b) Step 2: Form another two-foot bight with the running end. Place the rope at the two-foot bight along side on top of the original bight ensuring the running end is on the same side as the original bight.

(c) Step 3: Continue making two foot bights, laying them alternately into your palm until there is only six to eight feet remaining. At that point, begin wrapping the two pigtails horizontally four to six times at the mid way point of the ropes in a bight from bottom to top.

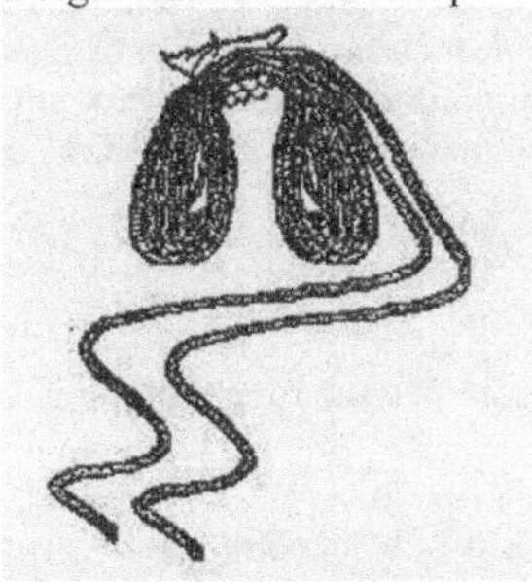

TWO FOOT BIGHTS ON BOTHS SIDES WITH 6 TO 8 FEET REMAINING

(d) Step 4: After completing your wraps, form a bight with the remaining pigtail and then thread it underneath your palm and upwards to one-foot above the coiled rope.

(e) Step 5: With the remaining pigtail, thread it through the one-foot bight in step four.

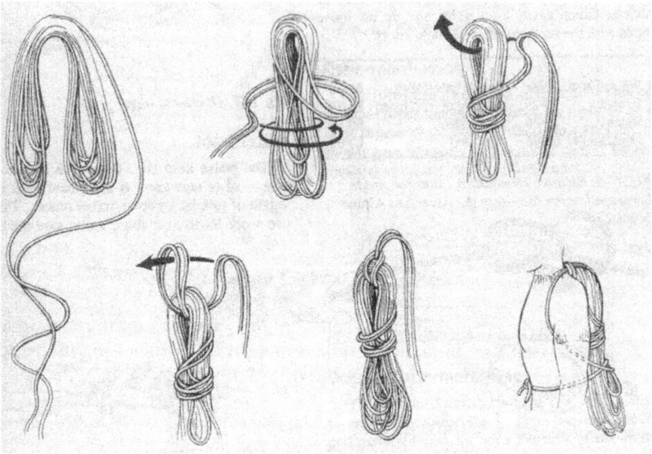

TYING AND CARRYING THE BUTTERFLY COIL

(2) Carrying the Butterfly coil. Separate the running ends, placing the coil in the center of the back of the carrier, then fun the two ends over his shoulders so as to form shoulder straps. The running ends are then brought under the arms, crossed in the back over the coil, brought around the body of the carrier and tied off with a square knot at his stomach.

6. **ROPE THROWING**. To insure that the rope will not get tangled when deployed, certain steps must be taken.

a. With a stacked rope, anchor off the standing end.

b. Take the opposite end of the rope and make 6-8 coils and place them in your strong arm. These wraps will serve as a throwing weight that you can aim.

c. 10-15 feet from the strong-arm coils create a second set of 6-8 wraps and place them in your weak arm.

d. From the edge of the cliff, sound off with the command "STAND-BY FOR ROPE". Just before you release the coils, sound off with the command "ROPE". At that time drop the weak arm coils from the cliff.

e. While taking aim, throw your strong-arm coils overhand or sidearm hard enough to hit your intended target.

f. If the throw was misdirected due to wind, tree, etc., reorganize and attempt to re-deploy the rope.

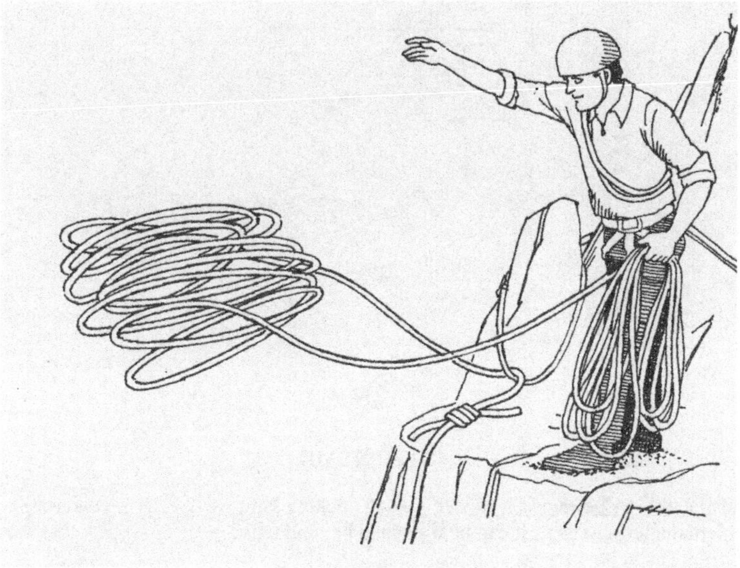

THROWING THE ROPE

6. **MOUNTAINEERING KNOTS.**

a. **Class I** – *End of the Rope Knots*

(1) Square Knot. Used to tie ends of two ropes of equal diameter together. It should be secured by overhand knots on both sides of the square knot.

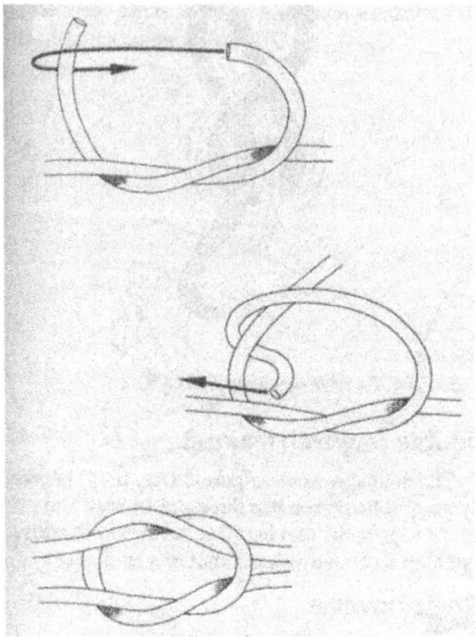

SQUARE KNOT

(2) Double Fisherman's Knot. It is a self-locking knot used for tying two ropes of equal diameter together. It can be tightened beyond untying.

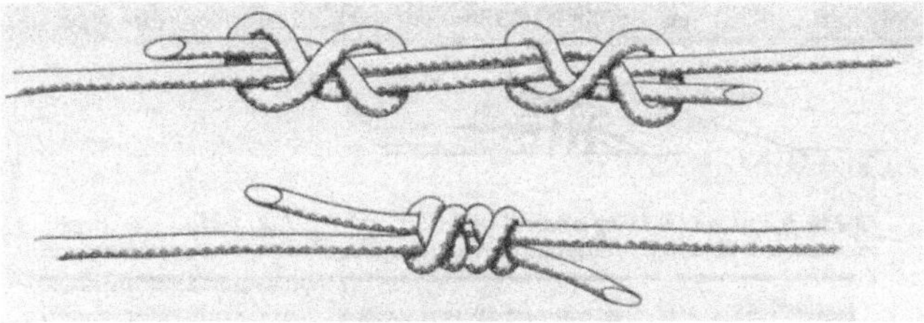

DOUBLE FISHERMAN'S KNOT

(3) Water/Tape Knot. Used to secure webbing or tape runners. It is constructed by tying an overhand knot (without twists) in one end of the tape, and threading the other end of the tape through the knot from the opposite direction. After the knot is dressed down, each pigtail should be a minimum of two inches long.

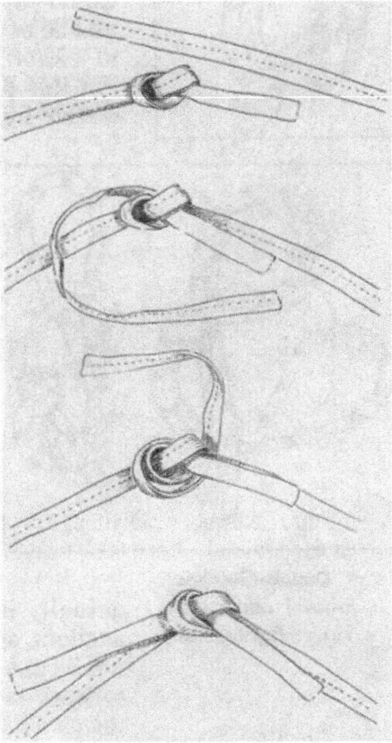

WATER TAPE KNOT

(4) Double Sheet Bend. Used to tie the ends of two or more ropes of equal or unequal diameter together.

SHEET BEND

b. **Class II** - *Anchor Knots*

(1) Bowline. Used to tie a fixed loop in the end of a rope. This knot is always tied with the pigtail on the inside and secured with an overhand knot.

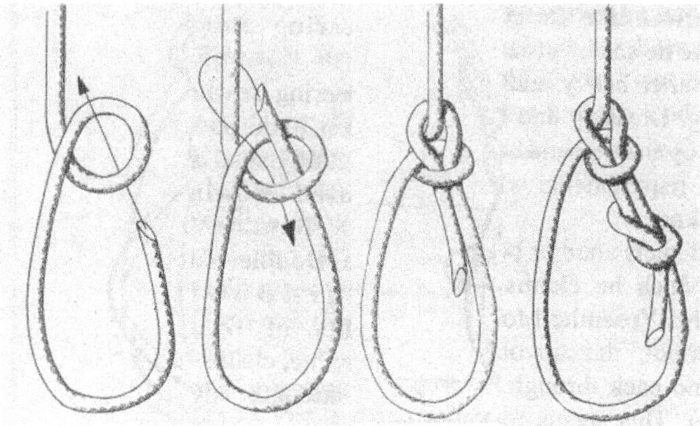

BOWLINE

(2) Round Turn with Two Half Hitches or a Bowline. A loop which runs around an object in such a manner as to provide 360 degree contact and may be used to distribute the load over a small diameter anchor. It will be secured with two half hitches or a bowline.

ROUND TURN AND TWO HALF HITCHES AND

ROUND TURN AND A BOWLINE

(3) Clove Hitch. This knot is an adjustable hitch. It could be considered a middle-of-the-rope anchor knot at the end-of-the-rope when used in conjunction with a bowline or round turn and two half hitches.

(1) Around-the-object Clove Hitch.

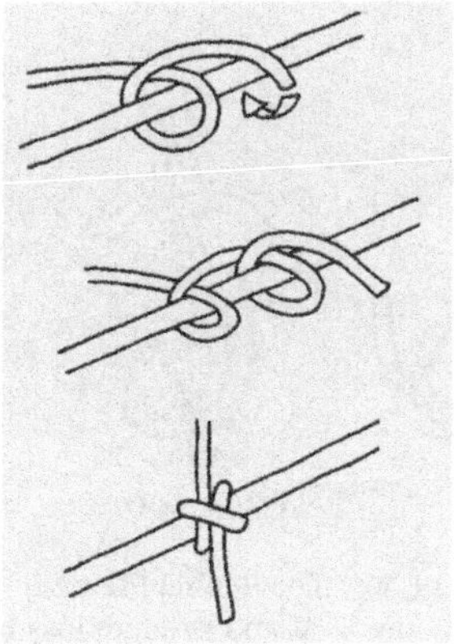

AROUND THE OBJECT CLOVE HITCH

(2) Over-the-object Clove Hitch.

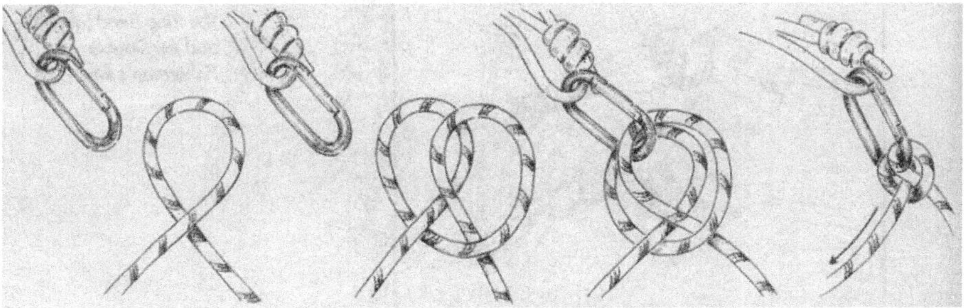

OVER THE OBJECT CLOVE HITCH

c. **Class III** - *Middle-of-the-rope*

(1) Figure-of-Eight-Loop. This is a strong knot that can be readily untied after being under load.

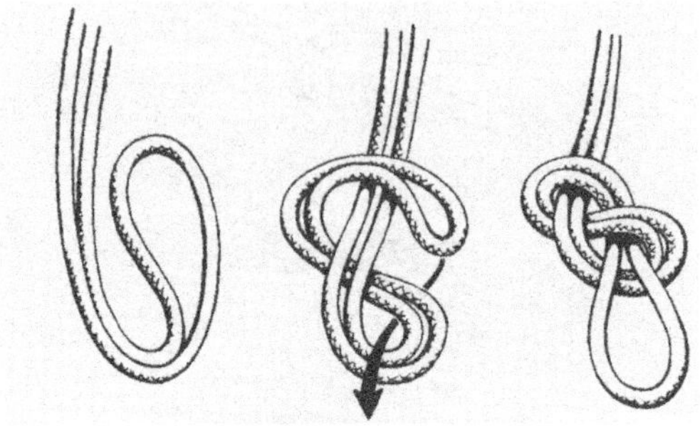

FIGURE 8 LOOP

(2) Double Figure of Eight Loop. The double Figure of Eight Loop is a strong knot and the double loop reduces the wear and strength loss from the rope bending around the carabiner by splitting the load between the two loops.

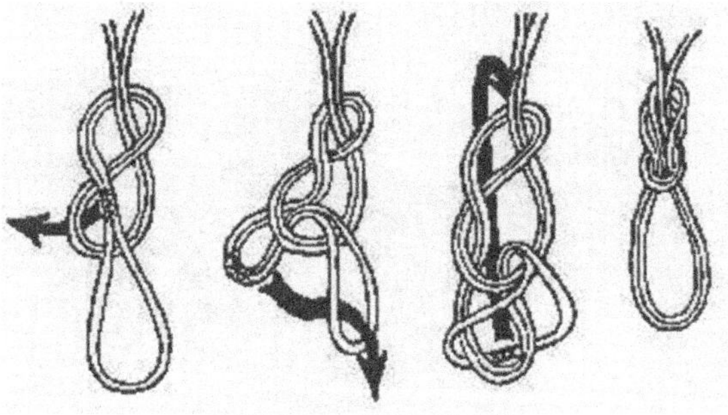

DOUBLE FIGURE 8 LOOP

(3) Two Loop Bowline. This knot can be used to construct a self-equalizing belay or to tie the middle person in on three people on a rope.

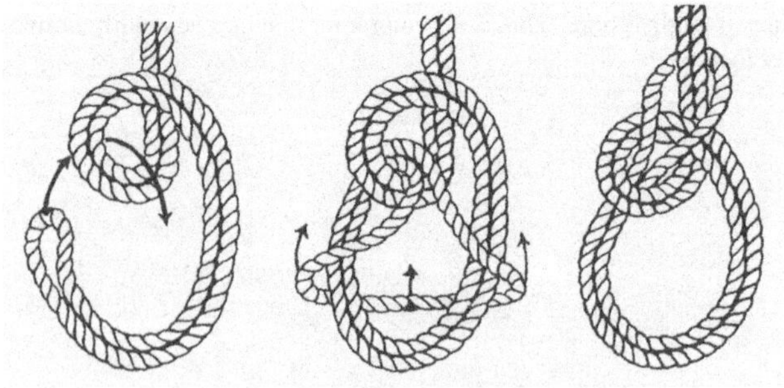

TWO LOOP BOWLINE

d. **Class IV** - *Special Knots*

(1) Prusik Knot. This knot functions by introducing friction that can be alternately set and released. For best results, tie the knot with a smaller diameter cord on a larger diameter cord. If slippage occurs, more wraps maybe used.

(a) Middle-of-the-Rope Prusik. This knot is created with an endless loop also known as a Prusik Cord. Do not tie this knot with tape due to less friction.

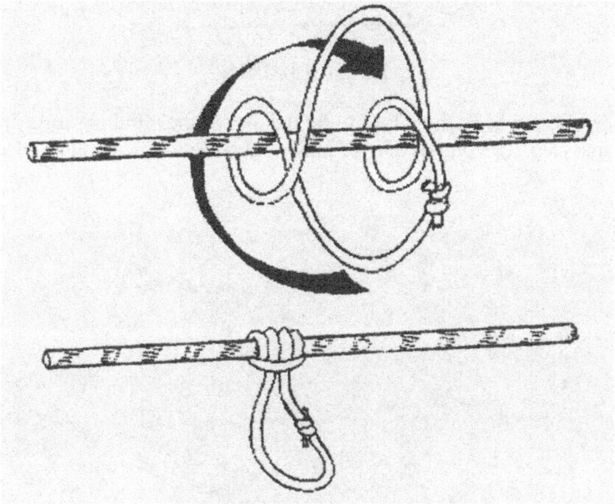

MIDDLE OF THE LINE PRUSIK

(b) End-of-the-Rope Prusik. This knot is always secured with a bowline. Do not tie this knot with tape due to less friction.

END OF THE LINE PRUSIK

(c) French Prusik. This can be constructed with a Prusik Cord or a single strand of cord with figure-of-eight loops on each end.

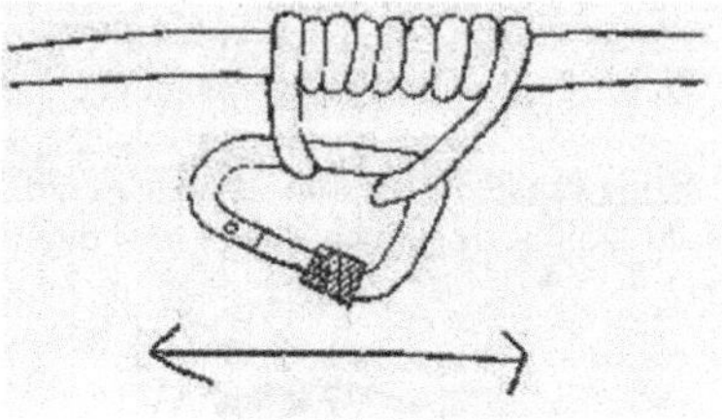

FRENCH PRUSIK

(2) Retraced Figure of Eight. Used to tie the end of the climbing rope into a harness or swammi wrap. The pigtail may or may not be secured with an overhand.

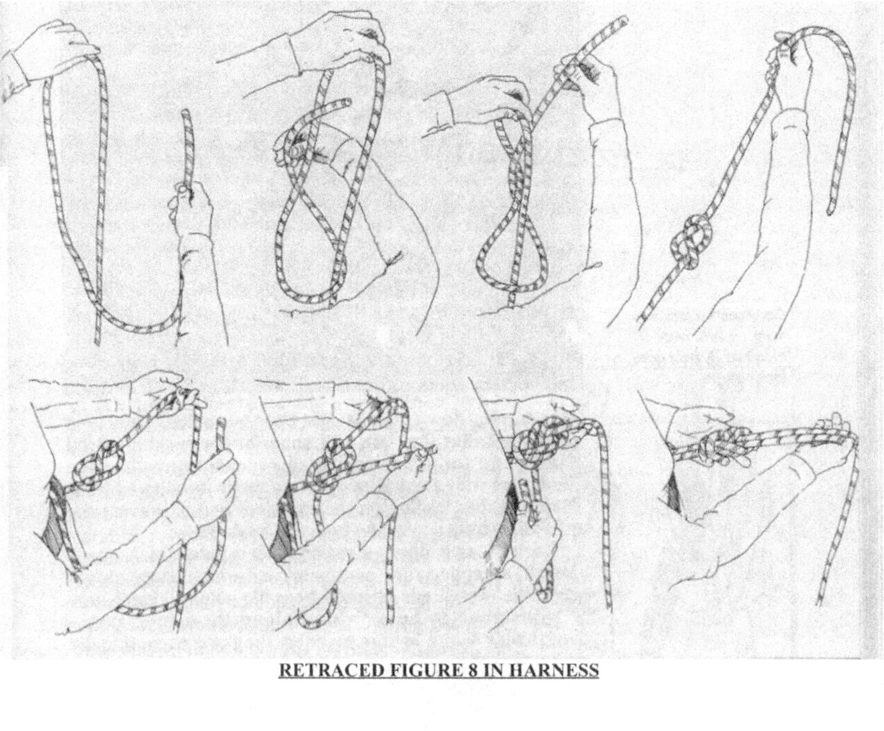

RETRACED FIGURE 8 IN HARNESS

(3) Directional Figure-Of-Eight. When tied and tensioned is applied to the standing and running ends of the rope, the knot will not pull apart. The loop will point toward the direction of pull.

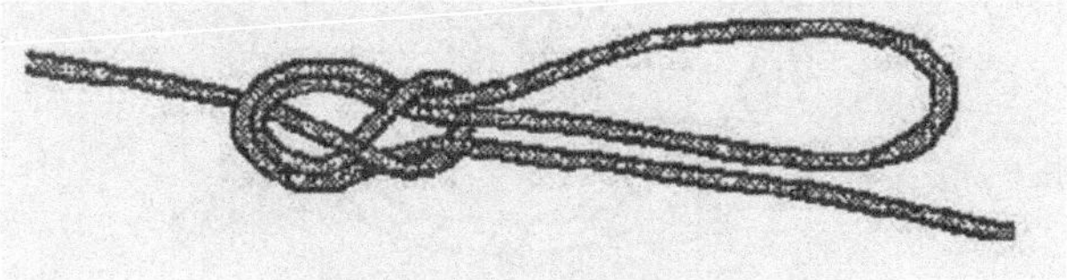

DIRECTIONAL FIGURE 8

(4) Slip Figure 8. This knot is used for retrievable anchors and fixed rope installations for the ease of untying the knot.

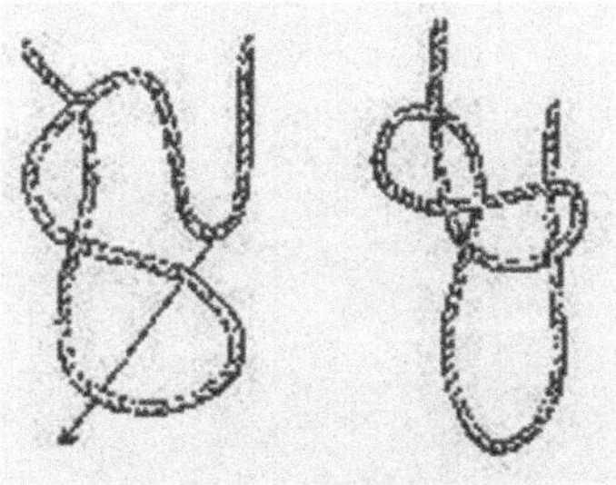

SLIP FIGURE 8

(5) Kliemheist. This is a friction knot.

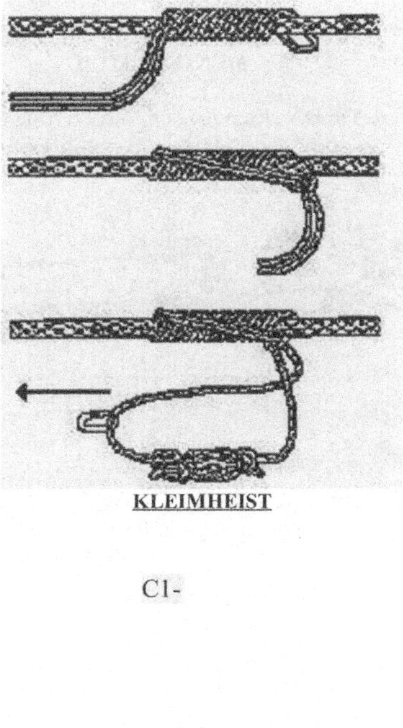

KLEIMHEIST

(6) Overhand Knot. Can be used to secure primary knots on itself.

OVERHAND

(7) M nter Hitch. This is a simple hitch in the rope that is clipped into a carabiner to put friction on the line.

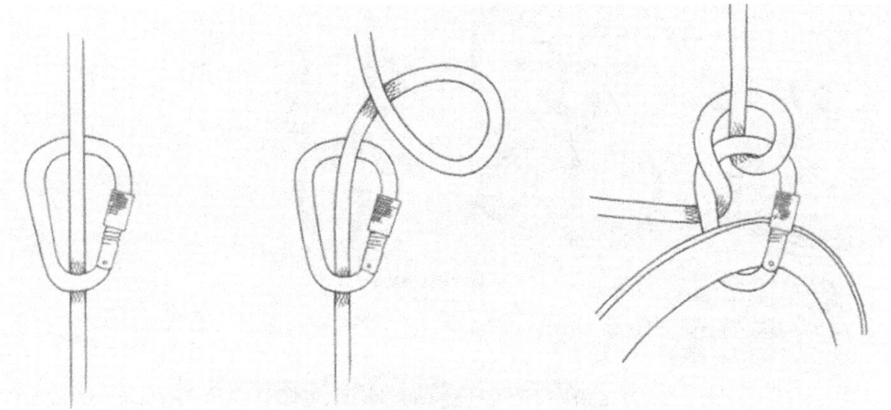

MÜNTER HITCH

(8) Timber Hitch. A Timber Hitch is used to fix a rope to a pole or equivalent for hoisting or towing purposes. It has the capability of casting off easily.

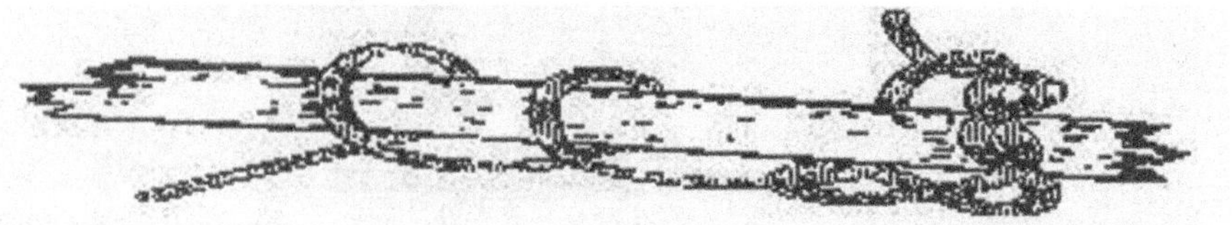

TIMBER HITCH

(9) Mariner's Hitch. This knot's advantage over the prusik is that it can be released under load.

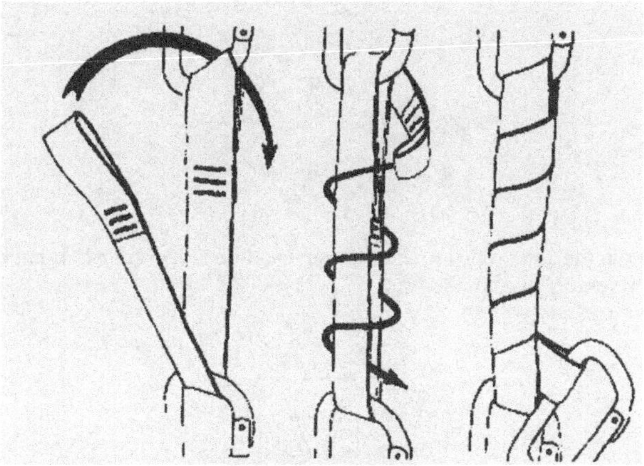

MARINER'S HITCH

(10) Krägur Knot. This friction knot is used to prevent unnecessary slippage when tied onto a rope of the same diameter.

KRÄGUR KNOT

(11) Rappel Seat. This is used as an expedient support harness for rappelling, crossing rope bridges, etc. It is constructed as follows:

(a) Center the sling rope on the left hip.

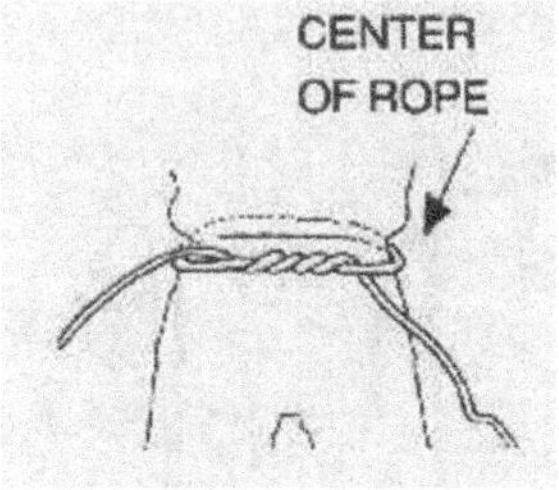

(b) Wrap the sling rope around the waist and tie at least one half hitch around itself (preferably two) in the front.

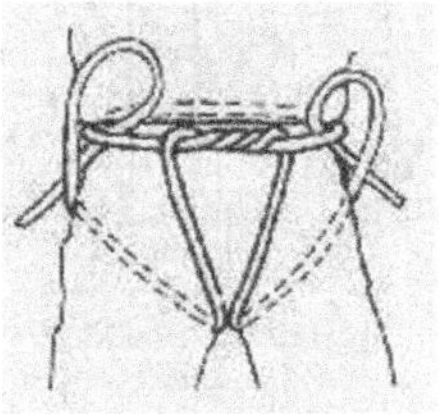

(c) Bring the running ends down through the legs, up over the buttocks, and over the original waist wrap, down between the waist wrap and the waist, and over itself, forming a bight. Cinch this up tightly.

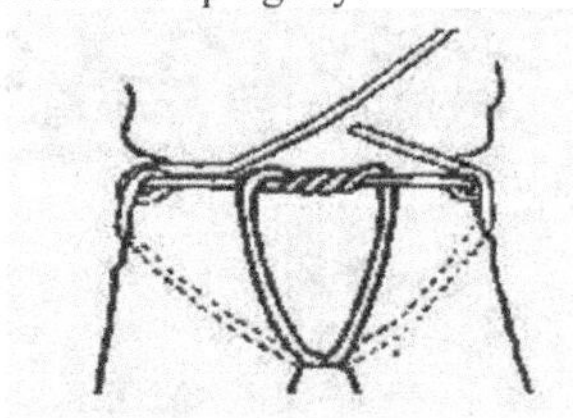

(d) Now tie a square knot with two overhand knots on the left hip ensuring that the overhand knots encompass all the wraps.

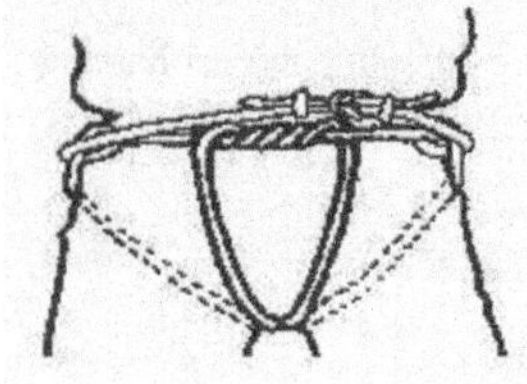

(e) Tuck any excess rope into a pocket.

(12) Bowline on a coil. Used by the first and last men on a climbing rope to tie into the rope. An overhand knot is used behind the knot. It distributes the force of a fall over a larger area of a climber's waist and is preferable to a single bowline around the waist. The bowline on a coil can also be used to take up excess rope. The bowline on a coil should have 4-6 wraps around the waist.

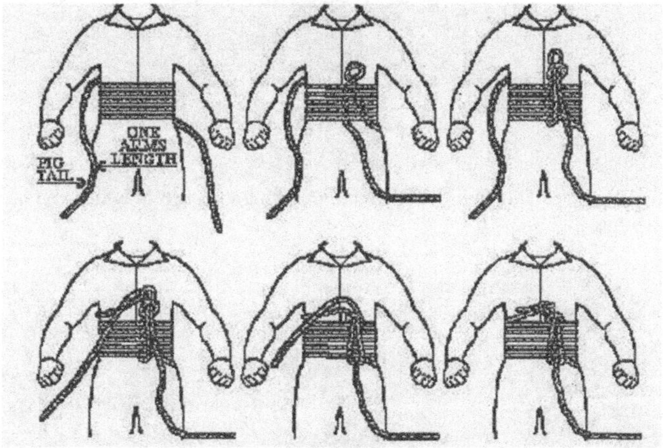

BOWLINE ON A COIL

(13) Swami wrap (Swami belt). The swami wrap (four to six wraps of rope or nylon webbing tied around the waist) popular with rock climbers who wanted to increase the working length of their climbing ropes. It also offers some improvement over the bowline on the coil in distributing the forces sustained in a fall over a larger area of the midsection of the body, but does not eliminate the problem of suffocation when hanging. The Swami wrap is constructed as follows:

(a) Take a sling rope and find the middle.

(b) Place the middle of the sling rope in the small of your back.

(c) Bring both pigtails to the front and continue to wrap around your body until both pigtails are about 15-20 inches long.

(d) Tie the pigtails on the left side with a square knot and two overhand knots.

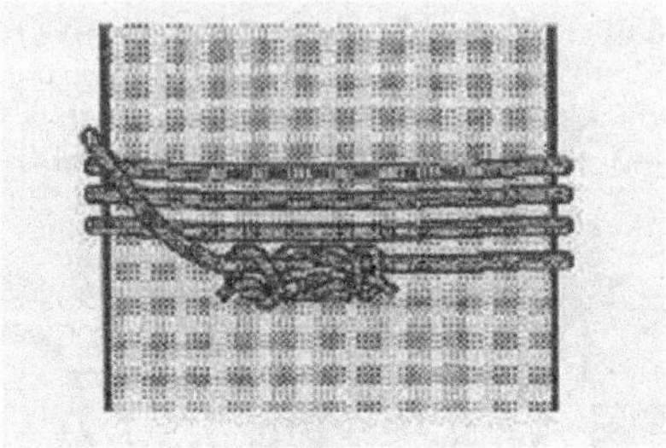

SWAMMI WRAP

(14) Three Loop Bowline. This knot will provide three bights, each of which can be adjusted against the others.

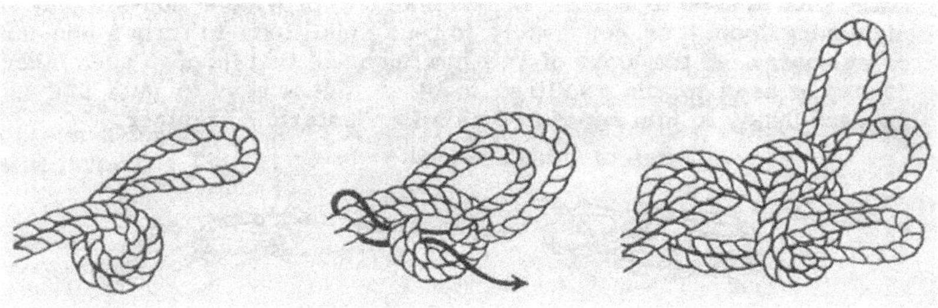

THREE LOOP BOWLINE

7 **KNOT STRENGTH.** Rope, cordage and webbing is strongest when loaded in a straight line. When bending a rope or web to create a knot, the strength of the rope is reduced. All knots should be dressed properly for maximum effective use.

KNOT	RELATIVE STRENGTH OF KNOT
No Knot	100%
Figure 8	75-80%
Bowline	70-75%
Double Bowline	70-75%

Double Fisherman	65-70%
Water Tape Knot	60-70%
Clove Hitch	60-65%
Overhand Knot	60-65%
Square Knot	45%

1. **KNOT TESTING TIME LIMIT.** The following times must be met to pass the knot tying portion of this course.

KNOT	UNIT OPERATIONS TIME LIMIT	SML, ACC TIME LIMIT (BLINDFOLDED)
Square knot	60 Seconds	30 Seconds
Double Fisherman's Knot	60 Seconds	30 Seconds
Water/Tape Knot	60 Seconds	30 Seconds
Round Turn and a Bowline	60 Seconds	30 Seconds
Round Turn and Two Half Hitches	60 Seconds	30 Seconds
Clove Hitch (around the object)	60 Seconds	30 Seconds
Munter Hitch	60 Seconds	30 Seconds
Slip Figure 8	60 Seconds	30 Seconds
Figure 8 Loop	60 Seconds	30 Seconds
Directional Figure 8	60 Seconds	30 Seconds
End of the Line Prusik	90 Seconds	45 Seconds
Retrace Figure 8	90 Seconds	45 Seconds
Rappel (Swiss) Seat	2 Minutes	60 Seconds

UNITED STATES MARINE CORPS
Mountain Warfare Training Center
Bridgeport, California 93517-5001

SMO
05/02/01

STUDENT OUTLINE

NOMENCLATURE AND CARE OF MOUNTAINEERING EOUIPMENT

LESSON PURPOSE. The purpose of this period of instruction is to introduce you to the types of equipment used here at MWTC and how to care for it so as to prevent its untimely failure. This lesson relates to all climbing and installation work that you perform here at MWTC.

OUTLINE

1. **ROPES.** All ropes used in the military must meet UIAA standards or U. S. Federal Test Standard 191A. Most ropes have a 5-year shelf life and maximum 2-year service life.

a. Static. Black in color.

(1) Construction. Kernmantle

(2) Minimum tensile strength. 7500 lbs.

(3) Maximum elongation. 1.5%

(4) Diameter. 11mm

(5) Sizes.

(a) 165 ft ± 5 ft

(b) 300 ft ± 10 ft

(6) Usage. Rescue operations and bridging where a low amount of elongation is desirable under a working load.

b. Dynamic. Olive Drab in color.

(1) Construction. Water-resistant treated Kernmantle to reduce friction.

(2) Minimum tensile strength. 6500 lbs.

(3) Maximum elongation. 6%

(4) Diameter. 10.5mm and 11mm

(5) Size. 165 ft ± 5 ft.

(6) Usage. For lead climbing/party climbing.

c. Maxim Dry Rope. Olive Drab or Multi-Colored.

(1) Construction. Water-repellent treated Kernmantle.

(2) Minimum tensile strength. 3472 lbs.

(3) Maximum elongation. 6%

(4) Diameter. 9mm

(5) Size. 150 ft ± 5 ft

(6) Usage. For glacier travel/ice climbing

d. Gold Line II.

(1) Construction. Eight strand braided nylon plymor.

(2) Minimum tensile strength. 4500 lbs.

(3) Maximum elongation. 20%

(4) Diameter. 11 mm

(5) Size. 300 ft or 600 ft spools

(6) Usage. Sling Ropes and litters only.

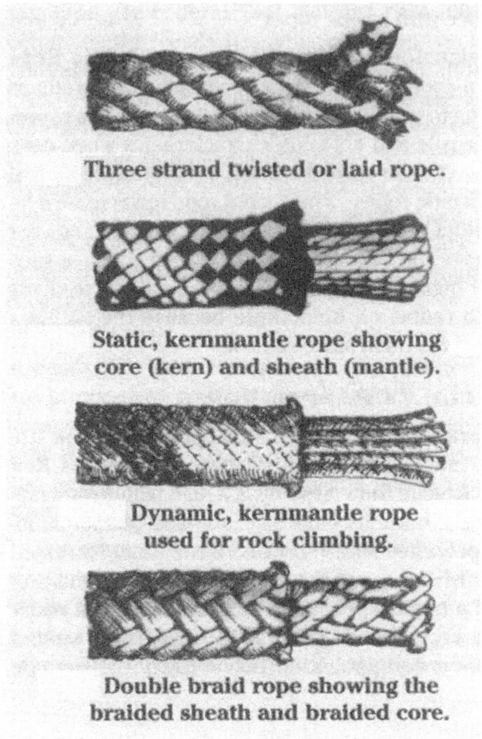

INSIDES AND OUTSIDE SHEATHS OF VARIOUS TYPES OF ROPES

NOTE: Sling ropes are made from 15 foot lengths of plymor or dynamic rope ONLY. Twenty-five foot practice coils should be constructed with static rope, but dynamic rope can be used.

2. ADVANTAGES/DISADVANTAGES

a. Advantages of Nylon Rope

(1) High strength to weight ratio.

(2) Good energy absorption in dynamic ropes.

(3) Flexible.

(4) Rot resistant, not affected by frost.

b. Disadvantages of Nylon Rope

(1) Low melting point. Nylon fuses at 400°F and melts at 480°F.

(2) Susceptible to abrasions and cuts.

(3) Affected by chemicals and light.

c. Advantages of Manila Rope

(1) Easily gripped.

(2) Hard wearing.

(3) Does not deteriorate in heat.

d. Disadvantages of Manila Rode

(1) Heavy, kinks, especially when wet. Absorbs water and swells.

(2) Burns at +300°F.

(3) Edible by rodents.

3. GENERAL INFORMATION

a. Nylon rope stretches under tension and will rupture at between 30% and 70% elongation, depending on construction.

b. Nylon rope loses as much as 30% strength when wet.

c. Temperatures as low as 250°F will damage a nylon rope.

4. NYLON WEBBING

a. The type of nylon webbing available is tubular. Tubular nylon webbing is very strong and flexible. All rules that apply to nylon rope apply to tubular nylon webbing. The size of nylon webbing used is:

(1) 1 inch tubular nylon. Tensile strength approximately 4,000 - 4,500 lbs., depending on the manufacturer.

b. Pre-sewn Spectra Runners. Tensile strength approximately 5,500 lbs.

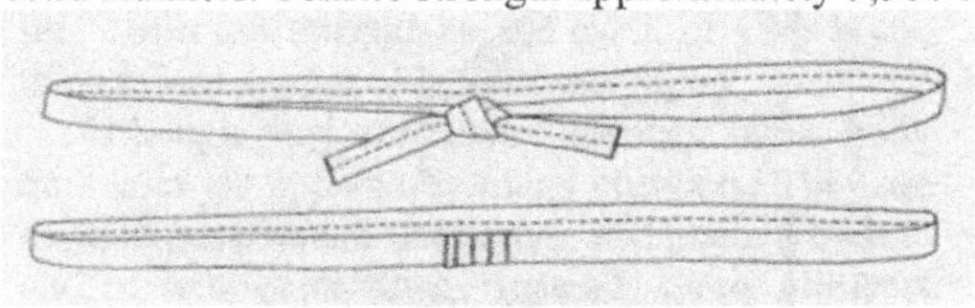

NYLON WEBBING

NOTE: These are minimum strengths. Some manufactures make even stronger webbing.

1. **CARABINERS**. Also commonly known as snap links. There are two different types of carabiners frequently used at MWTC, the steel locking carbine and the non-locking aluminum carabiners.

a. Steel Locking Carabiner. There are two different sizes of carabiners:

(1) Steel Locking Stubai 82.

(2) Large Locking "D" Stubai 85.

b. Aluminum Non-Locking Carabiner.

(1) The standard "D" non-locking carabiner with a tensile strength of 2000 lbs. or more.

2. **NOMENCLATURE OF A CARABINER**.

a. There are four parts to a non-locking carabiner. They are as follows:

(1) Gate.

(2) Gate pivot pin.

(3) Locking pin

(4) Body.

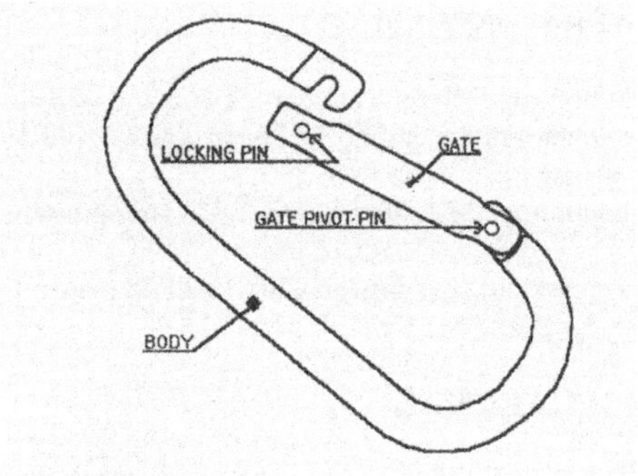

STEEL NON-LOCKING CARABINER

b. There are five parts to a locking carabiner. They are as follows:

(1) Gate

(2) Gate pivot pin

(3) Locking notch

(4) Locking nut

(5) Body

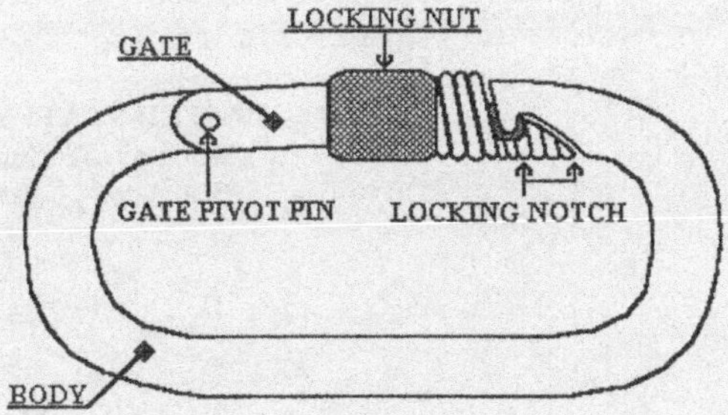

STEEL LOCKING CARABINER

3. **CARE, MAINTENANCE AND SERVICEABILITY OF A CARABINER**. The following inspections and maintenance should be performed on a routine basis:

a. Care.

(1) Avoid dropping of a carabiner onto a hard surface, hairline factures could occur.

(2) Keep the carabiners off of the bare ground because dirt and grit can get into the working parts.

(3) If the carabiner has fallen from a significant height, it should be retired.

b. Maintenance.

(1) Remove all dirt, moisture and grime after use.

(2) Lubricate with tri-flow graphite and wipe off with a clean cloth.

c. Serviceability Checks.

(1) The gate snaps shut with no friction and with no gap between the locking pin and locking notch.

(2) There is no excessive side to side movement of the gate

(3) The pivot pin is tight.

(4) That the spring-loaded gate functions properly.

(5) The locking pin is tight.

(6) The locking nut travels freely and locks securely.

(7) There are no cracks or flaws in the metal.

NOTE: The weakest part of a carabiner is the gate. If an engraver is used to mark a carabiner, it should be applied to the gate and not the load bearing side.

UNITED STATES MARINE CORPS
Mountain Warfare Training Center
Bridgeport, California 93517-5001

WMC
04/11/02

LESSON PLAN

BALANCE CLIMBING

1. LEARNING OBJECTIVES

a. TERMINAL LEARNING OBJECTIVE. In a summer mountainous environment, execute balance climbing, in accordance with the references. (WMC)

b. ENABLING LEARNING OBJECTIVES.

(1) With the aid of references, state orally the safety requirements for balance climbing, in accordance with the references. (WMC.C3a)

(2) With the aid of references, state orally the individual preparations for a balance climb, in accordance with the references. (WMC.C3b)

(3) In a summer mountainous environment, execute the duties of a spotter for a balance climb, in accordance with the references. (WMC.C3c)

(4) In a summer mountainous environment, execute the commands used between the climber and the spotter during a balance climb, in accordance with the references. (WMC.C3d)

(5) In a summer mountainous environment, execute each type of hold used in balance climbing, in accordance with the references. (WMC.C3e)

(6) Without the aid of references and given the acronym "CASHWORTH", describe orally the nine considerations for proper body position while climbing, in accordance with the references. (WMC.C3f)

OUTLINE

1. **SAFETY PRECAUTIONS**. There are two safety precautions that always apply to balance climbing. They are as follows:

a. Never climb more than 10 feet above the ground. By this it is meant that the climber's feet are never more than 10 feet above the ground.

b. A spotter is required for all balance climbs.

SPOTTER WATCHING THE CLIMBER

2. **INDIVIDUAL PREPARATIONS**. Prior to beginning a balance climb there are seven things that the climber must do to prepare himself. They are as follows:

a. Helmet with a serviceable chinstrap must be worn.

b. Sleeves rolled down_to give hand and arm freedom of movement. Blouse tucked in to your trousers. In case of fall, it may catch on a rock and cause you to flip over sideways.

c. All watches, rings, and jewelry must be removed before climbing.

d. Gloves will not be worn, as they can slip, and also give a false feel for the rock.

e. Unblouse trousers, if they restrict movement.

f. Soles of boots clean and dry.

g. Select route where vegetation is minimal. Never use vegetation for hand or foot holds.

3. **DUTIES OF THE SPOTTER**. The spotter is the balance climber's partner who, rather than climbing himself, acts as the safety man for the climber during the climb. The five duties of the spotter are as follows:

a. Positions himself directly behind the climber before the climb starts.

b. Maintains his position facing the cliff, directly below the climber and approximately 3-4 feet away from the base of the cliff, for the duration of the climb. He will move diagonally as necessary to remain below the climber.

c. The spotter will stand with his feet shoulder width apart and arms ready to stop the climber if he falls.

d. If the climber falls, the spotter will not "catch" him; he will prevent the climber from falling further down the hill. He will do this by pushing the climber towards the base of the cliff, thereby preventing him from tumbling backwards.

e. At no time will the spotter allow anyone to come between himself and the face of the cliff while a balance climb is taking place. He will require anyone who wants to pass by his position to go behind him.

4. **SPOTTING AND CLIMBING COMMANDS**. The following are the commands used by both the spotter and climber.

COMMAND	GIVEN BY	MEANING
"Last Name, climbing."	Climber	I am ready to climb.
"Climb Climber."	Spotter	I am ready to spot you.
"Last Name, off Climb."	Climber	I am off the climb.
"ROCK!"	Climber or Spotter.	A rock has been knocked off the rock face and is falling.
"FALLING!"	Climber	I am Falling.

a. If the command "ROCK" is given, all personnel in the vicinity will take the following action:

(1) If close to the cliff face, move against the cliff face with your face against the cliff face and your hands between you and the cliff face.

(2) If not close to the cliff face, look up to locate the rock and avoid it.

5. **ACTIONS IF FALLING**. If, while making a balance climb, the climber feels himself slipping and beginning to fall, he will take the following action:

a. Sound the command "falling".

b. Push himself away from the rock face.

c. Maintain proper body position as follows:

(1) Head up.

(2) Hands out toward the rock.

(3) Body relaxed.

(4) Feet kept below the body, slightly apart.

(5) Ensure you face the cliff face as you fall.

6. **TYPES OF HOLDS**. There are five basic holds that are used in balance climbing. They are as follows:

a. Push Holds.

(1) Most effective when hands are kept low.

(2) Often used in combination with a pull hold.

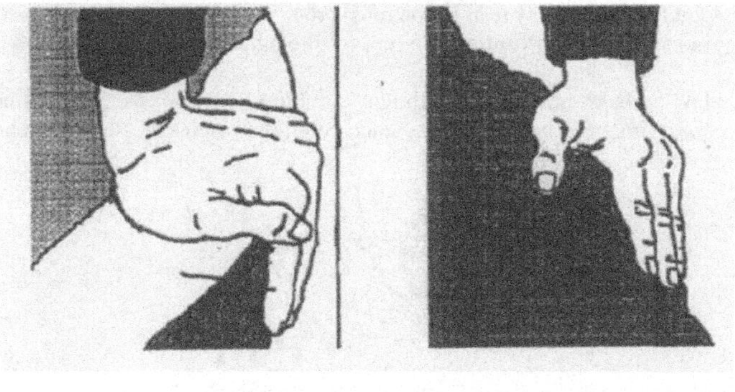

TYPES OF PUSH HOLDS

b. Pull Holds.

(1) The easiest hold to use and, consequently, often overused.

(2) Can be effective on small projections.

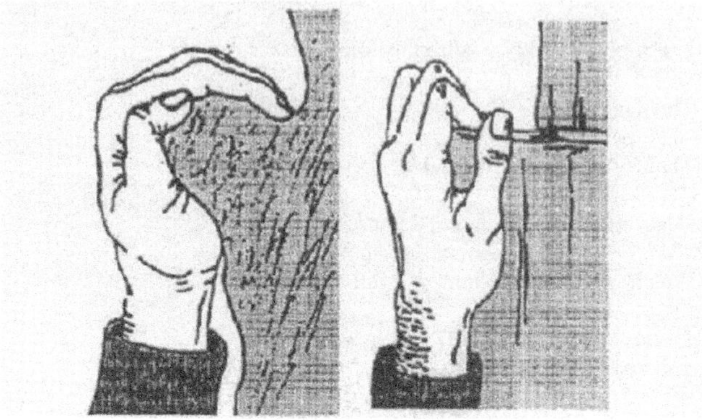

TYPES OF PULL HOLDS

c. Foot Holds

(1) Feet should be positioned with the inside of the foot to the rock.

(2) Use full sole contact as much as possible.

(3) Avoid crossing your feet. If you must cross your feet use a change step. A change step is a method of substituting one foot for the other foot on the same foothold.

(4) Making maximum use of footholds, climbing with your feet, is an effective means of conserving your body strength, since your leg muscles are stronger than your arm muscles.

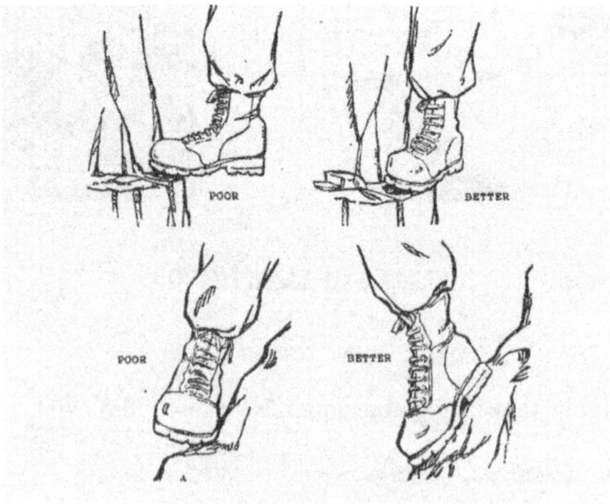

USE OF FOOT HOLDS

d. Friction Holds. A friction hold is anytime you are relying on the friction of your foot or hand against the face of the rock for traction, rather than pushing/pulling against a projection on the face of the rock.

(1) It is a type of hold that feels very insecure to an inexperienced climber.

(2) The effectiveness of this type of hold is dependent upon many things i.e., type, condition and angle of the rock face, type of boot soles, confidence, etc.

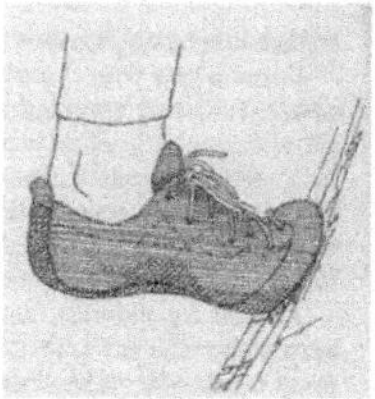

FRICTION USED IN SMEARING

e. Jam Holds. This type of hold involves jamming/wedging any part of your body or your entire body into a crack/opening in the rock.

(1) An important consideration is that you do not jam such that you cannot free that portion of your body after you complete the move. This sounds like a ridiculous statement; however, you must remember that after you complete your move, you may be withdrawing the portion of your body that you used from a different angle than you inserted it.

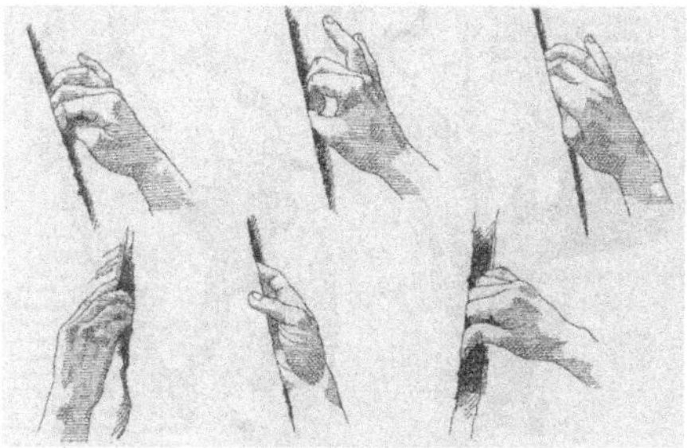

FINGER JAMS

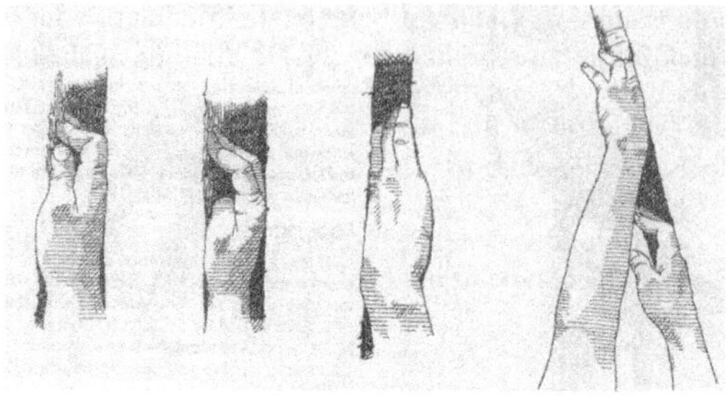

HAND JAMS

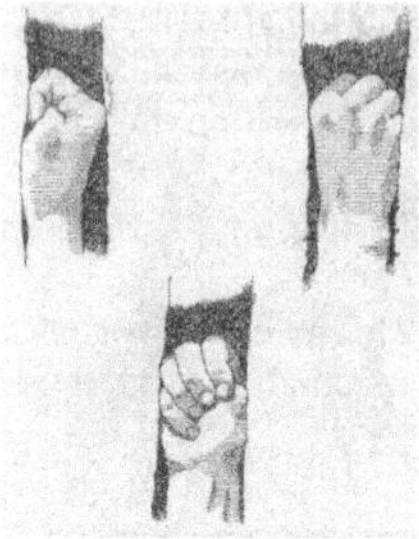

FIST JAMS

7.) **COMBINATION HOLDS.** The five types of holds just mentioned above are not just used individually. They are most often used in combinations with each other. Some examples are:

a. Chimney Climbing. This is when you insert your entire body into a crack in the rock and by using both sides of the opening, and possibly all five types of basic holds, move up' the crack.

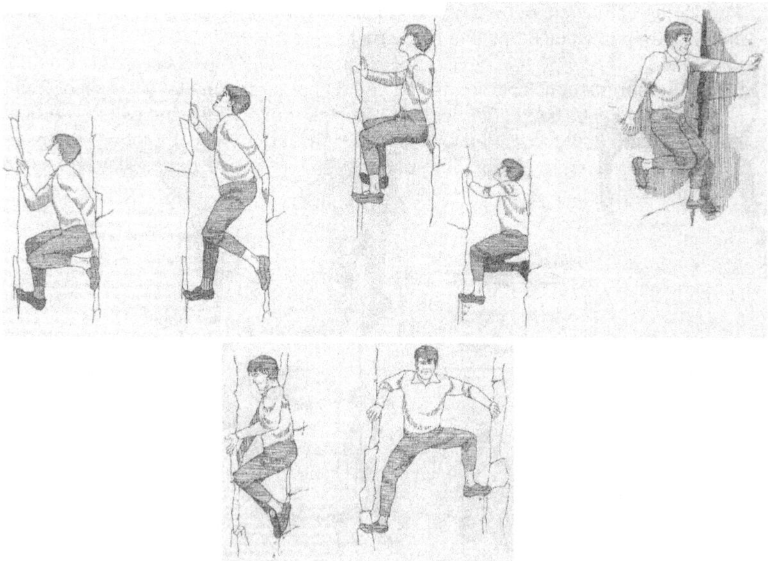

CHIMNEY CLIMBING TECHNIQUES

b. Lie-back. This is a combination of both pull holds with your hands and friction holds with your feet.

LIE BACK TECHNIQUE

c. Push-Pull. As the name implies, this is when you use a push hold and a pull hold together.

PUSH-PULL

d. Mantling. This is a technique where you continue to climb without moving your hands off a projection by pulling yourself up until your hands are at chest level and then invert your hands and push on the same projection.

MANTELING

MANTELING WITH FACE HOLD

e. Cross-Pressure in Cracks. This is a technique of putting both hands in the same crack and pulling your hands apart to hold/raise yourself.

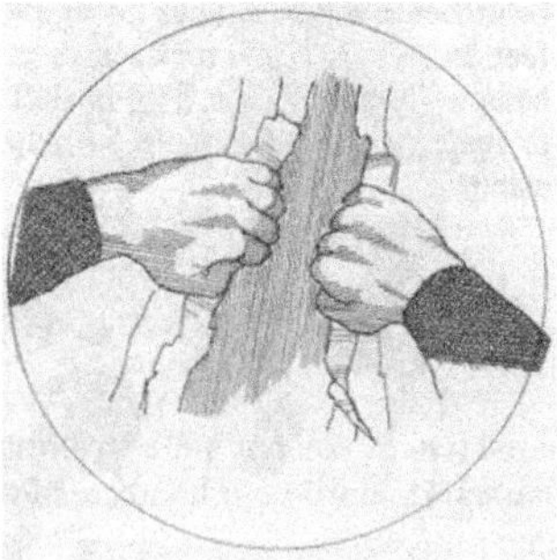

CROSS-PRESSURE METHOD

f. Inverted. Pull or push.

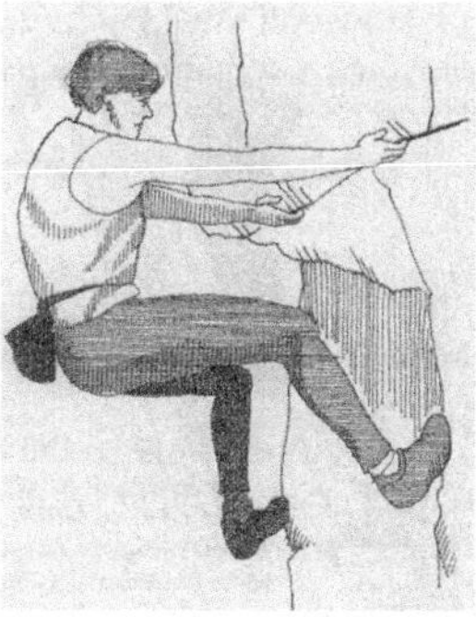

INVERTED TECHNIQUE USING COUNTERFORCE PRESSURE BETWEEN FEET AND HANDS

g. Pinch. As the name implies this is a grip used on tiny little nubbins.

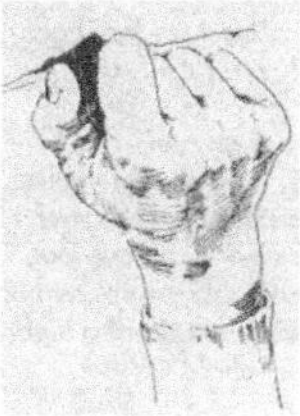

PINCH

h. Stemming. The spreading of arms or legs to maintain a proper body position. (i.e. usually used in a book or chimney.)

STEMMING

8. **GENERAL USE OF HOLDS**. How you use an individual hold is dependent on your experience level, or sometimes, your imagination. Here are some general guidelines.

a. Most handholds can be used as foot holds as you move up the rock.

b. Use all holds possible in order to conserve energy.

c. Even small projections may be used as holds.

d. Do not make use of your knees or elbows due to the reason that it is skin on bone and a slip could occur if pressure is exerted on them. Knees and elbows can be used with the extension of a limb jam.

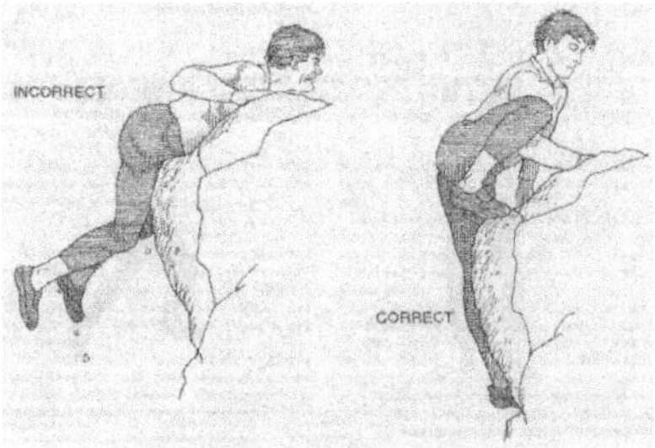

PROPER AND IMPROPER

9. **MOVEMENT ON SLAB**. Movement on slab is based on friction holds.

a. Use any and all irregularities in the slope to gain additional friction.

b. Traversing requires both hands and feet.

TRAVERSING

c. Descending steep slab may require turning inboard to face the slab and backing down.

d. The biggest mistake in slab climbing is leaning into the rock. Maintain maximum friction by keeping weight centered.

SLAB CLIMBING

10. BODY POSITION

a. The climber should climb with his body in balance by keeping his weight centered over and between his feet. Don't hug the rock. Don't over extend and become "spread-eagled". While climbing, keep in mind the acronym "CASHWORTH" for proper body position and movement. (WMC.C3F)

C - Conserve energy.

A - Always test holds.

S - Stand upright, on flexed joints.

H - Hands kept low; handholds should be waist to shoulder high.

W - Watch your feet.

O - On three points of contact.

R - Rhythmic movement.

T - Think ahead.

H - Heels kept low, lower than the toes.

UNITED STATES MARINE CORPS
Mountain Warfare Training Center
Bridgeport, California 93517-5001

WMC.C4
04/11/02

STUDENT OUTLINE

NATURAL AND ARTIFICIAL ANCHORS

TERMINAL LEARNING OBJECTIVE In a summer mountainous environment, establish an anchor system in accordance with the references. (WMC.C4)

ENABLING LEARNING OBJECTIVES.

(1) Without the aid of references and from a given list, choose the two types of anchors, in accordance with the references. (WMC.C4a)

(2) In a summer mountainous environment, establish a natural anchor point, in accordance with the references. (WMC.C4b)

OUTLINE

1. **TYPES OF ANCHORS.** (WMC.C4a)There are two types of anchors that we use. These two types are:

1) NATURAL
2) ARTIFICIAL

C4

2. **GENERAL CONSIDERATIONS FOR ANCHORS**. Anytime you employ natural or artificial anchors; there are special considerations that you must apply. These considerations apply both to the anchor itself and to the material being used to build the anchor. Some examples of these considerations are:

a. Whether using natural or artificial anchors, the installing unit must insure that the anchor is suitable for the load.

b. The anchor position must be relative to the direction of pull on the anchor.

c. The angle between the anchor points should not exceed 90 degrees. This is to ensure that no added stress is put upon the anchors, as well as the equipment being used to construct the anchor.

(1) To decrease the angle between anchor points, materials (i.e. sling ropes, web runners, Prusik cord, etc..) could be used to extend the anchor which will decrease the angle between anchor points.

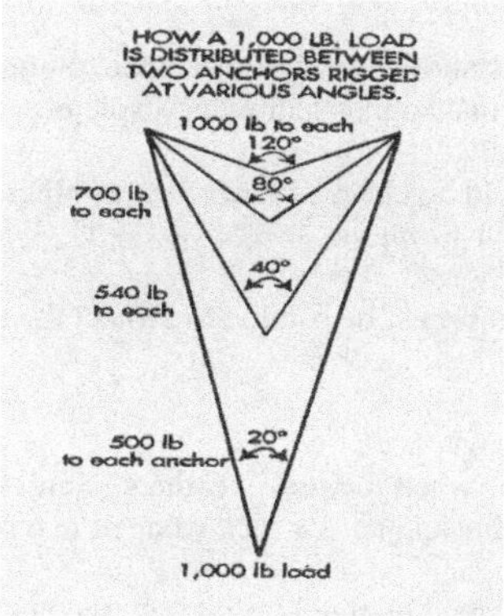

3. **NATURAL ANCHORS.** A natural feature is the preferred type of anchor point. Some examples and considerations are as followed:

a. Types of natural anchors.

(1) Trees.

(a) Select a tree that has not been chopped, burned or is rotten.

(b) The tree should be at least 6" in diameter and strong enough to support the intended load.

(c) Trees growing on rocky terrain should be treated with suspicion, since the roots normally are shallow and spread out along a relatively flat surface.

(2) Shrubs and Bushes.

(a) Select a shrub or bush that is alive and is not brittle, charred or loose.

(b) To avoid leverage, locate the central root and construct the anchor as near to the base as possible.

(3) Rocks and Boulders.

(a) Stability is of prime importance when considering a rock or boulder for an anchor. It must be strong enough and secure enough for the intended load.

(b) All surfaces of the rock or boulder should be inspected for any rough or sharp points. These areas must be padded to protect the rope from being abraded or cut.

(4) Spikes and Flakes.

(a) To check stability of a spike or flake, thump it with the heel of your hand. Anything that sounds hallow is suspicious.

(b) They should be checked for cracks or other signs of weathering that may impair their firmness.

(c) Sharp edges must be padded to protect the rope against cuts and abrasion.

(5) Threads and Chockstones.

(a) A thread is when the rock weathers or cracks to form a hole in the main wall. A chockstone is a rock wedged in a crack.

(b) Check a thread by thumping it with the heel of your hand. Make a visual inspection for cracks and weather. Common sense will prevail for choosing this type of anchor.

(c) When choosing a chockstone, ensure that it has substantial contact with the crack and that the stone's symmetry corresponds with the intended direction of pull.

b. Types of Natural High Tension Anchors. The following are types of Natural High Tension Anchor Systems used to construct installations. These can be tied on any suitable natural anchor point:

(1) Tree Wrap Anchor. The tree wrap is an anchor used to relieve tension on the actual knot itself. This system requires more rope when tied around a large anchor point.

(a) Tie a Figure 8 Loop on the standing end of the rope and wrap the 4-6 times around an anchor point.

(b) Attach a locking carabiner through the knot's loop and clip it onto the running end of the rope.

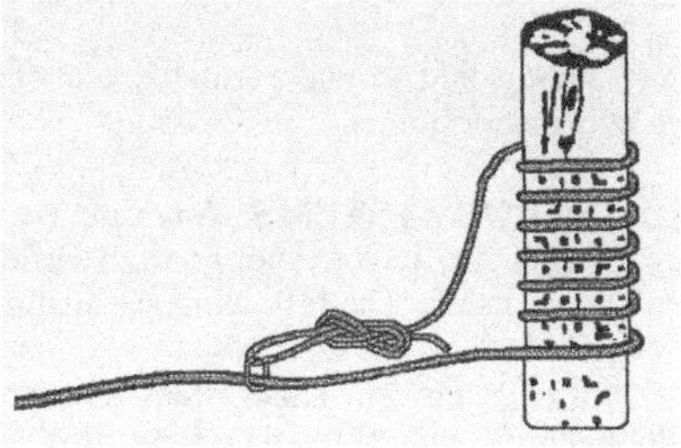

NOTE: If the anchor point to be wrapped is larger then 18' around, then three wraps will suffice.

(2) Swami Wrap Anchor. When tying the Swami Wrap, ensure that the joining knot is kept on the side of the anchor point. The Swami Wrap can be loosened under load if need be. This system can be substituted for a tree wrap when the installation's rope length is an issue.

(a) Select a suitable anchor point and tie a Swami Wrap around it.

(b) Clip a locking carabiner into as many wraps as possible. This will serve as the attachment point for the anchor system.

(3) Figure of 8 Anchor. The Figure of 8 Anchor is a quick and efficient system. This system cannot be loosened under load. It is tied in the following manner:

(a) Tie a Figure of 8 Loop on the standing end of the rope and wrap the knot around a suitable anchor point. Attach a locking carabiner through the knot's loop and clip it onto the running end of the rope.

(b) Before tensioning the anchor system, adjust the rope so the running end of the rope runs smoothly through the carabiner towards the direction of pull. This will prevent any lateral tension.

(b) If two ropes are used, the upper rope's carabiner has its gate upwards, and the lower rope's carabiner has it's gate downward.

c. Primary and Secondary Natural Anchor Point System. The two anchor points should be in line with the direction of pull. The primary anchor is the point nearest the running end while the secondary anchor point is directly behind the primary. It is constructed in the following manner:

(1) The primary anchor knot should be an around the object clove hitch. This is chosen for ease of untying the system after tension has been placed on the rope.

(2) The secondary knot is tied around a suitable anchor point ensuring that the rope is taut between the two anchor points.

4. **ARTIFICIAL ANCHOR SYSTEMS**. Any time we use anything other then a natural feature, we are using an artificial anchor point. Artificial anchors can be constructed in the ground, or on the rock itself. The following are artificial anchor systems:

a. Single Timber Deadman. This system is constructed in the ground and it requires considerable time and effort. The steps of it's construction are as follows:

(1) Dig a trench 6 feet long and 3 feet deep and wide enough to work in at a 90 degree angle to the direction of the pull.

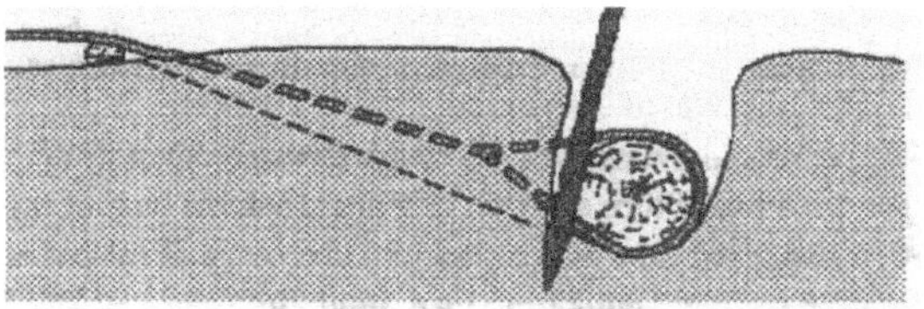

(2) Dig another trench about 12 inches wide. This trench is dug so that it intersects the main trench at a right angle in the middle. The bottom of this trench should be parallel to the direction of pull and should join the bottom of the main trench.

(3) Take an anchoring device (i.e. log, engineer stakes, bundled up branches etc.) that is strong enough to support the intended load. The anchor is placed into the main trench and covered with dirt with the exception of that part of the anchor that joins the second trench. Stakes approximately 3 feet long should be driven approximately 1-1/2 feet into the ground between the dead-man and the slanted side of the trench to assist in holding the dead-man in place if the soil is soft.

b. Picket Hold Fast. The picket hold fast is an easier anchor to construct than the Deadman, and can be used almost anywhere. The strength of the picket system depends on the pickets and the soil or snow conditions. The picket system can be used for both high and non-tension systems. Construction is as follows:

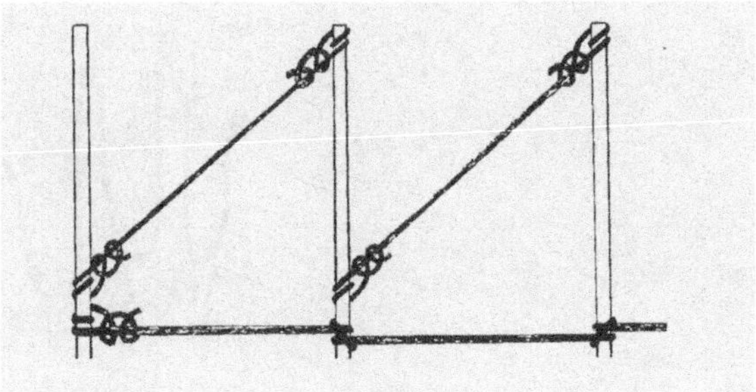

(1) Three stakes (i.e. logs, engineer stakes, etc.) are driven into the ground at a 30 degree angle against the direction of pull.

(2) The line of pickets should be driven in line with the direction of pull. The distance between the pickets can be anywhere from 3-12 feet apart depending on the terrain and soil conditions.

(3) Before tying any rope to the picket anchors, the base of the pickets must be buffed or padded.

(4) To tie a rope to the pickets, go to the furthest picket away from the cliff edge. Tie a round turn and two half hitches at the base of the picket. Take the running end of the rope and tie an over the object clove hitch to the base of the middle picket. Then with the running end of the rope tie an over the object clove hitch to the picket closest to the edge of the cliff face. Ensure that there is tension on the rope in between each picket.

(5) To tie off the pickets to themselves, go to the furthest picket from the cliff edge. And with a sling rope, tie a round turn and two half hitches at the base of that picket. With the running end of that sling rope tie a round turn and two half hitches at the top of the middle picket. Using a second sling rope tie a round turn and two half hitches to the base of the middle picket. Then with the running end of that sling rope tie a round turn and two half hitches to the top of the picket closest to the cliff edge.

c. <u>Equalized Anchor</u>. This system is built with a minimum of three pieces of protection. It can be tied with the standing end of a rope or by utilizing a practice coil.

(1) Construction with the Standing End of the Rope:

(a) Tie a Figure of 8 Loop in the standing end of the rope.

(b) Place a carabiner in each artificial anchor point and attach the knot in either of the outside carabiners.

(c) Clip the rope into the remaining carabiners.

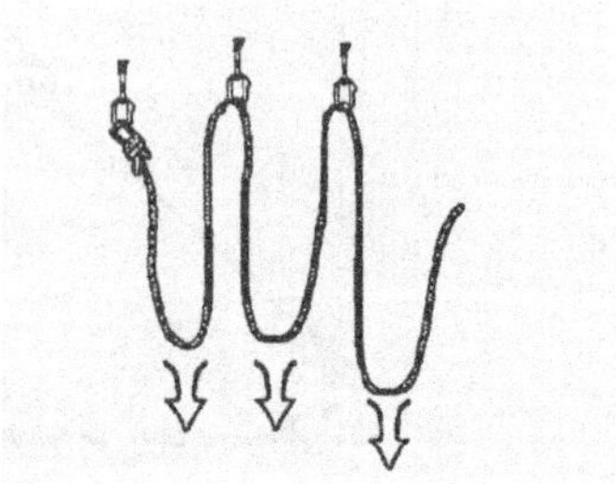

(d) A bight of rope is then pulled down after each carabiner into the anticipated direction of pull. With all three bights, tie an overhand knot around itself to include the running end of the rope.

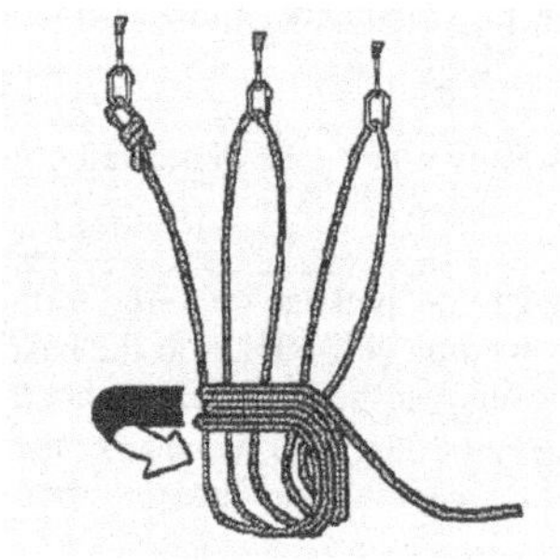

(e) With the running end of the rope coming from the overhand knot, tie a Figure of 8 Loop and attach it to the overhand knot's loop with a locking carabiner.

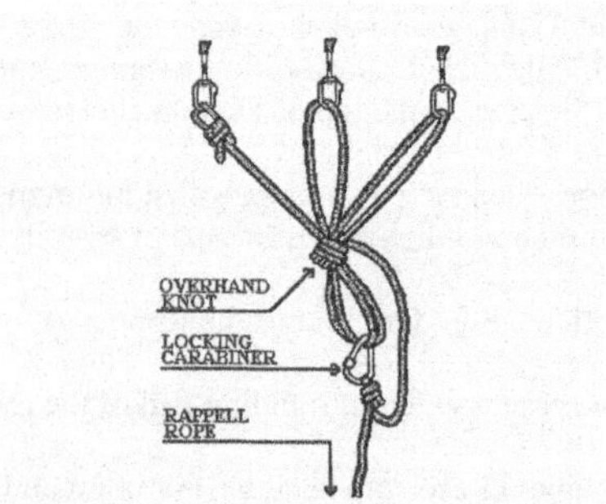

(f) Ensure that there is sufficient slack in the dead rope to prevent disfiguring the knot thus ruining the structural integrity of the knot.

(2) Construction with the Cordalette Method:

(a) Create an endless loop with a practice coil and clip it into each of the artificial anchor points.

(b) A bight of rope is then pulled down between each carabiner into the intended direction of pull.

(c) With all three bights, tie an overhand knot on itself.

(d) Place a locking carabiner into the overhand knot's loop. This will serve as the attachment point for the anchor system.

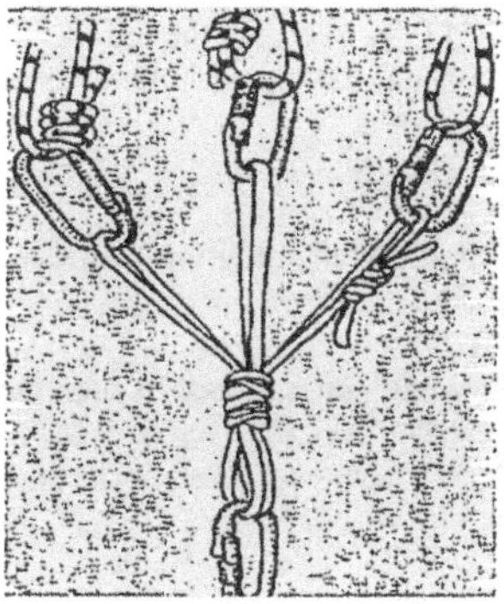

NOTE: If constructing a two-rope high-tension installation, a minimum of three artificial anchors per rope will be used.

UNITED STATES MARINE CORPS
Mountain Warfare Training Center
Bridgeport, California 93517-5001

SML
ACC
02/11/02

LESSON PLAN

SIT HARNESS

LEARNING OBJECTIVES

TERMINAL LEARNING OBJECTIVE: Without the aid of references, properly wear and maintain the MWTC sit harness, in accordance with the references.

ENABLING LEARNING OBJECTIVES

(1) Without the aid of references and given a diagram, name in writing the parts of a sit harness, in accordance with the references.

(2) Without the aid of references, wear a sit harness, in accordance with the references.

BODY

1. (5 Min) **NOMENCLATURE**

   a. Waist belt.

   b. Leg loops (adjustable).

   c. Buttocks straps (adjustable).

   d. Fastex buckle - 2.

   e. Doughnut.

   f. Equipment loops.

   g. D-ring (older models).

   h. Waist belt tie-in point.

   i. Crotch strap.

WEARING OF THE SIT

C5

HARNESS

a. First, disconnect the fastex buckle at the rear of the harness.

b. Hold the harness in front of you, put your feet through the leg loops ensuring that the buckles on the leg loops are outboard, on your thighs.

c. Fasten the waist belt into the buckle, ensuring that it is a tight but comfortable fit. **You must ensure that the waist belt is threaded back through the buckle, this action locks the waist belt to the buckle. Failure to do this will cause the waist belt to slip through the buckle when it is under load.**

d. Adjust the leg loops so that they are high on your thigh, once you have them adjusted, get your buddy to clip the fastex buckle male to the most comfortable female fastex. Then adjust the buttock straps so as the leg loops are held up.

e. If all the above are done correctly, the harness should now be a comfortable but snug fit, after you have fitted it for the first time there is no need to go through the same procedure each time you put it on. Simply hold the harness in front of you, step into it and attach the waist belt to the buckle in the approved manner.

2. TYING IN THE ROPE (END OF ROPE)

a. Pass the end of the climbing rope up through the crotch strap, then through the doughnut and through the waist belt tie-in point.

b. Tie the rope using a retraceable figure eight, adjusting the knot to get it as close as possible to the body.

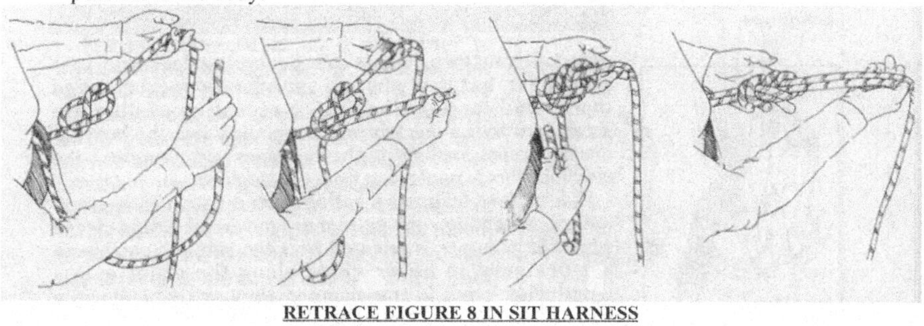

RETRACE FIGURE 8 IN SIT HARNESS

3. TYING IN THE ROPE (MIDDLE OF THE ROPE)

a. Take up a bight of rope and tie a figure eight loop, then take a steel locking carabiner and attach the carabiner to the harness by securing the crotch strap, doughnut, and the waist belt tie-in loop. Now attach the figure eight loop to the carabiner.

4. (5 Min) **CARE AND MAINTENANCE**

a. Avoid contact with chemicals, as this will damage the nylon.

b. Regularly inspect for signs of abrasions and normal wear. Pay particular attention to wear points such as the tie-in loops, buckles, and sewn joints.

c. Keep away from heat such as open flames, cigarettes, etc.

d. If soiled by grit and seawater, wash in lukewarm water with pure soap and allow to dry in a warm room away from direct heat.

e. Two to three years of life can be expected during normal climbing use.

f. It is recommended that a harness that has experienced a serious fall should be discarded.

g. Under no circumstances will you ever tie into the equipment rack of any sit harness as a belay, anchor point, etc.

UNITED STATES MARINE CORPS
Mountain Warfare Training Center
Bridgeport, California 93517-5001

FMST.07.33
05/02/01

STUDENT OUTLINE

ESTABLISHMENT OF RAPPEL POINTS AND RAPPELLING

TERMINAL LEARNING OBJECTIVE Given a mountainous environment, mountaineering equipment, helmet, sit harness, rappel gloves, 165 ft static rope, rescue 8, prusic cord, and one or more assistants, conduct rappelling operations to support mission requirements per the reference. (FMST.07.33)

ENABLING LEARNING OBJECTIVES.

(1) Without the aid of references, select from a given list the three criteria involved in site selection for a rappel site, in accordance with the references. (FMST.07.33a)

(2) In a summer mountainous environment, execute the commands used between a rappeller and his belay man when conducting a rappel, in accordance with the references. (FMST.07.33b)

(3) Without the aid of references, select from a given list three of the nine duties of the rappel point NCOIC, in accordance with the references. (FMST.07.33c)

(4) Without the aid of references, select from a given list three of the six duties of the first man down a rappel rope, in accordance with the references. (FMST.07.33d)

(5) Without the aid of references, select from a given list the three types of rappels, in accordance with the references. (FMST.07.33e)

(6) In a summer mountainous environment, rappel down a cliff face using a seat-hip rappel, in accordance with the references. (FMST.07.33g)

C6

OUTLINE

1. **INHERENT DANGER OF RAPPELLING**. Rappelling is inherently dangerous because rappellers rely totally on the equipment.

   a. To ensure a safe training evolution, two ropes will be utilized at MWTC.

   b. All the rappels taught at MWTC can be completed with a single rope if the situation arises.

   c. If utilizing a one-rope system with the carabiner wrap, the rope will be attached to the carabiner with two wraps vice one.

2. **SITE SELECTION**. (FMST.07.33a) when selecting a rappel site consider these three factors.

   a. There must be a good anchor. As previously taught in *NATURAL* and *ARTIFICIAL ANCHORS*, natural anchors are preferred.

   b. The rappel route down should be as free of obstacles (i.e., vegetation, debris) as possible.

   c. There must be suitable loading and unloading platforms.

NOTE: The evaluation of a site for the above factors should be made by the rappel point NCOIC, who should be the most experienced rappeller in the unit.

3. **DUTIES OF THE RAPPEL POINT NCOIC**. (FMST.07.33c)Once a rappelling site has been selected, one person will be appointed to each rappel lane as a rappel point NCOIC. These individuals should have experience as rappellers. The rappel point NCOIC has ten duties and responsibilities, which are:

   a. Ensures that the anchor points are sound and that the knots are properly tied.

   b. Ensures that loose rock and debris is cleared from the loading platform.

   c. Allows only one man on the loading platform at a time and ensures that the rappel point is run in an orderly manner.

   d. Ensures that each man is properly prepared for the particular rappel; i.e. gloves on, sleeves down, helmet secured, rappel seat tied correctly and secured properly.

   e. Attaches the rappeller to the rope and ensures the rappeller knows the proper braking position for that particular rappel.

   f. Ensures that the proper commands or signals are used.

   g. Dispatches each man down the rope.

h. The rappel point NCOIC will be the last man down the rope.

i. The rappel point NCOIC will ensure that the ropes are inspected after every 50 rappels.

j. The rappel point NCOIC will maintain a rope log.

4. **TYPES/USE**. (FMST.07.33e) The three types of rappels and when they are preferred to be used are:

a. Hasty Rappel: It is used when carrying loads down moderate slopes.

b. Seat-Shoulder Rappel: It is used for heavily laden troops over vertical to near vertical cliff faces.

c. Seat-Hip Rappel: It is used when carrying loads over vertical to near vertical faces.

5. **HASTY RAPPEL**. The hasty rappel is the easiest type of rappel to prepare for. It requires no equipment other than a rope and gloves.

a. Conduct. A hasty rappel is conducted in the following manner:

(1) Sleeves will be rolled down and gloves will be put on.

(2) Face slightly sideways.

(3) Place the rappel rope across your back, grasping it with both hands, palms forward, and arms extended.

(4) The hand nearest the anchor is the guide hand. The hand farthest from the anchor is the brake hand.

(5) Lean out at a moderate angle to the slope.

(6) Descend down the hill facing half-sideways, taking small steps and continually looking downhill while leading with the brake hand.

(7) Feet should not cross and the downhill foot should lead at all times.

b. Braking. The steps for braking during a hasty rappel are as follows:

(1) Bring the lower (brake) hand across the front of the chest to brake.

(2) At the same time, turn to face up toward the anchor point.

6. **SEAT-SHOULDER RAPPEL**. The seat-shoulder rappel relies on friction as the main effort of controlling the descent. It is very efficient for men with heavy packs because it provides support for heavy loads on the back.

a. Conduct. A seat-shoulder rappel is conducted in the following manner:

(1) Put on your rappel-seat, roll down your sleeves and put on your gloves.

NOTE: To avoid causing a possible injury, it is advised that the rappel seat be constructed at the rear of the rappel site loading platform. If the rappel seat is worn for too long, the rappel seat could loosen up enough to cause you to slip out of your seat while rappelling.

(2) The steel locking carabiner is placed on the rappel-seat so that the gate opens down and away, to prevent the gate from opening once the wraps are placed into the carabiner.

(3) Step-up to the rope with you're left shoulder facing the anchor.

(4) The rappel rope is attached to the rappeller's hard point carabiner as follows:

(a) Snap the rope into the locking carabiner.

(b) Taking slack from the standing (anchor) end of the rope, make one wrap with the rope around the body of the carabiner and back through the gate.

(c) Ensure that the locking nut of the carabiner is fastened to lock the carabiner closed.

NOTE: If you are using only one rope to rappel with due to the tactical situation or equipment availability, the procedures are the same, EXCEPT the individual will make two wraps around the body of the carabiner instead of one.

(5) Take the rope across your chest, over your left shoulder, diagonally across your pack and down to the right (brake) hand.

(6) Descend by walking down the cliff using the braking procedure to control the rate of descent. Look under your brake arm for possible obstacles to avoid.

b. Braking. The steps for braking during a seat-shoulder rappel are as follows:

(1) Lean back.

(2) Face directly uphill while bringing the brake hand across the chest.

7. **SEAT-HIP RAPPEL**. The seat-hip rappel is the most commonly used rappel.

a. Conduct. A seat-hip rappel is conducted in the following manner:

(1) Construct the rappel seat; roll down sleeves and put gloves on.

(2) The steel locking carabiner is placed on the rappel-seat so that the gate opens up and away.

(3) Step up to the rope with your left shoulder facing the anchor.

(4) The rappel rope is snapped into the carabiner as follows:

(a) Snap the rope into the locking carabiner.

(b) Taking slack from the standing (anchor) end of the rope, make one wrap with the rope around the body of the carabiner and through the gate again.

(c) Ensure that the locking nut of the carabiner is fastened to lock the carabiner closed.

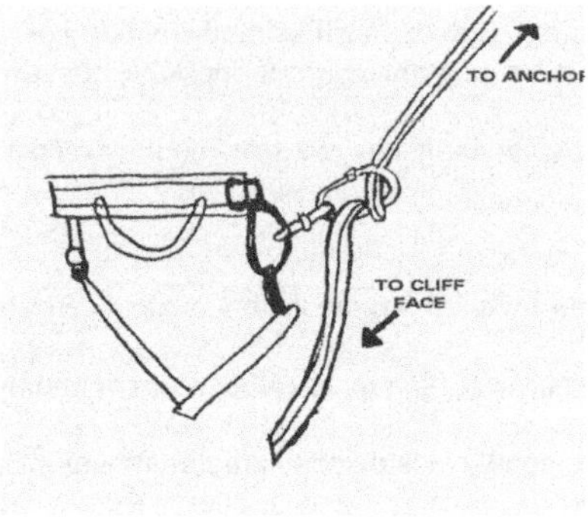

SEAT-HIP RAPPEL

NOTE: If you are using only one rope to rappel with due to the tactical situation or equipment availability, the procedures are the same, EXCEPT the individual will make two wraps around the body of the carabiner instead of one.

(5) The rappeller will grasp the running end of the rope with the brake (right) hand, palm down and turned slightly inboard, near the hip.

b. Braking. The steps in braking for a seat-hip rappel are as follows:

(1) Grasp the rope tightly with the brake hand.

(2) Take the brake hand and place it in the small of the back. This will create enough friction to stop all momentum.

NOTE: At no time will you bound or jump while you are descending. You "walk down" the cliff face using the proper braking procedure to control your rate of descent.

8. **SAFETY OF THE FIRST MAN DOWN**. To ensure the safety of the first man down, the rappeller will:

a. Before deploying the ropes, tie the two ends of the ropes together with an overhand knot.

b. Tie a friction knot on the standing end of the rappel rope with a Prusik cord.

c. Attach the rappel ropes to the harness's hard point with a locking carabiner.

d. Attach the Prusik cord to the harness's hard point with a locking carabiner. This cord is also referred to as a safety Prusik.

NOTE: If the Safety Prusik is too short, it can be extended with a web runner.

9. **DUTIES OF THE FIRST MAN DOWN**. (FMST.07.33d) The first man down the rope has specific duties. They are as follows:

a. Selects a smooth route down for the ropes.

b. Clears the route of loose rocks and debris.

c. The first man down will untie the overhand knot and straighten the ropes out once he reaches the bottom.

d. The first man down belays the next man down the rope.

(1) To belay correctly, the belay man will stand facing the cliff face, arms up with palms facing upward and over lapping. The rope will pass through the opening of the over lapping hand's thumb and index fingers.

(2) To stop a fallen rappeller, the belay man will grab the rope with closed fists and pull straight down bringing the forearms parallel to the deck.

e. Take charge of personnel as they arrive at the bottom to include appointing a belay man.

10. **RAPPELLING COMMANDS.** (FMST.07.33b) In order to conduct rappelling operations safely, it is essential that everyone understands the sequence of events. The following voice commands or rope tugs will be utilized:

VOICE COMMANDS	GIVEN BY	MEANING
'Lane # on Rappel"	Rappeller	I am ready to begin rappelling
'Lane # on Belay"	Belayer	I am on belay and you may begin to rappel
'Lane # off Rappel"	Rappeller	I have completed the rappel and am off the rope
"Lane # off Belay"	Belayer	I have completed the belaying of the rappeller

ROPE TUGS	GIVEN BY	MEANING
Three Tugs	Rappel Point NCOIC	The rappeller is ready to begin rappelling
Three Tugs	Belayer	I am on belay and the rappeller may begin to rappel
Three Tugs	Belayer	The rappeller is off the rappel rope

11. **TYING-OFF**. Occasionally, it may be necessary to stop during a rappel before reaching the bottom of a cliff. The following sequence is used:

a. The commands used in tying-off are as listed below:

COMMAND	GIVEN BY	MEANING
"Lane # Tying-off"	Rappeller	I am ready to tie-off, give me some slack.
"Lane # Tying-off"	Belayer	I have given you enough slack and you may tie-off.
"Lane # On rappel"	Rappeller	I have completed tying-off and I am ready to resume rappelling.
"Lane # On rappel"	Belayer	I am on belay, you may rappel.

b. The procedure used when tying-off is as follows:

(1) The rappeller gives the command, "Lane #, tying-off."

(2) The Belayer gives the rappeller slack and gives the command "Lane # Tying-off". He continues to hold the rope in the belay position, remaining alert and watching the rappeller.

(3) The rappeller quickly brings his brake hand to the twelve o'clock position, so that the running end and standing end are parallel.

(4) The rappeller grasps all ropes with his guide hand as close to the carabiner as possible.

(5) The rappeller steps over the rope so that the running end is going between his legs.

(6) The rappeller releases the rope with his brake hand then reaches under the running end, over the standing end. He then takes up a bight from the running end about two feet long and pulls it over the standing end and under the running end, forming a half hitch.

(7) He pulls the half hitch tightly against the guide hand.

(8) He works the half hitch down snugly against the carabiner while maintaining contact with the guide hand as long as possible.

NOTE: A loose half hitch could bind into the carabiner causing difficulties in clearing the knot out. Make sure that the first half hitch is dressed down tightly before moving it against the carabiner.

(9) Place another half hitch above the one already tied.

(10) To untie, reverse the steps. Remove the safety half hitch, then shrink the first half hitch down to a small loop by grabbing the running end of both ropes and pulling them straight out to the left. Place your right hand in the middle of the two bights until they are snug on all four fingers. Then remove one finger and make the bights snug on three fingers. Repeat this process until it's down to one finger and both the bights are equal.

(11) Now, grasp the running end with both hands and smartly jerk the running end of the ropes upward to pop the small loop out from the first half hitch. From this position, keep the guide hand around both of the ropes next to the carabiner; step back over the rope so that the running end is to your right side. Grab just the running end with your brake hand and quickly set the brake behind the small of your back then, readjust you guide hand onto the standing end. You now should be in the seat-hip rappel position.

NOTE: The wraps may bind up some after untying, where no further movement down is possible. To alleviate this, keep the brake on and force your body weight down to pop the wraps in the carabiner back to their intended position and then continue on with the rappel.

(12) The rappeller gives the command "Lane #, on rappel."

(13) The Belayer takes up the slack and gives the command "Lane #, on belay."

(14) The rappeller resumes to rappel down.

12. **RESCUE TECHNIQUES**. When conducting rappelling operations, the possibility of a rappeller getting caught on the rope due to either his clothing or equipment can occur. There are two types of rescues that can be performed to free the rappeller from this situation.

a. The Self-Rescue Technique.

(1) After realizing that you are caught up on the rope, check with your belay man to ensure that he has a solid brake set.

(2) Using a safety Prusik, tie a friction knot onto the rappel rope approximately an arms length above the area that is fouled. Anchor the other end of your safety Prusik to your rappel seats hard point. This is the same technique used as discussed earlier with the first man down the rope.

(3) Work the friction knot up until there is no tension on the rappel device. This will give you the necessary slack to free the malfunction.

(4) Once you have corrected the problem, continue to rappel down using the first man down method.

b. The Buddy Rescue Technique. Himself uses this method when the rappeller is unable to correct the problem.

(1) A rescuer will rappel down on another rope to the disabled rappeller and tie-off.

(2) The rescuer will then establish a safety Prusik onto the victim as taught in the self-rescue technique.

(3) After clearing the malfunction, continue with rappelling operations.

c. Rappelling Casualty Rescue. There are three different methods in which to get an injured rappeller to the bottom.

(1) The first method is to allow the belay man to lower the casualty by slowly releasing the tension on the rappel rope.

(2) The second method is used for critical injuries or when the belay man cannot properly control the casualty's descent.

(a) The rescuer will rappel down using another rope. Once the rescuer gets to the casualty, the rescuer will have his belay man brake him off, enabling the rescuer to use both hands.

(b) If necessary, the rescuer will perform the basic first aid needed. Once the casualty is ready to be lowered, the rescuer will call down to both of the belay men to simultaneously lower both men. The rescuer will hold onto the casualty the entire way down so that he doesn't bounce off the rock face.

(3) The third method is called a tandem rappel. Its used to rescue a casualty who has sustained an injury serious enough that he cannot operate the rappel device himself and requires the assistance of a second rappeller. This type of rescue can be conducted from the top of the cliff face or while on rappel, depending on where the injury occurred.

(a) If the injury occurs while on top of the cliff face, the following steps should be taken:

1. Take either a long sling rope or Prusik cord and tie a figure-eight knot offset so that one length of the cord is longer then the other.

2. Tie a figure-eight knot at each end.

3. Take a rappel device and attach it to the offset figure-eight loop with a carabiner and attach it to the rappel rope.

4. Next, attach the casualty's hard point to the short end of the sling rope/Prusik cord and the rescuer to the other end.

5. The rescuer will then utilize a safety Prusik in the same method as the first man down technique, except that the friction knot will be tied below the rappelling device and controlled with the brake hand.

6. The rescuer will then maneuver himself under the casualty to provide assistance and support.

7. At this point you are ready to rappel both rescuer and casualty at the same time.

NOTE: Your rappel device should be far enough in front of you so that the casualty will not be able to reach it.

(b) If the injury occurs during the rappel, the following steps are taken:

1. The rescuer will preset the same system on a separate rappel rope and rappel down to the casualty.

2. The rescuer will then take the short end of the sling rope and attach it to the casualty's hard point.

3. With a knife, the rescuer will cut the casualty's rappel rope away.

4. Rappel down to the bottom of the cliff face.

NOTE: THIS IS THE ONLY TIME THAT A KNIFE WILL BE USED DURING A RAPPEL. EXTREME CAUTION SHOULD BE USED. IT IS NOT TO BE DONE IN TRAINING, BUT ONLY DURING AN ACTUAL RESCUE AND AS A LAST RESORT.

13. **RETRIEVABLE RAPPELS**. Once a unit has rappelled down a vertical obstacle, it may be necessary to retrieve the rope(s). Depending on the height of the obstacle either one or two ropes will used to construct the rappel lane.

a. One-Rope Retrievable Rappel:

(1) Find the middle of the rope and place it directly behind a suitable anchor point.

(2) Join the pigtails of the rope with an overhand knot and deploy the rope down the obstacle.

(3) On one side of the rope in front of the anchor point, tie an over-the-object clove hitch onto a locking carabiner.

(4) On the other side of the rope, tie another over-the-object clove hitch and attach it to the same carabiner.

(5) The first man down will utilize a safety Prusik and untie the overhand knot after he reaches the bottom.

(6) All others conduct proper rappel procedures.

(7) The last man down will disconnect the carabiner from the rope, and ensuring that the middle of the rope is directly behind the anchor point, rappel down the rope.

(8) Pulling on either end of the rope then retrieves the rope.

b. Two-Rope Retrievable Rappel. This type of system is identical to the one rope retrievable rappel except for a few considerations.

(1) The reason for using two ropes vice one rope is that the height of the rappel is greater.

(2) When two ropes are used they should be joined together using a square knot. This knot will be placed directly behind the anchor point.

(3) All else remains the same as the one rope retrievable rappel except:

(a) The last man down will disconnect the carabiner from the rope and will move the joining knot as close to the vertical obstacle's edge as possible. This will prevent the knot from possibly getting caught up while retrieving.

(b) The last man down will then place a carabiner on the rope below the knot. This will enable him to know which line to pull for retrieval.

14. **EQUIPMENT**. Equipment should be worn in accordance with unit SOP. The unit has the responsibility to determine which methods it feels are most beneficial to the mission. Weapons should be worn across shoulder; muzzle down and away from the brake hand with a tight sling securely attached to the weapon. Weapons should also be dummy corded to the individuals.

UNITED STATES MARINE CORPS
Mountain Warfare Training Center
Bridgeport California, 93517-5001

WMC
04/11/02

STUDENT OUTLINE

TOP ROPING

LEARNING OBJECTIVES.

TERMINAL LEARNING OBJECTIVE. In a summer mountainous environment, and given a rock face and a belay man, conduct top roping with cartridge belt and rifle, in accordance with the references. (WMC)

ENABLING LEARNING OBJECTIVE

(1) In a summer mountainous environment, establish a belay point from the top of a cliff for use in a top rope climb, in accordance with the references. (WMC.a)

(2) In a summer mountainous environment, demonstrate the use of climbing commands while conducting a top rope, in accordance with the references. (WMC.b)

OUTLINE

1. **ESTABLISHING A BELAY STANCE FROM THE TOP**.

a. The Belayer will establish a sitting belay stance on the cliff head by:

(1) Constructing a suitable anchor with the standing end of the rope.

(2) With the direction of pull away from the anchor, tie a directional figure 8 loop near enough to the cliff's edge so that the climber can be observed, if possible.

(3) Tying a swami Wrap around oneself, clip a locking carabiner through all the rear wraps and clip a large locking carabiner through all the front wraps.

(4) Secure the rear locking carabiner into the directional figure of 8 loop.

(5) Secure the running end of the rope to the large locking carabiner utilizing a suitable friction belay (i.e., Munter hitch, stitch plate, etc.).

NOTE: Gloves **will not** be worn while belaying a climber during a top rope.

2. **SECURING THE CLIMBER TO THE ROPE**.

a. The climber will tie into the end of the top rope by:

(1) Constructing a Swami wrap around oneself.

(2) Tying a retrace figure 8 loop through all of the Swami's wraps or;

(3) A figure of 8 loop clipped into a locking carabiner secured through all the Swami's wraps.

b. Alternative methods.

(1) A bowline on a coil as discussed in ROPE MANAGEMENT or;

(2) Tie into the hard point of the harness as discussed in SIT HARNESS.

3. **ESTABLISHING A BELAY STANCE FROM THE BOTTOM**.

a. Belaying from the bottom is commonly referred to as a "yo-yo" or "Sling Shot" belay. This system can be constructed for either a direct or indirect belay.

(1) Construct a suitable top anchor utilizing either a pulley or one steel locking or two non-locking carabiners as the attachment point for the anchor.

(2) The rope will travel up from the bottom and pass through the attachment point then return to the bottom.

(3) The climber will be secured to one end of the rope and the Belayer from the other end.

4. **CLIMBER'S RESPONSIBILITIES**.

a. The climber ensures that the Belayer is anchored and on belay, by use of commands or prearranged signals prior to beginning the climb.

b. The climber will not out climb the belayer; this will cause slack in the rope between the belayer and the climber.

c. Avoid placing excess pressure/weight on belay man.

d. Weapons will be worn across shoulder, muzzle down and to the left, with a tight sling securely attached to the weapon.

5. **<u>CLIMBING COMMANDS AND SIGNALS</u>.** In order to conduct top rope operations safely, it is essential that everyone understands the sequence of events. The following voice commands or rope tugs will be utilized:

VOICE COMMANDS	GIVEN BY	MEANING
"Lane #, Up Rope"	Climber	Belayer needs to take in the slack.
"That's Me"	Climber	Excess slack has been taken up between us.
"Lane #, On Belay"	Belayer	I am on belay.
"Lane #, On Climb"	Climber	I am ready to climb.
"Climb Climber"	Belayer	You may begin to climb.

ROPE TUGS	GIVEN BY	MEANING
Three Tugs	Climber	Belayer needs to take in the slack.
Three Tugs	Belayer	I have taken up the slack and you may begin to climb.
Three Tugs	Climber	Climbing.

UNITED STATES MARINE CORPS
Mountain Warfare Training Center
Bridgeport, California 93517-5001

FMST.07.40
03/02/02

STUDENT OUTLINE

MOUNTAIN CASUALTY EVACUATIONS

TERMINAL LEARNING OBJECTIVES. In a summer mountainous environment and given a simulated/actual casualty, evacuate casualties, in accordance with the references. (FMST.07.40)

ENABLING LEARNING OBJECTIVES

(1) With the aid of references, state orally the general considerations for medevac procedures, in accordance with the references. (FMST.07.40a)

(2) In a summer mountainous environment, and with the aid of references, construct an expedient litter, in accordance with the references. (FMST.07.40b)

(3) In a summer mountainous environment, and with the aid of references, secure a simulated/actual casualty in a SKEDCO litter, in accordance with the references. (FMST.07.40c)

(4) In a summer mountainous environment, and with the aid of references, attach a belay line to a SKEDCO litter for movement up/down moderate slopes, in accordance with the references. (FMST.07.40d)

(5) In a summer mountainous environment, with the aid of references, and given a simulated/actual casualty, conduct a medevac using the barrow-boy method, in accordance with the references. (FMST.07.40e)

C8

OUTLINE

1. **GENERAL CONSIDERATIONS** (FMST.07.40a) the general considerations are a set of guidelines that can be used no matter how serious the casualty is. They are remembered by a simple acronym, **APASSNGG.**

   a. Apply Essential First Aid. (i.e. splints, pressure bandage, etc.)

   b. Protect the Patient from the Elements. Provide the casualty with proper insulation from the ground. Ensure that he is warm and dry. If there are any natural hazards (i.e. rock fall, lighting, etc.) either move the casualty as quickly as possible or ensure that he is well protected.

   c. Avoid Unnecessary Handling of Patient.

   d. Select Easiest Route. Send scouts ahead if possible, to break trails.

   e. Set Up Relay Points and Warming Stations. If the route is long and arduous, set up relay points and warming stations with minimum amount of medical personnel at warming stations to:

      (1) Permit emergency treatment. Treat for shock, hemorrhage, or other conditions that may arise.

      (2) Reevaluate the patient constantly. If patient develops increased signs of shock or other symptoms during the evacuation, he may be retained at an emergency station until stable.

   f. Normal Litter Teams Must Be Augmented in Arduous Terrain.

   g. Give Litter Teams Specific Goals. This litter teams job is extremely tiring, both physically and mentally. The litter teams must be given realistic goals to work towards.

   h. Gear. Ensure all of the patient's gear is kept with him throughout the evacuation.

2. **ONE -MAN CARRIES**

   a. Sling Rope Carry. The sling rope carry requires two men and a 15-foot sling rope. One as the bearer, and an assistant to help in securing the casualty to the bearer. Conscious or unconscious casualties my be transported in this manner:

(1) Bearer kneels on all fours and the assistant places casualty face down on bearer's back ensuring the casualty's armpits are even with the bearer's shoulders.

(2) He then finds the middle of the sling rope and places it between the casualty's shoulders and the ends of the sling rope are run under the casualties armpits, crossed and over the bearer's shoulders and under his arms.

(3) Then the ropes are run between the casualty's legs, around his thighs, and tied with a square knot with two overhands just above the bearer's belt buckle.

(4) Ensure the rope is tight. Padding, when available, should be placed where the rope passes over the bearer's shoulders and under the casualties thighs.

b. <u>The rope coil carry.</u> This requires a bearer and a rope coil. It can be used to carry a conscious or unconscious casualty.

(1) Position the casualty on his back.

(2) Separate the loops of the mountain coil into two approximately equal groups.

(3) Slip ½ of the coil over the casualty's left leg and ½ over his right leg so that the wraps holding the coil are in the casualty's crotch, the loops extending upward the armpits.

(4) The bearer lies on his back between the casualty's leg and slips his arms the loops. He then moves forward until the coil is extended. When using the rope coil the bearer ties the coil to himself vice slipping his arms through the loops.

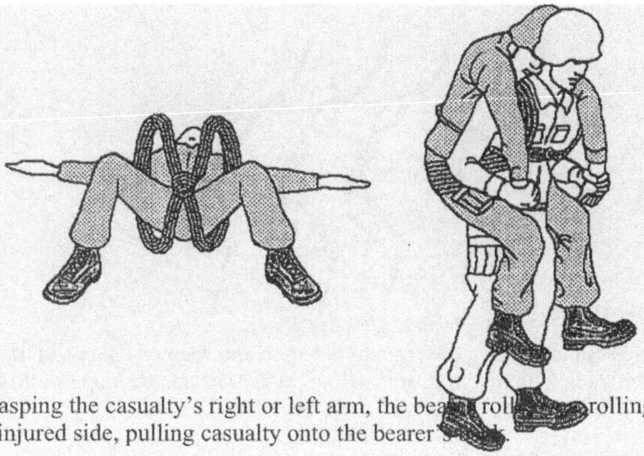

(5) Grasping the casualty's right or left arm, the bearer rolls, rolling to the casualty's uninjured side, pulling casualty onto the bearer's back.

(6) Holding the casualty's wrists, the bearer carefully stands, using his legs to lift up and keeping his back as straight as possible.

NOTE: The length of the coils on the rope coil and the height of the bearer are to be considered. If the coils are too long and the bearer happens to be a shorter person, it will require the coils to be uncoiled and shortened. If this is not done, then the casualty will hang too low on the bearer's back and make it a very cumbersome evacuation. A sling rope harness can be used around the victim's back and the bearer's chest, which will free the bearer's hands.

3. **TWO-MAN CARRY**

a. Pole Carry. The pole carry method is a field expedient method and should be considered as a last resort only, when narrow ledges must be traversed; vegetation limits the bearers to a narrow trail. This method is difficult for the bearers and uncomfortable for the casualty. Two bearers, four sling ropes and a 12 foot pole - 3 inches in diameter, are required for this carry.

(1) The casualty is placed on his back in a sleeping bag or wrapped in a poncho or blanket, then placed on an insulated pad.

(2) One sling rope is placed under the casualty below the armpits and tied with a square knot across the casualty's chest.

(3) The second sling rope is tied in the same manner at the casualty's waist.

(4) The third sling rope is placed at the casualty's legs below the knee.

(5) The fourth sling rope is tied around the ankles.

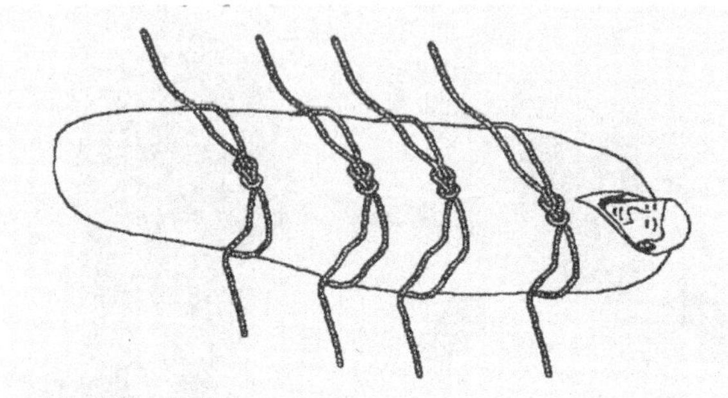

(6) The pole is placed along the casualty's length and secured using square knots with two overhands with the ends of the sling ropes. The square knots should be so tight that the overhands are tied onto themselves

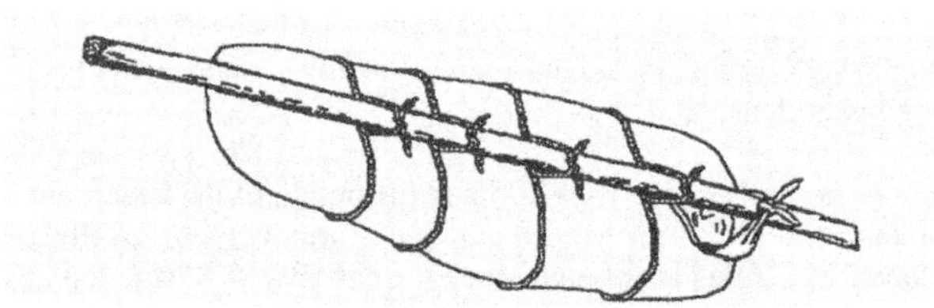

(7) The casualty should hang below the pole, as close to the pole as possible, to prevent swinging during movement.

(8) The casualty's head may be supported using a triangular bandage or a cartridge belt passed around the pole.

(9) For additional support and of movement, two additional bearers may be required, as well as a mountain coil.

(a) Mountain coil is split into two equal coils.

(b) Place knot of mountain coil under casualty's lower back.

(c) Additional bearers slip into each half of the hasty coil one on each side of casualty, aiding in support and movement of the casualty.

4. **EXPEDIENT LITTERS.** (FMST.07.40b) The two types of expedient litters that we will talk about are the alpine basket and the poncho litter.

a. Alpine Basket. To belay the alpine basket, the pre-rigs are attached to the bights formed coming through the loops.

(1) If barrow boy is to be used, the procedure previously discussed will be adhered to.

(2) If barrow boy is not used, then a tag line from the bottom must be implemented to keep the casualty away from the cliff face on the decent.

(3) Construction of the Alpine Basket:

(a) Start by making the same amount of bights as in the rope litter, but start from one end and tie a figure-of-eight loop to run the first bight through.

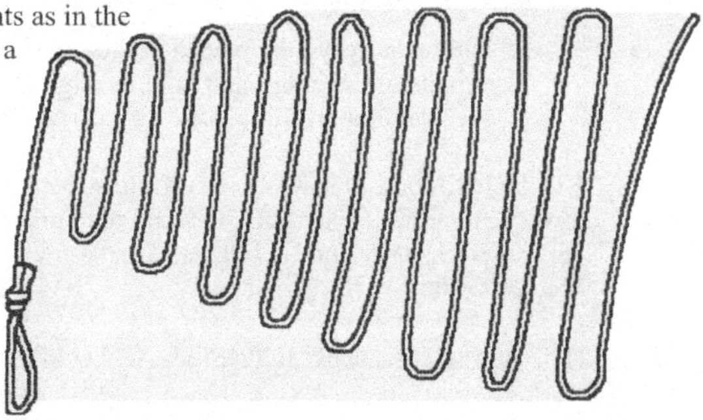

(b) Place padding i.e., isopor mat, on top of the bights and then lay the casualty on the padding and bights.

(c) Start at the casualty's feet and pull the first bight up around the casualty's ankles and through the figure-of-eight loop tied into the starting end of the rope.

(d) Go to the opposite side of the casualty and pull up the second bight and pull it through the loop formed by the bight that was pulled through the figure-of-eight.

(e) Continue until you get to the casualty's armpits, bring the second to the last bight up over the casualty's shoulder and into the bight and then bring the last bight up over the casualty's other shoulder and into the last bight formed.

(f) Secure the end of the bight with an appropriate anchor knot.

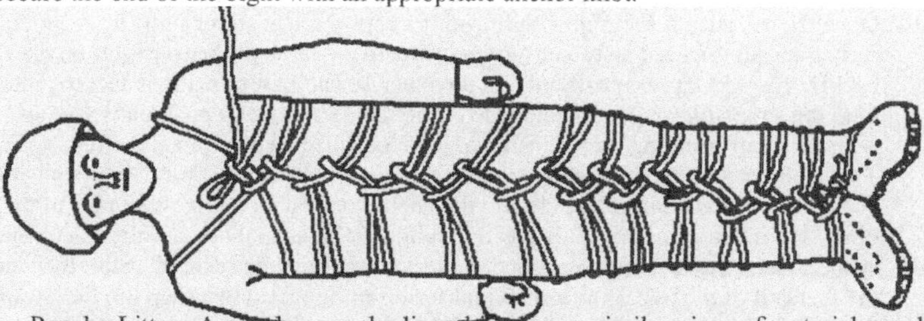

b. Poncho Litter. A poncho, poncho liner, bivy bag or similar piece of material may be used. In addition you will need six individuals with sling ropes.

(1) Lay poncho litter flat on the ground.

(2) Select six rocks about the size of a golf ball. Place one rock in each of the corners and one in the middle on each side in the middle of the litter. The rocks are placed on the under side of the poncho or like material. If a bivy bag is used the casualty should be zipped inside the bivy bag. The rocks should then be arranged in the same manner only on the inside below the zipper.

(3) Tie the sling rope together with an overhand knot. Take the middle of the rope and secure it around the rock with a clove hitch.

(4) An isomat may be laid on the poncho to help make the litter firmer.

(5) The casualty is then placed in the litter. The sling ropes are adjusted by feeding the pigtails of the over hand knot through itself to adjust for length. The loop is then put over the inboard shoulder of the carriers. Insure that the casualty is carried level.

5. **LITTERS.** There are two kinds of litters that are used for casualty evacuations in moderate to vertical terrain. The SKEDCO and the Stokes litter. Each has a variation of procedures for securing a casualty and rigging the litters for either raising or lowering. Let's first look at the SKEDCO litter.

(1) Securing a Casualty to the SKEDCO (FMST.07.40c)

(a) First unroll the litter. The litter must be re-rolled the opposite way to allow the litter to lay flat. Then lay the litter next to the casualty.

(b) If the casualty has any possibility of a spinal injury the Oregon Spine Splint must be used. Secure the splint to the casualty by use of the color-coded buckles. Experienced medical personal are recommended if spinal immobilization is necessary.

(c) Once the casualty is on the SKEDCO use the four body straps to secure the casualty to the litter. Unless injuries prevent, the casualty's arms should be at his sides to prevent further injuries to himself or the rescuers.

(d) Once the casualty is secured with the body straps the feet straps must be secured. The feet straps are secured last to ensure the casualty is in the proper position on the SKED. The feet can be positioned in three ways. The first position is feet together with the straps running on the outside of the feet. The second position is feet apart with the straps running on the inside of the feet. The last position is the feet stacked. This is the most uncomfortable position and not recommended for casualties with possible spinal cord injuries. Placing the heel of one foot on top of the toes of the other forms this position. This position will only be used for a casualty in confined spaces. Once the feet are positioned the feet straps must be secured. Start by bending the feet end of the SKED to form a plat form for the feet. Then loop the feet straps through the second grommets on each side.

(e) The last thing to do is to form the head end to protect the casualty's head. If possible the casualty should wear a helmet. Form the head end tying the pull strap up and secure it to the first body straps.

(2) Medevacing a Casualty. There are many ways to move a casualty once in the SKEDCO. However the medevac team must keep the general considerations in mind. The two methods that we will talk about next are the simplest and require the least amount of additional rigging

(a) The first way is to drag the casualty by the drag strap located at the head end of the SKED. You can also use the SKEDCO's carrying bag as a harness in conjunction with the pull strap and towing harness. If additional people are required, cordage can be added to the pull strap or the front carrying handles.

(b) The second way is to carry the casualty using the carrying handles. By using the set of four removable webbing the litter team can be augmented. To do this each pieces of webbing is tied to make an endless loop. Then pass a bight of the loop through one of the grommets to create additional handles.

(3) Rigging the SKED for Vertical Terrain and Helicopter Lift. In this type of terrain special requirements must be taken to raise or lower a casualty in the SKEDCO.

(a) Rigging the SKED for a Vertical Employment. A vertical raise or lower is used when moving a casualty on steep earth to avoid any further injury to the casualty. On vertical terrain the vertical raise or lower can be used if the terrain is not uniformed or there is a chance of rock fall. Ensure that the casualty's head is always above his feet.

(1) Identify the 30-foot piece of cordage that comes with the SKEDCO. Then tie a figure eight in the middle of the rope. If the rope is worn or missing, the same process can be done with two sling ropes.

(2) Next pass each end of the rope through the grommets at the head of the SKEDCO. Leaving approximately 1-2 feet of rope between the stretcher and the knot.

(3) Continue to feed each end through the grommets and the carrying handles towards the foot end of the SKEDCO. Pass the ends of the rope through the last grommets at the foot end and secure the two ends with a square knot without over hands.

(4) Bring the pigtails up and over the casualty's feet and pass the ends through the
a carrying handles towards the middle of the casualty. Then tie a square knot
b with two over hands.

c

(4) Rigging the SKEDCO for a Horizontal Employment. A horizontal raise or lower is preferred on uniformed vertical terrain. The horizontal employment allows the rescuer to assist the casualty on either a raise or lower. It also allows the rescuer the ability to monitor the casualty's condition and can easily treat the casualty if the need arises.

(a) Identify the two 4 inch nylon straps. They should be two lengths; one four inches shorter than the other. The shorter strap should be marked HEAD STRAPS.

(b) Insert one end of the head strap into a slot near the head of the litter. Then wrap the rest of the straps under the SKEDCO and pass the other end through the opposing slot. Do the same at the foot end of the SKEDCO with the other strap. Ensure that the strap runs smoothly under the SKEDCO.

(c) Connect the strap ends with the large locking caribiner that comes with the SKEDCO. If the caribiner is worn or missing, opposing Stubai 85 locking carabiners will suffice.

6. **STOKES LITTER**. The stokes litter is a litter that is constructed of metal tubing with a plastic covering. The litter is formed in a rectangular basket shape with mesh attached to the frame. Using the stokes for an evacuation (as with any evacuation) it should be padded for the casualty.

a. Securing the casualty to the stokes litter. In the event that the "seat belts" are missing from the stokes litter, sling ropes can be used to the lash the casualty. The steps involved are:

(1) Tie two sling ropes together using square knots and two overhands.

(2) Tie a stirrup hitch around ankles and feet, feed the two pigtails through the right angles of the stokes. Do not cross the ropes at the ankles.

(3) Lace the sling rope towards the casualty's head by passing the rope through the right angles (not over the top of the rails of the stokes).

(4) Secure the ends of the sling ropes by tying a clove hitch with two half hitches on the thick vertical bar located by the victim's shoulder.

7. **YOSEMITE PRE-RIG.** The title of this method, pre-rig implies the meaning ready to use, but in this case the rig must be constructed. Its purpose is for attaching a belay line to a litter and to make the litter easily adjusted. This is the one method that is used to secure a litter to a belay line.

a. Construction: This method normally requires four sling ropes. The steps are: (1) using one sling rope, tie a figure-of-eight loop with one tail.

(1) Take the remaining tail and run it through the window of the stokes litter or in the stirrups of the collapsible litter.

(2) Tie a Kragur knot onto the same sling rope.

(3) Repeat steps (1), (2), and (3) with the three remaining sling ropes.

(4) Suspend the litter to ensure that the comers are balanced.

8. **ASCENTS OR DESCENT OVER STEEP TO MODERATE SLOPES.** When the litter team is ascending or descending a slope they must consider the potential for further injury to the casualty or to themselves. If the risk of injury is high a belay line may be used to prevent injury to the casualty and the rescuers.

a. Preparing Casualty for Ascents or Decent over Steep to Moderate Terrain. This procedure will be depending on several things. Initially, site selection should contain the following features.

(1) Suitable anchor points.
(2) Clearance for casualty along the route
(3) Loading and unloading points.

b. Additional considerations.

(1) The casualty will always be rigged for vertical employment when on steep to moderate terrain.
(2) The smoothest possible route must be selected.
(3) Ensure that the casualty's head is above his feet.

c. Rescuers positions. There are two methods that can be used for the rescuers for moving a casualty in steep to moderate terrain.

(1) Two to four men will position themselves on each side of the litter. They can then carry the litter by the carrying handles. In steep terrain a second belay line may be used to assist the rescuers. We will discuses the belay line later in the chapter.

(2) The Caterpillar method will require as many personnel as possible. The personnel will split in half and position themselves on each side of the litter forming a tunnel. As the litter is raised or lowered each member will hand the litter to the next member in the tunnel. As the litter passes each person in the tunnel he will peel off and assume the lead either at the top or bottom of the tunnel. This will continue until the litter reaches its desired destination.

d. Belay Line. (FMST.07.40d) For belaying of a casualty, one rope will be used and from the top using one of two methods depending on the application.

(1) Body Belay. This method should only be used over moderate terrain. The belay man will establish a sitting position behind a suitable anchor (i.e., rock, tree, etc.) and pass the standing end of the rope behind his back. The running end of the rope will feed out from the belay man's right side. A figure of eight loop is tied to the end of the running end of the rope. It is then attached to the litter's figure eight loop with a locking carabiner. The belay man will then remove all of the slack between himself and the litter. The standing end of the rope should be stacked on the belay man's left side and run through his left side. As the casualty is lowered, the belay man will feed the rope from behind his back allowing it to run through his right hand. If the belay man needs to stop the casualty, he will clench the rope in his left hand, and

bring the rope to the center of his chest.

(2) Direct Belay. This method is the safest for either raising or lowering a casualty in either moderate to steep terrain.

(a) To Lower a Casualty. First a swami wrap will be tied around a suitable anchor point. Two locking carabiners will be clipped into all of the wraps of the swami wraps, gates up. A figure of eight loop is tied into the end of the static rope and attached to the litter with a locking carabiner. After all the slack has been taken up between the litter and the anchor, the rope must be tied through an appropriate belay device. The belay device is attached to the anchor through one of the two locking carabiners on the anchor. A safety (French) prussic will be tied to the running end of the rope and clipped into the second locking carabiner on the anchor. While the casualty is being lowered, one person will control the rope running through the belay device. The safety prussic will be controlled by a second person. Should the primary belay man lose control, the person operating the safety prussic simply lets go and the prussic will bind onto the rope, stopping the casualty.

(b) To Raise a Casualty. The anchors will be established in the same manner as discussed in lowering the casualty with one minor change. The one change is that instead of running the rope through a belay device, the rope will only run through a locking carabiner. The load will be raised by the use of a mule team. The mule must consist of as many people as possible. The mule team will raise the load in as straight a line from the anchor as possible. If the space does not permit a ninety-degree angle away from the anchor is also an option. The mule team will walk backward until the last man reaches his limit of advance. Once he reaches that limit he will peel off the end and return to the front of the mule team. This process is continued until the casualty reaches the top. If the load becomes unmanageable, the safety prussic will be allowed to bind on the rope while the mule team repositions themselves. If the person operating the safety prussic can not see the casualty a Point NCO will be in charge of communicating with the mule team.

(c) To Belay the Rescuers. If the route is too steep or the footing is poor the rescuers may need some assistance either on the raising or lowering of a casualty. If this is the case a separate belay line will be established for the rescuers. The anchor and the belay line are established in the same manner. The same anchor can be used if it is suitable for the load. The rescuers will then tie either around the chest bowlines or swami wraps. A figure eight loop will be tied into the end of the static rope and connected to the bottom rescuer with a locking carabiner. The other rescuers will connect themselves to the same rope with middle of the line prussic. They will be connected in this manner so that they can adjust their position to the casualty.

9. **BARROW BOY.** (FMST.07.40e) A barrow boy is no more than an assistant to the litter on vertical to near vertical cliff faces. The barrow boy can be used for either the Stokes or the

SKED litters. For this situation the Stokes litter should only be used in the horizontal position. However the SKED can be employed in either the horizontal or the vertical positions.

NOTE: Static load, lowered slowly, little friction, so don't need 2 ropes.

a. Rigging the Barrow Boy.

(1) First the rescuer must ensure that a suitable anchor has been established, a proper belay has been constructed, and that a safety prussic has been constructed.

(2) Then the rescuer must ensure that if an A-frame is used that it has been constructed and anchored properly.

(3) Next the rescuer will tie a rappel seat on. (a sit harness can also be used) Then he will ensure that an around the body bowline is tied onto the casualty. A figure eight will be tied on to the end to act as the casualty's safety.

(4) Then after running the running end of the rope though the carabiners or pulley of the A-frame he will tie a figure eight loop at the end of the static rope. He will then attach the figure eight loop to his hard point with a locking carabiner. Then he will tie a middle of the line prussic above the figure eight and attach it to the same locking carabiner in his hard point. This is called the adjustment prussic. It is used to adjust the position of the Barrow Boy in relation to the litter. Next he will take six to eight feet of slack from the end of the line figure eight and tie a directional figure of eight with direction of pull down. (Note: the prussic should be between the end of the line figure of eight and the directional figure of eight.) The directional figure of eight is the attaching point for the litter and the casualty's safety.

(5) Once the rescuer and the litter are secured, the belay man must take all the slack out of the system. The rescuers will the maneuver the litter through the apex of the A-frame with the help of the point NCO.

(6) Once onto the cliff face the rescuer will then position him self with his adjustment prussic so that he can be of most assistance to the litter on the raise or lower. The rescuer will pull the litter out away from the cliff face so that the casualty rides smoothly up or down the cliff face.

(7) The Point NCO will be in charge of the belay men or the mule team. He will also communicate with the rescuer about the rate of speed, if the rescuer needs to be stopped along the route, and when he reaches the top or bottom of the cliff.

10. **<u>TANDEM LOWERING.</u>** The tandem lowering system can be used for the walking wounded, POW's, or more serious casualties when situation would not permit using the barrow boy.

a. The assistant to the casualty should first tie a rappel seat on himself and then assist the casualty with his.

b. The assistant will take only one belay line and tie a end of the line figure 8 loop, and clip this into his rappel seat.

c. A directional figure 8 will be tied approximately 12 inches up the rope from the figure 8 loop (with the loop pointed down) and this will be clipped to the casualty's rappel seat.

d. If needed, safety prussic cords should be tied above the casualty's directional figure 8. These safety cords should be attached in the same manner as the first man down a rappel.

e. The casualty and the assistant will lower as one, with the assistant helping on the way down the cliff.

11. **OTHER CONSIDERATIONS.** All of the techniques we have discussed for the evacuation of a casualty from top to bottom can also be used on a suspension traverse or rope bridge, with a slight variation in the belay line. Two belay lines may be used for rope bridges and the suspension traverse, if they are available. No matter what type of litter is used, the individuals involved in the evacuation must ensure that the head is always uphill or not lower than the feet.

UNITED STATES MARINE CORPS
Mountain Warfare Training Center
Bridgeport, California 93517-5001

FMST.0
7.34
04/12/02

STUDENT HANDOUT

ONE-ROPE BRIDGE

LEARNING OBJECTIVES

TERMINAL LEARNING OBJECTIVES. In a summer mountainous environment, conduct bridging, in accordance with the references. (FMST.07.34)

ENABLING LEARNING OBJECTIVES

(1) Without the aid of references, list in writing the criteria for site selection for a one-rope bridge, in accordance with the reference. (FMST.07.34a)

(2) In a summer mountainous environment with an assault load , cross a one-rope bridge using the rappel seat method, in accordance with the reference. (FMST.07.34b)

(3) Without the aid of references, list in writing the rescue techniques, in accordance with the reference. (FMST.07.34c)

(4) In a summer mountainous environment transport a SKED stretcher across a one rope bridge, in accordance with the reference. (FMST.07.34d)

BODY

1. **SITE SELECTION**

a. The two criteria for site selection for a one-rope bridge are: (FMST.07.34a)

1) There must be suitable anchors on both sides of the stream.

2) The anchors must offer good loading and unloading platforms.

b. Other considerations involved are:

1) The site chosen for the initial crossing does not have to be at the location for the construction of the bridge, just as long as the rope can be taken to the selected site the

C9

crossing.

2) The site chosen for the lead swimmer to cross should be as free as possible from obstacles in the water, such as large boulders, stumps or logs as discussed in *STREAM CROSSING*.

3) The anchors must be close enough for the 150 foot coil to reach both near side and far side anchors. Keep in mind that it will take approximately 1/3 of the 150 foot rope for tightening and anchoring of the bridge.

2. **DISTANCE ESTIMATION**. The follow methods can be used to determine the distance between anchor points:

a. Azimuth Method. Shoot an azimuth to a point on the far side of the intended obstacle to cross. Then move LEFT or RIGHT (perpendicular to the azimuth) until you get a 15 degree offset of your previous azimuth. Next, measure the distance that you paced in feet from your first azimuth to your second azimuth, and multiply that distance by three. This total will give you the approximate distance across the obstacle in feet.

b. Unit Average Method. Get three Marines to judge (best guess) the distance across the intended obstacle. Add up the total accumulated distance, then divide by three. This will give an estimation of the distance across the obstacle.

Example:	Marine 1	240 ft
	Marine 2	230 ft
	Marine 3	250 ft
		720 ft

Divided by 3 = 240 ft

3. **ORGANIZATION**. The organization of construction is broken down into four groups. They are as follow:

a. The Bridging Team - consisting of the bridge NCO and another Marine.

b. The Safety Line Team - consisting of the lead swimmer and his belay man.

c. The Mule Team - consisting 3 Marines for a 9:1, and 6 Marines for a 3:1.

d. The Security Team – consisting of the rest of the party.

4. **CONSTRUCTION**. This bridge is ideal for squad and platoon sized units, for its quick construction and minimal amount of equipment required.

a. Once the site for the bridge has been designated by the bridge NCO, the Safety Line Team will move up stream from this site to enter the water. Their first step is to flake out the rope. Once this is accomplished, the lead swimmer's will take a bight of the rope and tie a figure 8 loop at the end that will be going to the far side, ensuring the knot has a 18"-24" loop. The lead swimmer's upstream arm will go in it, with the belay man tending his rope the lead swimmer will cross the never using the flow to assist him. (Ferry angle as taught in STREAM CROSSING)

b. At the same time as the lead swimmer is crossing, the bridging team will be preparing the near side anchor. The bridging team will tie a swammi wrap around the anchor, using a sling rope or practice coil, and ensuring the square knot is behind or on the side of the anchor. Once the anchor is secure, flake out the bridging line in order to send it across.

c. Once the lead swimmer is across he will move to the site where the far side anchor will be established and secure his rope. The belay man will then move down to the near side anchor, and as close to the waters edge as possible. He will attach the bridging line to the safety by tying a middle of the rope in a figure 8 knot to the safety line and an end of the rope figure 8 knot to the bridging line connecting the two ropes with an 85 carabiner. The lead swimmer will then pull the rope across. The belay man with the help of another Marine will belay the bridging line across keeping tension on the rope to keep the rope out of the water.

d. After the bridging line is across the lead swimmer will detach the bridging line from the safety line and secure it to the far side anchor using a tree wrap, wrapping from right to left, ensuring there are 3-5 wraps.

e. Once the bridging line is secure. The lead swimmer and the belay man will secure the safety line on both banks, ensuring the rope is creating a ferry angle (minimum 45 degrees). This will be the safety line for the Marines crossing the bridge.

NOTE: If a fall from a rope bridge would result in death or serious injury, then utilize a safety line that will be anchored parallel with the bridge and hand tightened. The Marine crossing will clip into both ropes prior crossing.

f. On the near side bank:

1. The bridging team will need the following gear to build the mechanical advantage:

(a) Four steel locking carabiners & 3 -Three foot prusiks

2. Once the bridging line is secured on the far side, the bridging team will take one 85 and clip it into the near side anchor (swammi wrap). Then they will take the bridging line and clip into that 85, then using one three foot long prusik (16" loop) the bridging line will be secured to the 85 by tying a French Prusik, which will act as a braking knot. The bridging team will now pull the rope taut, and begin construction of the mechanical advantage.

3. The first step is to take one three foot prusik (16" loop) and tie a French prusik on the bridging line as far away from the anchor as possible. Then take one 85 and clip it to the loops of the French prusik.

4. Taking a bight from the running end of the rope clip it into the 85 hanging from the tails of the French prusik.

5. Then bring a bight back from the clipped in rope to the anchor, and taking one 85 clip it into the 85 created in step (2), take the bight of rope and clip into this 85.

6. Take one three foot prusik and tie a French prusik onto the bottom end of the bight clipped into the 85 created in step (3), take one 85 and clip it into the tail loops of the French prusik.

7. From the anchor, take a bight of rope from the 85 created in step (6) and clip it into the 85 created in step (6).

8. You have made approximately a 9:1 mechanical advantage.

NOTE: If the manpower is available the squad/platoon has the option of building a 3:1 mechanical advantage vise a 9: 1, the mule team will just have to add another body having no more then 3 bodies on a 9:1 and 6 bodies on a 3:1

g. Tensioning of the Bridge

1. Now that the bridge is built, the bridge NCO can call up the mule team to tension the bridge.

2. This procedure is facilitated by the mechanical advantage system that was just put into place. The braking knot (French prusik) is used to hold tension on the bridging line while the bridge is being tensioned.

3. The mule team will begin to pull on the running end of the rope coming out of the mechanical advantage. They will pull the rope straight back, trying to keep it in line with the bridge as best as possible.

4. The mule team will tension the rope as much as possible, the bridge NCO will monitor the system. Once the mule team cannot pull anymore tension, the bridge NCO will then have them hold the rope in place, and will reset the braking knot. Once the braking knot is set the bridge NCO will ask for slack from the mule team and cycle the system out. This process will continue until the bridge is tight.

5. On the last cycle the bridge NCO will set the brake knot and with the help of another Marine they will make a bight out of the running end and bring it around the tree while keeping tension.

6. Take a bight and make a complete round turn on the body of the 85 created in step (2). Last, the bridge NCO will tie two half hitches encompassing all the ropes just behind the anchor.

7. At this stage the bridge is tight and secured, the bridge NCO will now call up the remainder of the squad/platoon to cross. The bridge NCO will monitor the crossing of all Marines, and will be the last one to cross.

TRANSITION: Now that we have discussed the construction of the bridge, are there any questions over the material? Next we will discuss the crossing of the bridge.

5. **<u>CROSSING</u>.** (FMST.07.34b) The method used to cross is known as a horizontal traverse. This traverse can be accomplished in the following ways.

a. <u>Rappel Seat Method</u>

1.) The Marine ties himself into a rappel seat and inserts a large steel locking carabiner with the gate facing down and away.

2.) The Marine faces the bridge with his right hand towards the near anchor to snap into the bridge, once the 85 is locked onto the bridge rope, a helper flips the 85 over so that the locking nut screws down. A carrier rope with a girth hitch may be added to assist the Marine in mounting the rope.

3.) The Marine hangs below the bridge from his rappel seat with his head pointing in the direction of the far anchor and allows his legs to hang free.

4) Pulling with his arms makes progress.

5) This method is the safest and therefore the preferred method.

6) If the Marine must take a pack across, he may wear it with the waistband secured: However, the preferred method is to have another carabiner (aluminum non-locking is sufficient) attached to the pack frame at the top and attach this to the bridge behind the Marine putting his legs through the shoulder straps and pulling the pack across with him.

7) One man at a time will cross, although one can load and another can unload concurrently.

8) Weapons will be worn across the shoulder; muzzle down with a tight sling securely attached to the weapon.

b. Pulley Method. This method is used when the one rope bridge is long, uphill, or speed is vital and the Marines crossing it have a lot of heavy personal equipment; M24OG's, radios, etc.

1) Equipment required. One pulley, four steel locking carabiners, and a hauling line twice the length of the obstacle.

2) Setting up the system. Construction of the first suspension point is done by attaching a pulley to the one rope bridge, one 85 carabiner is attached to the pulley, gate down, and a second 85 is attached to the first 85 with the gate facing the near side bank. Into this second 85, attach the hauling rope with a figure-8 loop. The figure-8 is placed halfway along the hauling rope and attached to the 85, which is then locked. Four feet down the line another figure-8 on a bight is placed into the second suspension point.

3) Connection. The Marine clips his 85 into the lower 85, his equipment is clipped into the second 85. The mule team starts to pull.

c. Helmets will be worn for all methods of crossing the bridge. Gloves are optional.

NOTE: There are additional methods of crossing a one rope bridge such as the Commando crawl, Monkey crawl and hand over hand techniques, however, they are not taught at MWTC due to safety reasons.

6. **RESCUE TECHNIQUES** If an individual is unable to complete the crossing on the one rope bridge using the rappel seat method of crossing (i.e. from a fall or exhaustion), a rescue will have to be made in the following manner: (FMST.07.34c)

a. Reach. First try to reach the victim by using an object such as a pole, your hand, if the victim is close enough, etc ...

b. Throw. If reaching the victim is impossible, try throwing a rope to the victim and have him attach the rope to himself, preferably his seat, and pull him back to the desired side of the installation.

c. Tow. If the victim is unable to catch or reach a rope being thrown to him or the victim is unconscious, tie a figure-of-eight loop into the middle of the safety line and then connect the rope to the rope bridge with a steel locking carabiner. The mule team will pull the carabiner up against the victim's seat and begin towing the victim to the desired side of the installation. The rope used to tow the victim should be twice the length of the span of the rope bridge (If necessary, ropes can be tied together to accommodate the span). This will allow towing from either side without having to throw or have a Marine carry the safety line back across the bridge each time a towing rescue is performed.

d. Go. If all else fails, the last option will be to go after the victim. The rescuer will move out onto the rope bridge with a safety line attached to himself at approximately eight feet from the end of the safety line. In the end of the safety line tie a figure-of-eight loop and insert a steel locking carabiner in the end of it. Once the rescuer has made contact with the victim. He will attach the steel locking carabiner to either the victim's seat (preferably the victim's carabiner that is attached to the rope bridge), or to the bridge itself. To ensuring that the carabiner is placed so that it pulls against the victim's carabiner. The mule team now starts to pull both the victim and the rescuer to the desired side of the installation.

e. Cut. If while crossing the rope over water the individual goes underwater and no other rescue technique can be employed, the rope will be cut.

7. RETRIEVING THE ONE ROPE BRIDGE

a. Before the bridge NCO sends the last Marine to cross, they must make the bridge retrievable.

b. The first step is to break down the mechanical advantage, ensuring the braking knot is set before doing so. The system will be broken down until the bridging line is attached only to the brake knot.

c. They will now take the rope around to the back of the anchor, and tie a slip figure 8, with the loop of the slip 8, take it and attach it to the bridging line just in back of the braking knot with a 85.

d. Now pull the running end coming out of the slip 8 until it tightens, and bites on the rope. Once the rope bites, tie a thumb knot behind the slip 8 as close to the slip figure 8 as possible.

e. The bridge NCO, with the help of another Marine, will now untie the swammi wrap, until all of the tension is on the bridging line.

f. The safety line will now be brought up, and attached to the end of the bridging line with an overhand knot.

g. The bridge NCO and the other. Marine will now cross the bridge.

NOTE: It is possible that the bridge may lose tension after breaking the bridge down. The packs of the last two Marines across can be sent over with other Marines, prior to the slip figure being tied.

h. After both Marines have crossed, the safely line will be brought up from down stream.

i. The tree wrap will now be taken off. Once the tree wrap is off, all knots and carabiners will be taken out.

j. The safety line will now be pulled, the bridge is retrieved.

UNITED STATES MARINE CORPS
Mountain Warfare Training Center
Bridgeport, California 93517-5001

WMC
04/12/02

STUDENT OUTLINE

SUMMER MOUNTAIN WARFIGHTING LOAD REQUIREMENTS

LESSON PURPOSE. The purpose of this period of instruction is to introduce the student to the different levels of gear requirements in accordance with MWTC's uniform codes. This lesson relates to mountain walking and tactical evolutions

OUTLINE

1. **BASIC UNIFORM REQUIREMENTS.**

a. Marines operating in mountainous terrain will generally wear the standard issued utility uniform. However, due to the wide range of temperatures, and sudden changes in weather in the mountains, Marines will find the need to be continually adding and removing layers. The amount of clothing worn may vary depending upon the severity of the weather, the activity level of the Marine, and the individual metabolism of the Marine. Every person in that unit should still maintain the same outer camouflage layer.

b. As part of the basic uniform, each man will be required to have in his possession, at all times, seven required pocket items. These seven items should be carried in the pockets of your utility uniform:

(1) Pocketknife

(2) Whistle

(3) Pressure Bandage

(4) Chapstick and sunscreen

(5) Sunglasses

(6) Survival Kit and rations (fire starting material, signaling material, food gathering material, water procuring material, 1st aid, shelter material, and some high energy, lightweight snacks.)

(7) Notebook with pen/pencil

c. Some additional items that should be carried in your pockets at all times:

(1) Contact gloves

(2) Paracord (10 meters)

(3) Flashlight with tactical lens and spare batteries

(4) Chemlights

2. **ASSAULT LOAD.**

a. The Assault Load is equipment in addition to the basic uniform requirements. It is carried in the load-bearing vest (LBV), butt pack and the pack system. This is the equipment carried for short duration missions such as security patrols or during the assault. It is also carried at all times when away from the bivouac site.

(1) An extra insulating layer (Field jacket liner, wooly pully, etc.)

(2) Protective layer (ECWCS parka and trousers)

(3) LBV with 2 quarts of water and first aid kit

(4) Helmet

(5) Rations for the time away from your bivouac site

(6) Cold weather hat or balaclava

(7) Individual mountaineering gear

a. Sling rope

b. Carabiners

c. Rappelling Gloves

(8) Specialized mountaineering equipment

(9) Mission essential gear

3. **COMBAT LOAD**. The Combat Load is the equipment carried for longer duration missions.

The following items are in addition to the items already being carried in the Assault Load:

a. Sleeping bag with bivy bag

b. Isopor mat

c. Stove

d. Fuel bottle

e. Poncho (for expedient shelters or medevac purposes)

4. **EXISTENCE LOAD**. The Existence Load is to be packed for longer duration combat missions, where it is necessary to replace worn out items. These items could be uniforms, boots, personal hygiene gear, etc. The Existence Load should be brought to the forward elements once the situation allows.

5. **PACKING CONSIDERATIONS.** Because most Marines are familiar with how to pack a pack, these are general guidelines only.

a. When not wearing your protective layer or insulating layers, keep it handy at the top of the pack. When taking a break during a movement, you will be able to quickly don a layer to prevent getting chilled.

b. Keep your stove and fuel bottle in the outside pockets of your pack. They may leak and soak your equipment with fuel if stored inside your pack.

c. Keep the climbing rope in the rope bag. This will reduce the rope's exposure to petroleum products, ultra-violet light, cuts, and abrasions.

d. Ice ax is placed through the ice ax loops with spike up, adze outboard, and secured in place.

e. Crampons are attached to the outside of the pack with the tips covered and facing outboard.

Desert Operations

Department of Defense

Pentagon
Publishing

* FM 90-3
FMFM 7-27

Headquarters
Department of the Army
The Marine Corps
Washington, DC
24 August 1993

FM 90-3
DESERT OPERATIONS

Table of Contents

DISTRIBUTION RESTRICTION: Approved for public release; distribution is unlimited.

* This publication supersedes FM 90-3, 19 August 1977.
U.S. Marine Corps: PCN 139 000670 00

INTRODUCTION

Arid regions make up about one-third of the earth's land surface, a higher percentage then that of any other type of climate. As we have seen in the recent past, some of these regions—because of diverse and conflicting culturs, strategic importance, and natural resources—have become centers of conflict.

Military leaders have long recognized the potential for US involvement in conflict In these regions. Exercises at the Array's National Training Center, Fort Irwin and the marine Corps' Marine Air Ground Combat Center, Twentynine Palms. California, have provided an opportunity for virtually all our ground forces to experience desert conditions, The success of Operation Desert Storm can be directly attributed to this realistic training.

Desert operations demand adaptation to the environment and to the limitations imposed by terrain and climate, Success depends on an appreciation of the effects of arid conditions on soldiers (both physically and psychologically), on equipment and facilities, and on eombot and support operations. Leaders and soldiers must continually evaluate the situation and be ready to react to changing conditions. Equipment and tactics must be modified and adapted to a dusty, rugged landscape where temperatures vary from extreme highs to freezing lows and where visibility can clange from 30 miles to 30 feet in a matter of minutes.

The key to success in desert operations is mobility. This was clearly evident in the ground operations of Desert Storm, The tactics employed to achieve victory over Iraq were wide, rapid flanking movements similar to those executed by Montgomery and Rommel during World War II, During Desert Storm, however, new technologies increased higher-echelon headquarter' ability to target, attack, and fight deep operations simultaneously. Modem weapon systems like the

Ml Al Abrams tank, Bradley fighting vehicle, light armored vehicle, and assault amphibious vehicle, coupled with newly developed navigation and targeting devices, contributed im measurably. Taclictl units were able to fight battles with minimal direction; leaders were able to exercise initiative based on a clear understanding of their commanders' intent. Current doctrine—Focused on improving mobility and implemented through, the planning, preparation, and execution processes, battle drills, and tactical SOPs, paved the way for the overwhelming triumph.

Arid regions create ***bold*** opportunities and resurants **for** soldiers and marines at all levels. The US military's performance in Desert Storm shows it understands these factors and has successfully addressed the effects of desert warfare TO troops, equipment, and operations. As they prepare for the future, leaders, soldiers, and marines must study past carnpaigns and use the lessons they learn **to** reduce casualties, use the environment to their advantage, and ensure victory on the desert battlefield.

CHAPTER 1

THE ENVIRONMENT AND ITS EFFECTS ON PERSONNEL AND EQUIPMENT

This chapter describes the desert environment and how it affects personnel and equipment.

Section I. The Environment

Successful desert operations require adaptation to the environment and to the limitations its terrain and climate impose. Equipment and tactics must be modified and adapted to a dusty and rugged landscape where temperatures vary from extreme highs down to freezing and where visibility may change from 30 miles to 30 feet in a matter of minutes. Deserts are arid, barren regions of the earth incapable of supporting normal life due to lack of water. See Figure 1-1 for arid regions of the world. Temperatures vary according to latitude and season, from over 136 degrees Fahrenheit in the deserts of Mexico and Libya to the bitter cold of winter in the Gobi (East Asia). In some deserts, day-to-night temperature fluctuation exceeds 70 degrees Fahrenheit. Some species of animal and plant life have adapted successfully to desert conditions where annual rainfall may vary from 0 to 10 inches.

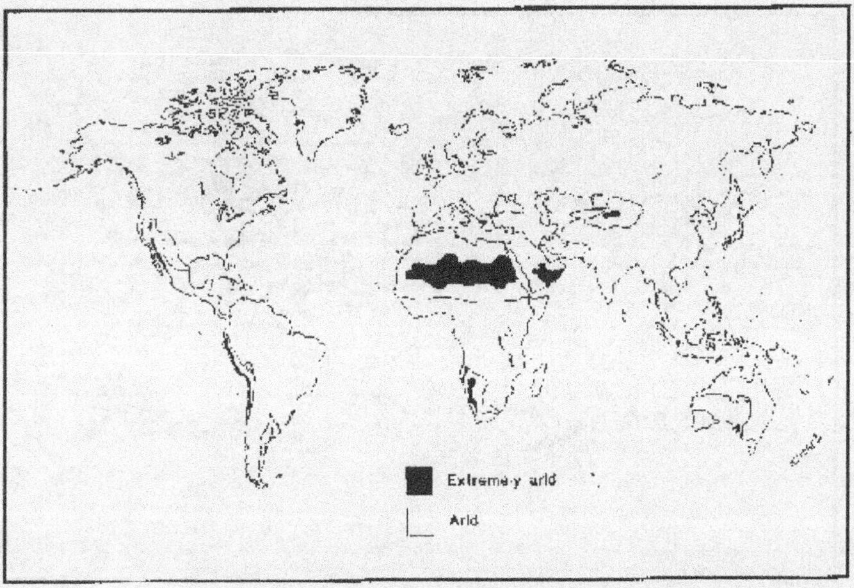

Figure 1-1. Desert of the world.

Desert terrain also varies considerably from place to place, the sole common denominator being lack of water with its consequent environmental effects, such as sparse, if any, vegetation. The basic land forms are similar to those in other parts of the world, but the topsoil has been eroded due to a combination of lack of water, heat, and wind to give deserts their characteristic barren appearance. The bedrock may be covered by a flat layer of sand, or gravel, or may have been exposed by erosion. Other common features are sand dunes, escarpments, wadis, and depressions. This environment can profoundly affect military operations. See Figure 1-2 for locations of major deserts of the world, and Appendix A for additional information on desert countries of the world.

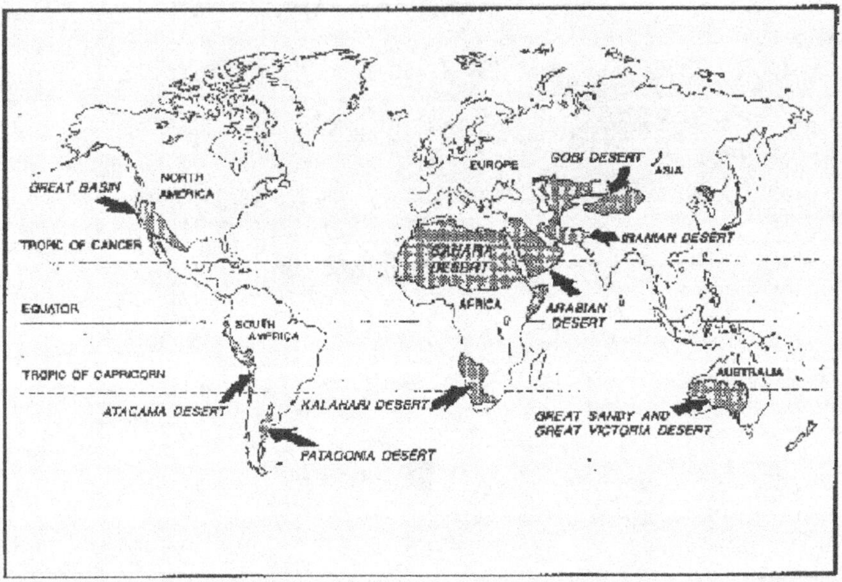

It is important to realize that deserts are affected by seasons. Those in the Southern Hemisphere have summer between 21 December and 21 March. This 6-month difference from the United States is important when considering equipping and training nonacclimatized soldiers/marines for desert operations south of the equator.

TERRAIN

Key terrain in the desert is largely dependent on the restrictions to movement that are present. If the desert floor will not support wheeled vehicle traffic, the few roads and desert tracks become key terrain. Crossroads are vital as they control military operations in a large area. Desert warfare is often a battle for control of the lines of communication (LOC). The side that can protect its own LOC while interdicting those of the enemy will prevail. Water sources are vital, especially if a force is incapable of long distance resupply of its water requirements. Defiles play an important role, where they exist. In the Western Desert of Libya, an escarpment that paralleled the coast was a barrier to movement except through a few passes. Control of these passes was vital. Similar escarpments are found in Saudi Arabia and Kuwait.

Types of Desert Terrain

There are three types of desert terrain: mountain, rocky plateau, and sandy or dune terrain. The following paragraphs discuss these types of terrain.

Mountain Deserts

Mountain deserts are characterized by scattered ranges or areas of barren hills or mountains, separated by dry, flat basins. See Figure 1-3 for an example of mountain desert terrain. High ground may rise gradually or abruptly from flat areas, to a height of several thousand feet above sea level. Most of the infrequent rainfall occurs on high ground and runs off in the form of flash floods, eroding deep gullies and ravines and depositing sand and gravel around the edges of the basins. Water evaporates rapidly, leaving the land as barren as before, although there may be short-lived vegetation. If sufficient water enters the basin to compensate for the rate of evaporation, shallow lakes may develop, such as the Great Salt Lake in Utah or the Dead Sea; most of these have a high salt content.

Figure 1-3, Example of mountain desert terrian,

Rocky Plateau Deserts

Rocky plateau deserts are extensive flat areas with quantities of solid or broken rock at or near the surface. See Figure 1-4 for an example of a rocky plateau desert. They may be wet or dry, steep-walled eroded valleys, known as wadis, gulches, or canyons. Narrow valleys can be extremely dangerous to men and materiel due to flash flooding after rains; although their flat bottoms may be superficially attractive as assembly areas. The National Training Center and the Golan Heights are examples of rocky plateau deserts.

figure1-4, Example of rocky plateau desert terrain.

Sandy or Dune Deserts

Sandy or dune deserts are extensive flat areas covered with sand or gravel, the product of ancient deposits or modern wind erosion. "Flat" is relative in this case, as some areas may contain sand dunes that are over 1,000 feet high and 10-15 miles long; trafficability on this type of terrain will depend on windward/leeward gradients of the dunes and the texture of the sand. See Figure 1-5 for an example of a sandy desert. Other areas, however, may be totally flat for distances of 3,000 meters and beyond. Plant life may vary from none to scrub, reaching over 6 feet high. Examples of this type of desert include the ergs of the Sahara, the Empty Quarter of the Arabian desert, areas of California and New Mexico, and the Kalahari in South Africa. See Figure 1-6 for an example of a dune desert.

figure 1 -3. Example of sandy desert terrain,

Figure 1-6. Example of dune desert terrain.

Trafficability

Roads and trails are rare in the open desert. Complex road systems beyond simple commercial links are not needed. Road systems have been used for centuries to connect centers of commerce, or important religious shrines such as Mecca and Medina in Saudi Arabia. These road systems are supplemented by routes joining oil or other mineral deposits to collection outlet points. Some surfaces, such as lava beds or salt marshes, preclude any form of routine vehicular movement, but generally ground movement is possible in all directions. Speed of movement varies depending on surface texture. Rudimentary trails are used by minor caravans and nomadic tribesmen, with wells or oases approximately every 20 to 40 miles; although there are some waterless stretches which extend over 100 miles. Trails vary in width from a few meters to over 800 meters.

Vehicle travel in mountainous desert country may be severely restricted. Available mutes can be easily blinked by the enemy or by climatic conditions. Hairpin turns are common on the edges of precipitous mountain gorges, and the higher passes may be blocked by snow in the winter.

Natural Factors

The following terrain features require special considerations regarding trafficability.

Wadis or dried water courses, vary from wide, but barely perceptible depressions of soft sand, dotted with bushes, to deep, steep-sided ravines. There frequently is a passable route through the bottom of a dried wadi. Wadis can provide cover from ground observation and camouflage from visual air reconnaissance. The threat of flash floods after heavy rains poses a significant danger to troops and equipment downstream. Flooding may occur in these areas even if it is not raining in the immediate area. See Figure 1-7 for an example of a wadi.

Figure 1-7. Example of a wadi

Salt marsh (sebkha) terrain is impassable to tracks and wheels when wet. When dry it has a brittle, crusty surface, negotiable by light wheel vehicles only. Salt marshes develop at points where the water in the subsoil of the desert rose to the surface. Because of the constant evaporation in the desert, the salts carried by the water are deposited, and results in a hard, brittle crust.

Salt marshes are normally impassable, the worst type being those with a dry crust of silt on top. Marsh mud used on desert sand will, however, produce an excellent temporary road. Many desert areas have salt marshes either in the center of a drainage basin or near the sea coast. Old trails or paths may cross the marsh, which are visible during the dry season but not in the wet season. In the wet season trails are indicated by standing water due to the crust being too hard or too thick for it to penetrate. However, such routes should not be tried by load-carrying vehicles without prior reconnaissance and marking. Vehicles may become mired so severely as to render equipment and units combat ineffective. Heavier track-laying vehicles, like tanks, are especially susceptible to these areas, therefore reconnaissance is critical.

Man-made Factors

The ruins of earlier civilizations, scattered across the deserts of the world, often are sited along important avenues of approach and frequently dominate the only available passes in difficult terrain. Control of these positions maybe imperative for any force intending to dominate the immediate area. Currently occupied dwellings have little impact on trafficability except that they are normally located near roads and trails. Apart from nomadic tribesmen who live in tents (see Figure 1-8 for an example of desert nomads), the population lives in thick-walled structures with small windows, usually built of masonry or a mud and straw (adobe) mixture. Figure 1-9 shows common man-made desert structures.

Figure 1-8. Example of desert nomads

Figure 1-9. Common man_made desert.

Because of exploration for and production of oil and other resources, wells, pipelines, refineries, quarries, and crushing plants may be of strategic importance in the desert. Pipelines are often raised 1 meter off the ground-where this is the case, pipelines will inhibit movement. Subsurface pipelines can also be an obstacle. In Southwest Asia, the subsurface pipelines were indicated on maps. Often they were buried at such a shallow depth that they could be damaged by heavy vehicles traversing them. Furthermore, if a pipeline is ruptured, not only is the spill of oil a consideration, but the fumes maybe hazardous as well.

Agriculture in desert areas has little effect on trafficability except that canals limit surface mobility. Destruction of an irrigation system, which may be a result of military operations, could have a devastating effect on the local population and should be an important consideration in operational estimates. Figure 1-10 shows an irrigation ditch.

Figure 1-10. Irrigation ditch

TEMPERATURE

The highest known ambient temperature recorded in a desert was 136 degrees Fahrenheit (58 degrees Celsius). Lower temperatures than this produced internal tank temperatures approaching 160 degrees Fahrenheit (71 degrees Celsius) in the Sahara Desert during the Second World War. Winter temperatures in Siberian deserts and in the Gobi reach minus 50 degrees Fahrenheit (minus 45 degrees Celsius). Low temperatures are aggravated by very strong winds producing high windchill factors. The cloudless sky of the desert permits the earth to heat during sunlit hours, yet cool to near freezing at night. In the inland Sinai, for example, day-to-night temperature fluctuations are as much as 72 degrees Fahrenheit.

WINDS

Desert winds can achieve velocities of near hurricane force; dust and sand suspended within them make life intolerable, maintenance very difficult, and restrict visibility to a few meters. The Sahara "Khamseen", for example, lasts for days at a time; although it normally only occurs in the spring and summer. The deserts of Iran are equally well known for the "wind of 120 days," with sand blowing almost constantly from the north at wind velocities of up to 75 miles per hour.

Although there is no danger of a man being buried alive by a sandstorm, individuals can become separated from their units. In all deserts, rapid temperature changes invariably follow strong winds. Even without wind, the telltale clouds raised by wheels, tracks, and marching troops give away movement. Wind aggravates the problem. As the day gets warmer the wind increases and the dust signatures of vehicles may drift downwind for several hundred meters.

In the evening the wind normally settles down. In many deserts a prevailing wind blows steadily from one cardinal direction for most of the year, and eventually switches to another direction for the remaining months. The equinoctial gales raise huge sandstorms that rise to several thousand feet and may last for several days. Gales and sandstorms in the winter months can be bitterly cold. See Figure 1-11 for an example of wind erosion.

Figure 1-11. Example of wind crosion

Sandstorms are likely to form suddenly and stop just as suddenly. In a severe sandstorm, sand permeates everything making movement nearly impossible, not only because of limited visibility, but also because blowing sand damages moving parts of machinery.

WATER

The lack of water is the most important single characteristic of the desert. The population, if any, varies directly with local water supply. A Sahara oasis may, for its size, be one of the most densely occupied places on earth (see Figure 1-12 for a typical oasis).

Figure 1-12. Typical oasis

Desert rainfall varies from one day in the year to intermittent showers throughout the winter. Severe thunderstorms bring heavy rain, and usually far too much rain falls far too quickly to organize collection on a systematic basis. The water soon soaks into the ground and may result in flash floods. In some cases the rain binds the sand much like a beach after the tide ebbs allowing easy maneuver however, it also turns loam into an impassable quagmire obstacle. Rainstorms tend to be localized, affecting only a few square kilometers at a time. Whenever possible, as storms approach, vehicles should move to rocky areas or high ground to avoid flash floods and becoming mired.

Permanent rivers such as the Nile, the Colorado, or the Kuiseb in the Namib Desert of Southwest Africa are fed by heavy precipitation outside the desert so the river survives despite a high evaporation rate.

Subsurface water may be so far below the surface, or so limited, that wells are normally inadequate to support any great number of people. Because potable water is absolutely vital, a large natural supply may be both tactically and strategically important. Destruction of a water supply system may become a political rather than military decision, because of its lasting effects on the resident civilian population.

Finding Water

When there is no surface water, tap into the earth's water table for ground water. Access to this table and its supply of generally pure water depends on the contour of the land and the type of soil. See Figure 1-13 for water tables.

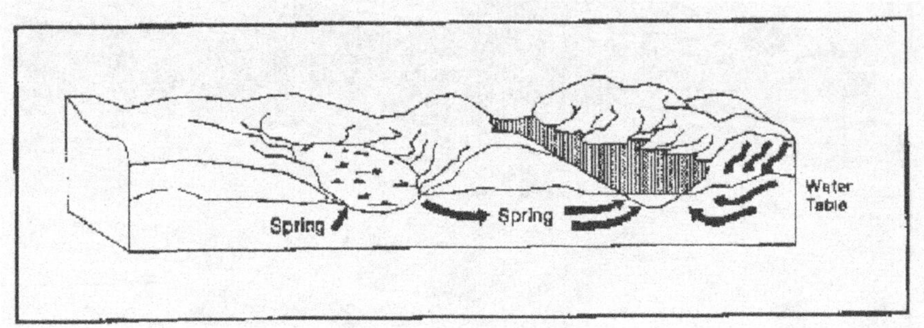

Figure 1-13. Water tables.

From Rocky Soil

Look for springs and seepages. Limestone has more and larger springs than any other type rock. Because limestone is easily dissolved, caverns are readily etched in it by ground water. Look in these caverns for springs. Lava rock is a good source of seeping ground water because it is porous. Look for springs along the walls of valleys that cross the lava flow. Look for seepage where a dry canyon cuts through a layer of porous sandstone.

Watch for water indicators in desert environments. Some signs to look for are the direction in which certain birds fly, the location of plants, and the convergence of game trails. Asian sand grouse, crested larks, and zebra birds visit water holes at least once a day. Parrots and pigeons must live within reach of water. Cattails, greasewoods, willows, elderberry, rushes, and salt grass grow only where ground water is near the surface. Look for these signs and dig. If you do not have a bayonet or entrenching tool, dig with a flat rock or sharp stick.

Desert natives often know of lingering surface pools in low places. They cover their surface pools, so look under brush heaps or in sheltered nooks, especially in semiarid and brush country.

Places that are visibly damp, where animals have scratched, or where flies hover, indicate recent surface water. Dig in such places for water. Collect dew on clear nights by sponging it up with a handkerchief. During a heavy dew you should be able to collect about a pint an hour.

Dig in dry stream beds because water may be found under the gravel. When in snow fields, put in a water container and place it in the sun out of the wind.

From Plants

If unsuccessful in your search for ground or runoff water, or if you do not have time to purify the questionable water, a water-yielding plant may be the best source. Clear sap from many plants is easily obtained. This sap is pure and is mostly water.

Plant tissues. Many plants with fleshy leaves or stems store drinkable water. Try them wherever you find them. The barrel cactus of the southwestern United States is a possible source of water (see Figure 1-14). Use it only as a last resort and only if you have the energy to cut through the tough, spine-studded outer rind. Cut off the top of the cactus and smash the pulp within the plant. Catch the liquid in a container. Chunks may be carried as an emergency water source. A barrel cactus 3-1/2 feet high will yield about a quart of milky juice and is an exception to the rule that milky or colored sap-bearing plants should not be eaten.

figure 1-14. Barrel cactus of possible source of water.

Roots of desert plants. Desert plants often have their roots near the surface. The Australian water tree, desert oak, and bloodwood are some examples. Pry these roots out of the ground, cut them into 24-36 inch lengths, remove the bark, and suck the water.

Vines. Not all vines yield palatable water, but try any vine found. Use the following method for tapping a vine--it will work on any species:

Step 1. Cut a deep notch in the vine as high up as you can reach.

Step 2. Cut the vine off close to the ground and let the water drip into your mouth or into a container.

Step 3. When the water ceases to drip, cut another section off the top. Repeat this until the supply of fluid is exhausted.

Palms. Burl, coconut, sugar and nipa palms contain a drinkable sugary fluid. To start the flow in coconut palm, bend the flower stalk downward and cut off the top. If a thin slice is cut off the stalk every 12 hours, you can renew the flow and collect up to a quart a day.

Coconut. Select green coconuts. They can be opened easily with a knife and they have more milk than ripe coconuts. The juice of a ripe coconut is extremely laxative; therefore, do not drink more than three or four cups a day.

The milk of a coconut can be obtained by piercing two eyes of the coconut with a sharp object such as a stick or a nail. To break off the outer fibrous covering of the coconut without a knife, slam the coconut forcefully on the point of a rock or protruding stump.

Survival Water Still

You can build a cheap and simple survival still that will produce drinking water in a dry desert. Basic materials for setting up this still are-

- 6-foot square sheet of clean plastic.
- A 2- to 4-quart capacity container.
- A 5-foot piece of flexible plastic tubing.

Pick an unshaded spot for the still, and dig a hole. If no shovel is available, use a stick or even your hands. The hole should be about 3 feet across for a few inches down, then slope the hole toward the bottom as shown in Figure 1-15 which depicts a cross section of a survival still. The hole should be deep enough so the point of the plastic cone will be about 18 inches below ground and will still clear the top of the container. Once the hole is properly dug, tape one end of the plastic drinking tube inside the container and center the container in the bottom of the hole. Leave the top end of the drinking tube free, lay the plastic sheet over the hole, and pile enough dirt around the edge of the plastic to hold it securely. Use a fist-size rock to weight down the center of the plastic; adjust the plastic as necessary to bring it within a couple of inches of the top of the container. Heat from the sun vaporizes the ground water. This vapor condenses under the plastic, trickles down, and drops into the container.

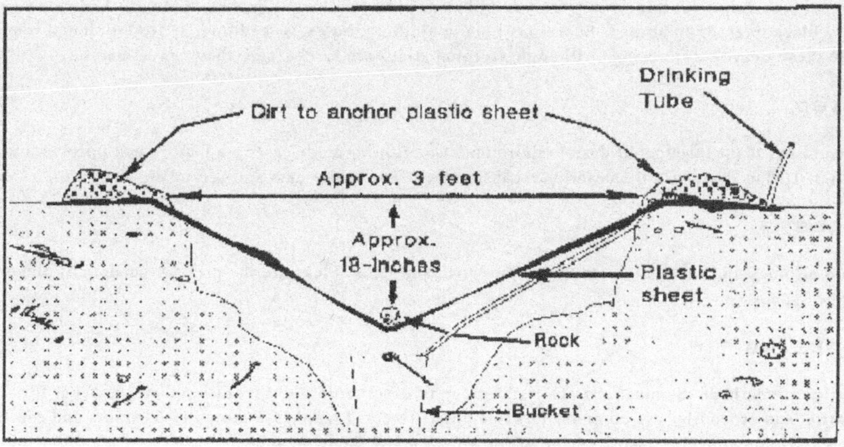

Figure 1-15. Cross-section of a survival still.

VEGETATION

The indigenous vegetation and wildlife of a desert have physiologically adapted to the conditions of the desert environment. For example, the cacti of the American desert store moisture in enlarged stems. Some plants have drought-resistant seeds that may lie dormant for years, followed by a brief, but colorful display of growth after a rainstorm. The available vegetation is usually inadequate to provide much shade, shelter, or concealment, especially from the air. Some plants, like the desert gourd, have vines which grow to 4.5 meters (15 feet). Others have wide lateral roots just below the surface to take advantage of rain and dew, while still others grow deep roots to tap subsurface water. Presence of palm trees usually indicates water within a meter of the surface, salt grass within 2 meters, Cottonwood and willows up to 4 meters. In addition to indicating the presence of water, some plants are edible.

WILDLIFE

Invertebrates such as ground-dwelling spiders, scorpions, and centipedes, together with insects of almost every type, are in the desert. Drawn to man as a source of moisture or food, lice, mites, and flies can be extremely unpleasant and carry diseases such as scrub typhus and dysentery. The stings of scorpions and the bites of centipedes and spiders are extremely painful, though seldom fatal. Some species of scorpion, as well as black widow and recluse spiders, can cause death. The following paragraphs describe some of the wildlife that are encountered in desert areas and the hazards they may pose to man.

Scorpions

Scorpions are prevalent in desert regions, particularly active at night. Scorpions are They prefer damp locations and are easily recognizable by their crab-like appearance, and by their long tail which ends in a sharp stinger. Adult scorpions vary from less than an inch to almost 8 inches in length. Colors range from nearly black to straw to striped. Scorpions hide in clothing, boots, or bedding, so troops should routinely shake these items before using. Although scorpion stings are rarely fatal, they can be painful.

Flies

Flies are abundant throughout desert environments. Filth-borne disease is a major health problem posed by flies. Dirt or insects in the desert can cause infection in minor cuts and scratches.

Fleas

Avoid all dogs and rats which are the major carriers of fleas. Fleas are the primary carriers of plague and murine typhus.

Reptiles

Reptiles are perhaps the most characteristic group of desert animals. Lizards and snakes occur in quantity, and crocodiles are common in some desert rivers. Lizards are normally harmless and can be ignored; although exceptions occur in North America and Saudi Arabia.

Snakes, ranging from the totally harmless to the lethal, abound in the desert. A bite from a poisonous snake under two feet long can easily become infected. Snakes seek shade (cool areas) under bushes, rocks, trees, and shrubs. These areas should be checked before sitting or resting. Troops should always check clothing and boots before putting them on. Vehicle operators should look for snakes when initially conducting before-operations maintenance. Look for snakes in and around suspension components and engine compartments as snakes may seek the warm areas on recently parked vehicles to avoid the cool night temperatures.

Sand vipers have two long and distinctive fangs that may be covered with a curtain of flesh or folded back into the mouth. Sand vipers usually are aggressive and dangerous in spite of their size. A sand viper usually buries itself in the sand and may strike at a passing man; its presence is alerted by a characteristic coiling pattern left on the sand.

The Egyptian cobra can be identified by its characteristic cobra combative posture. In this posture, the upper portion of the body is raised vertically and the head tilted sharply forward. The neck is usually flattened to form a hood. The Egyptian cobra is often found around rocky places and ruins and is fairly common. The distance the cobra can strike in a forward direction is equal to the distance the head is raised above the ground. Poking around in holes and rock piles is particularly dangerous because of the likelihood of encountering a cobra. See Figure 1-16 for an example of a viper and cobra.

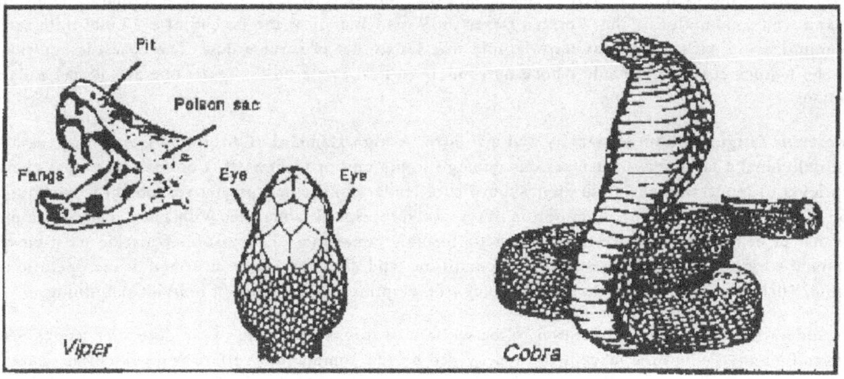

Figure 1-6. Sand viper and cobra

Mammals

The camel is the best known desert mammal. The urine of the camel is very concentrated to reduce water loss, allowing it to lose 30 percent of its body weight without undue distress. A proportionate loss would be fatal to man. The camel regains this weight by drinking up to 27 gallons (120 liters) of water at a time. It cannot, however, live indefinitely without water and will die of dehydration as readily as man in equivalent circumstances. Other mammals, such as gazelles, obtain most of their required water supply from the vegetation they eat and live in areas where there is no open water. Smaller animals, including rodents, conserve their moisture by burrowing underground away from the direct heat of the sun, only emerging for foraging at night. All these living things have adapted to the environment over a period of thousands of years; however, man has not made this adaptation and must carry his food and water with him and must also adapt to this severe environment.

Dogs are often found near mess facilities and tend to be in packs of 8 or 10. Dogs are carriers of rabies and should be avoided. Commanders must decide how to deal with packs of dogs; extermination and avoidance are two options. Dogs also carry fleas which may be transferred upon bodily contact. Rabies is present in most desert mammal populations. Do not take any chances of contracting fleas or rabies from any animal by adopting pets.

Rats are carriers of various parasites and gastrointestinal diseases due to their presence in unsanitary locations.

Section II. Environmental Effects on Personnel

There is no reason to fear the desert environment, and it should not adversely affect the morale of a soldier/marine who is prepared for it. Lack of natural concealment has been known to induce temporary agoraphobia (fear of open spaces) in some troops new to desert conditions, but this fear normally disappears with acclimatization. Remember that there is nothing unique about either living or fighting in deserts; native tribesmen have lived in the Sahara for thousands of years. The British maintained a field army and won a campaign in the Western Desert in World War II at the far end of a 12,000-mile sea line of communication with equipment considerably inferior to that in service now. The desert is neutral, and affects both sides equally; the side whose personnel are best prepared for desert operations has a distinct advantage.

The desert is fatiguing, both physically and mentally. A high standard of discipline is essential, as a single individual's lapse may cause serious damage to his unit or to himself. Commanders must exercise a high level of leadership and train their subordinate leaders to assume greater responsibilities required by the wide dispersion of units common in desert warfare. Soldiers/marines with good leaders are more apt to accept heavy physical exertion and uncomfortable conditions. Every soldier/marine must clearly understand why he is fighting in such harsh conditions and should be kept informed of the operational situation. Ultimately, however, the maintenance of discipline will depend on individual training.

Commanders must pay special attention to the welfare of troops operating in the desert, as troops are unable to find any "comforts" except those provided by the command. Welfare is an essential factor in the maintenance of morale in a harsh environment, especially to the inexperienced. There is more to welfare than the provision of mail and clean clothing. Troops must be kept healthy and physically fit; they must have adequate, palatable, regular food, and be allowed periods of rest and sleep. These things will not always be possible and discomfort is inevitable, but if troops know that their commanders are doing everything they can to make life tolerable, they will more readily accept the extremes brought on by the environment.

HEAT

The extreme heat of the desert can cause heat exhaustion and heatstroke and puts troops at risk of degraded performance. For optimum mental and physical performance, body temperatures must be maintained within narrow limits. Thus, it is important that the body lose the heat it gains during work. The amount of heat accumulation in the human body depends upon the amount of physical activity, level of hydration, and the state of personal heat acclimatization. Unit leaders must monitor their troops carefully for signs of heat distress and adjust schedules, work rates, rest, and water consumption according to conditions.

Normally, several physical and physiological mechanisms (e.g., convection and evaporation) assure transfer of excess body heat to the air. But when air temperature is above skin temperature (around 92 degrees Fahrenheit) the evaporation of sweat is the only operative mechanism. Following the loss of sweat, water must be consumed to replace the body's lost fluids. If the body fluid lost through sweating is not replaced, dehydration will follow. This will hamper heat dissipation and can lead to heat illness. When humidity is high, evaporation of sweat is inhibited and there is a greater risk of dehydration or heat stress. Consider the following to help prevent dehydration:

- Heat, wind, and dry air combine to produce a higher individual water requirement, primarily through loss of body water as sweat. Sweat rates can be high even when the skin looks and feels dry.
- Dehydration nullifies the benefits of heat acclimatization and physical fitness, it increases the susceptibility to heat injury, reduces the capacity to work, and decreases appetite and alertness. A lack of alertness can indicate early stages of dehydration.

- Thirst is not an adequate indicator of dehydration. The soldier/marine will not sense when he is dehydrated and will fail to replace body water losses, even when drinking water is available. The universal experience in the desert is that troops exhibit "voluntary dehydration" that is, they maintain their hydration status at about 2 percent of body weight (1.5 quarts) below their ideal hydration status without any sense of thirst.

Chronic dehydration increases the incidence of several medical problems: constipation (already an issue in any field situation), piles (hemorrhoids), kidney stones, and urinary infections. The likelihood of these problems occurring can be reduced by enforcing mandatory drinking schedules.

Resting on hot sand will increase heat stress-the more a body surface is in contact with the sand, the greater the heat stress. Ground or sand in full sun is hot, usually 30-45 degrees hotter than the air, and may reach 150 degrees Fahrenheit when the air temperature is 120 degrees Fahrenheit. Cooler sand is just inches below the surface; a shaded trench will provide a cool resting spot.
At the first evidence of heat illness, have the troops stop work, get into shade, and rehydrate. Early intervention is important. Soldiers/ marines who are not taken care of can become more serious casualties.

WARNING

One heat casually Is usually followed by Others and Is a warning that the entire unit may be at risk, This is the "Week Link Rule." The slalus Of the whole unit Is assessed at this point.

ACCLIMATIZATION

Acclimatization to heat is necessary to permit the body to reach and maintain efficiency in its cooling process. A period of approximately 2 weeks should be allowed for acclimatization, with progressive increases in heat exposure and physical exertion. Significant acclimatization can be attained in 4-5 days, but full acclimatization takes 7-14 days, with 2-3 hours per day of exercise in the heat. Gradually increase physical activity until full acclimatization is achieved.

Acclimatization does not reduce, and may increase, water requirements. Although this strengthens heat resistance, there is no such thing as total protection against the debilitating effects of heat. Situations may arise where it is not possible for men to become fully acclimatized before being required to do heavy labor. When this happens heavy activity should be limited to cooler hours and troops should be allowed to rest frequently. Check the weather daily. Day-to-day and region-to-region variations in temperatures, wind, and humidity can be substantial.

CLIMATIC STRESS

Climatic stress on the human body in hot deserts can be caused by any combination of air temperature, humidity, air movement, and radiant heat. The body is also adversely affected by such factors as lack of acclimatization, being overweight, dehydration, alcohol consumption, lack of sleep, old age, and poor health.

The body maintains its optimum temperature of 98.6 degrees Fahrenheit by conduction/convection, radiation, and evaporation (sweat). The most important of these in the daytime desert is evaporation, as air temperature alone is probably already above skin temperature. If, however, relative humidity is high, air will not easily evaporate sweat and the cooling effect is reduced. The following paragraphs describe the effects of radiant light, wind, and sand on personnel in desert areas.

Radiant Light

Radiant light comes from all directions. The sun's rays, either direct or reflected off the ground, affect the skin and can also produce eyestrain and temporarily impaired vision. Not only does glare damage the eyes but it is very tiring; therefore, dark glasses or goggles should be worn.

Overexposure to the sun can cause sunburn. Persons with fair skin, freckled skin, ruddy complexions, or red hair are more susceptible to sunburn than others, but all personnel are susceptible to some degree. Personnel with darker complexions can also sunburn. This is difficult to monitor due to skin pigmentation, so leaders must be ever vigilant to watch for possible sunburn victims. Sunburn is characterized by painful reddened skin, and can result in blistering and lead to other forms of heat illness.

Soldier/marines should acquire a suntan in gradual stages (preferably in the early morning or late afternoon) to gain some protection against sunburn. They should not be permitted to expose bare skin to the sun for longer than five minutes on the first day, increasing exposure gradually at the rate of five minutes per day. They should be fully clothed in loose garments in all operational situations. This will also reduce sweat loss. It is important to remember that—

- The sun is as dangerous on cloudy days as it is on sunny days.
- Sunburn ointment is not designed to give complete protection against excessive exposure.
- Sunbathing or dozing in the desert sun can be fatal.

Wind

The wind can be as physically demanding as the heat, burning the face, arms, and any exposed skin with blown sand. Sand gets into eyes, nose, mouth, throat, lungs, ears, and hair, and reaches every part of the body. Even speaking and listening can be difficult. Continual exposure to blown sand is exhausting and demoralizing. Technical work spaces that are protected from dust and sand are likely to be very hot. Work/rest cycles and enforced water consumption will be required.

The combination of wind and dust or sand can cause extreme irritation to mucous membranes, chap the lips and other exposed skin surfaces, and can cause nosebleed. Cracked, chapped lips make eating difficult and cause communication problems. Irritative conjunctivitis, caused when fine particles enter the eyes, is a frequent complaint of vehicle crews, even those wearing goggles. Lip balm and skin and eye ointments must be used by all personnel. Constant wind noise is tiresome and increases soldier/marine fatigue, thus affecting alertness.

When visibility is reduced by sandstorms to the extent that military operations are impossible, soldiers/marines should not be allowed to leave their group for any purpose unless secured by lines for recovery.

The following are special considerations when performing operations in dust or sand:

- Contact lenses are very difficult to maintain in the dry dusty environment of the desert and should not be worn except by military personnel operating in air conditioned environments, under command guidance.
- Mucous membranes can be protected by breathing through a wet face cloth, snuffing small amounts of water into nostrils (native water is not safe for this purpose) or coating the nostrils with

a small amount of petroleum jelly. Lips should be protected by lip balm.

- Moving vehicles create their own sandstorms and troops traveling in open vehicles should be protected.
- Scarves and bandannas can be used to protect the head and face.
- The face should be washed as often as possible. The eyelids should be cleaned daily.

BASIC HEAT INJURY PREVENTION

The temperature of the body is regulated within very narrow limits. Too little salt causes heat cramps; too little salt and insufficient water causes heat exhaustion. Heat exhaustion will cause a general collapse of the body's cooling mechanism. This condition is heatstroke, and is potentially fatal. To avoid these illnesses, troops should maintain their physical fitness by eating adequately, drinking sufficient water, and consuming adequate salt. If soldiers/marines expend more calories than they take in, they will be more prone to heat illnesses. Since troops may lose their desire for food in hot climates, they must be encouraged to eat, with the heavier meal of the day scheduled during the cooler hours.

It is necessary to recognize heat stress symptoms quickly. When suffering from heatstroke, the most dangerous condition, there is a tendency for a soldier/marine to creep away from his comrades and attempt to hide in a shady and secluded spot; if not found and treated, he will die. When shade is required during the day, it can best be provided by tarpaulins or camouflage nets, preferably doubled to allow air circulation between layers and dampened with any surplus water.

WARNING

Drink water even if not thirsty. Thirst Is not an adequate warning of dehydration.

Approximately 75 percent of the human body is fluid. All chemical activities in the body occur in a water solution, which assists in the removal of toxic body wastes and plays a vital part in the maintenance of an even body temperature. A loss of 2 quarts of body fluid (2.5 percent of body weight) decreases efficiency by 25 percent and a loss of fluid equal to 15 percent of body weight is usually fatal. The following are some considerations when operating in a desert environment:

- Consider water a tactical weapon. Reduce heat injury by forcing water consumption. Soldiers/marines in armored vehicles, MOPP gear, and in body armor need to increase their water intake.
- When possible, drink Cool (50-55 degrees Fahrenheit) water.
- Drink one quart of water in the morning, at each meal, and before strenuous work. In hot climates drink at least one quart of water each hour. At higher temperatures hourly water requirements increase to over two quarts.
- Take frequent drinks since they are more often effective than drinking the same amount all at once. Larger soldiers/marines need more water.
- Replace salt loss through eating meals.
- When possible, work loads and/or duration of physical activity should be less during the first days of exposure to heat, and then should gradually be increased to follow acclimatization.

- Modify activities when conditions that increase the risk of heat injury (fatigue/loss of sleep, previous heat exhaustion, taking medication) are present.
- Take frequent rest periods in the shade, if possible. Lower the work rate and work loads as the heat condition increases.
- Perform heavy work in the cooler hours of the day such as early morning or late evening, if possible.

A description of the symptoms and treatment for heat illnesses follows:

- Heat cramps.

  - Symptoms: Muscle cramps of arms, legs, and/or stomach. Heavy sweating (wet skin) and extreme thirst.

  - First aid: Move soldier/marine to a shady area and loosen clothing. Slowly give large amounts of cool water. Watch the soldier/marine and continue to give him water, if he accepts it. Get medical help if cramps continue.

- Heat exhaustion.

  - Symptoms: Heavy sweating with pale, moist, cool skin; headache, weakness, dizziness, and/or loss of appetite; heat cramps, nausea (with or without vomiting), rapid breathing, confusion, and tingling of the hands and/or feet.

  - First aid: Move the soldier/marine to a cool, shady area and loosen/remove clothing. Pour water on the soldier/marine and fan him to increase the cooling effect. Have the soldier/marine slowly drink at least one full canteen of water. Elevate the soldier's/marine's legs. Get medical help if symptoms continue; watch the soldier/marine until the symptoms are gone or medical aid arrives.

- Heatstroke.

  - Symptoms: Sweating stops (red, flushed, hot dry skin).

  - First aid: Evacuate to a medical facility immediately. Move the soldier/marine to a cool, shady area and loosen or remove clothing if the situation permits. Start cooling him immediately. Immerse him in water and fan him. Massage his extremities and skin and elevate his legs. If conscious, have the soldier/marine slowly drink one full canteen of water.

DANGER

Heatstroke is a medical emergency. Seek medical attention immediately.

WATER SUPPLY

Maintaining safe, clean, water supplies is critical. The best containers for small quantities of water (5 gallons) are plastic water cans or coolers. Water in plastic cans will be good for up to 72 hours; storage in metal containers is safe only for 24 hours. Water trailers, if kept cool, will keep water fresh up to five days. If the air temperature exceeds 100 degrees Fahrenheit, the water temperature must be monitored. When the temperature exceeds 92 degrees Fahrenheit, the water should be changed, as bacteria will multiply. If the water is not changed the water can become a source of sickness, such as diarrhea. Ice in containers keeps water cool. If ice is put in water trailers, the ice must be removed prior to moving the trailer to prevent damage to the inner lining of the trailer.

Potable drinking water is the single most important need in the desert. Ensure nonpotable water is never mistaken for drinking water. Water that is not fit to drink but is not otherwise dangerous (it may be merely oversalinated) may be used to aid cooling. It can be used to wet clothing, for example, so the body does not use too much of its internal store of water.

Use only government-issued water containers for drinking water. Carry enough water on a vehicle to last the crew until the next planned resupply. It is wise to provide a small reserve. Carry water containers in positions that-

- Prevent vibration by clamping them firmly to the vehicle body.
- Are in the shade and benefit from an air draft.
- Are protected from puncture by shell splinters.
- Are easily dismounted in case of vehicle evacuation.

Troops must be trained not to waste water. Water that has been used for washing socks, for example, is perfectly adequate for a vehicle cooling system.

Obtain drinking water only from approved sources to avoid disease or water that may have been deliberately polluted. Be careful to guard against pollution of water sources. If rationing is in effect, water should be issued under the close supervision of officers and noncommissioned officers.

Humans cannot perform to maximum efficiency on a decreased water intake. An acclimatized soldier/marine will need as much (if not more) water as the nonacclimatized soldier/marine, as he sweats more readily. If the ration water is not sufficient, there is no alternative but to reduce physical activity or restrict it to the cooler parts of the day.

In very hot conditions it is better to drink smaller quantities of water often rather than large quantities occasionally. Drinking large quantities causes excessive sweating and may induce heat cramps. Use of alcohol lessens resistance to heat due to its dehydrating effect. As activities increase or conditions become more severe, increase water intake accordingly.

The optimum water drinking temperature is between 10 degrees Celsius and 15.5 degrees Celsius (50-60 degrees Fahrenheit). Use lister bags or even wet cloth around metal containers to help cool water.

Units performing heavy activities on a sustained basis, such as a forced march or digging in, at 80 degrees wet bulb globe temperature index, may require more than 3 gallons of drinking water per man. Any increase in the heat stress will increase this need. In high temperatures, the average soldier/marine will require 9 quarts of water per day to survive, but 5 gallons are recommended. Details on water consumption and planning factors are contained in Appendix G.

HINT: Taking a military issue wool sock, soaking it in water, and wrapping it around a canteen or placing it on a **water** bottle, will cool the water through evaporation.

While working in high desert temperatures, a man at rest may lose as much as a pint of water per hour from sweating. In very high temperatures and low humidity, sweating is not noticeable as it evaporates so fast the skin will appear dry. Whenever possible, sweat should be retained on the skin to improve the cooling process; however, the only way to do this is to avoid direct sun on the skin. This is the most important reason why desert troops must remain fully clothed. If a soldier/marine is working, his water loss through sweating (and subsequent requirement for replenishment) increases in proportion to the amount of work done (movement). Troops will not always drink their required amount of liquid readily and will need to be encouraged or ordered to drink more than they think is necessary as the sensation of thirst is not felt until there is a body deficit of 1 to 2 quarts of water. This is particularly true during the period of acclimatization. Packets of artificial fruit flavoring encourages consumption due to the variety of pleasant tastes.

All unit leaders must understand the critical importance of maintaining the proper hydration status. Almost any contingency of military operations will act to interfere with the maintenance of hydration. Urine provides the best indicator of proper hydration. The following are considerations for proper hydration during desert operations:

- Water is the key to your health and survival. Drink before you become thirsty and drink often, When you become thirsty you will be about a "quart and a half low".
- Carry as much water as possible when away from approved sources of drinking water. Man can live longer without food than without water.
- Drink before you work; carry water in your belly, do not "save" it in your canteen. Learn to drink a quart or more of water at one time and drink frequently to replace sweat losses.
- Ensure troops have at least one canteen of water in reserve, and know where and when water resupply will be available.
- Carbohydrate/electrolyte beverages (e.g., Gatorade) are not required, and if used, should not be the only source of water. They are too concentrated to be used alone. Many athletes prefer to dilute these 1:1 with water. Gaseous drinks, sodas, beer, and milk are not good substitutes for water because of their dehydrating effects.
- If urine is more colored than diluted lemonade, or the last urination cannot be remembered, there is probably insufficient water intake. Collect urine samples in field expedient containers and spot check the color as a guide to ensuring proper hydration. Very dark urine warns of dehydration. Soldiers/marines should observe their own urine, and use the buddy system to watch for signs of dehydration in others.
- Diseases, especially diarrheal diseases, will complicate and often prevent maintenance of proper hydration.

Salt, in correct proportions, is vital to the human body; however, the more a man sweats, the more salt he loses. The issue ration has enough salt for a soldier/marine drinking up to 4 quarts of water per day. Unacclimatized troops need additional salt during their first few days of exposure and all soldiers/marines need additional salt when sweating heavily. If the water demand to balance sweat loss rises, extra salt must be taken under medical direction. Salt, in excess of body requirements, may cause increased thirst and a feeling of sickness, and can be dangerous. Water must be tested before adding salt as some sources are already saline, especially those close to the sea.

COLD

The desert can be combine to produce dangerously cold. The dry air, wind, and clear sky can bone-chilling discomfort and even injury. The ability of the body to maintain body temperature within a narrow range is as important in the cold as in the heat. Loss of body heat to the environment can lead to cold injury; a genera] lowering of the body temperature can result in hypothermia, and local freezing of body tissues can lead to frostbite. Hypothermia is the major threat from the cold in the desert, but frostbite also occurs.

Troops must have enough clothing and shelter to keep warm. Remember, wood is difficult to find; any that is available is probably already in use. Troops maybe tempted to leave clothing and equipment behind that seems unnecessary (and burdensome) during the heat of the day. Cold-wet injuries (immersion foot or trench foot) may be a problem for dismounted troops operating in the coastal marshes of the Persian Gulf during the winter. Some guidelines to follow when operating in the cold are—

- Anticipate an increased risk of cold-wet injuries if a proposed operation includes lowland or marshes. Prolonged exposure of the feet in cold water causes immersion foot injury, which is completely disabling.
- Check the weather-know what conditions you will be confronting. The daytime temperature is no guide to the nighttime temperature; 90-degree-Fahrenheit days can turn into 30-degree-Fahrenheit nights.
- The effects of the wind on the perception of cold is well known. Windchill charts contained in FM 21-10 allow estimation of the combined cooling power of air temperature and wind speed compared to the effects of an equally cooling still-air temperature.

CLOTHING

Uniforms should be worn to protect against sunlight and wind. Wear the uniform loosely. Use hats, goggles, and sunscreen. Standard lightweight clothing is suitable for desert operations but should be camouflaged in desert colors, not green. Wear nonstarched long-sleeved shirts, and full-length trousers tucked into combat boots. Wear a scarf or triangular bandanna loosely around the neck (as a sweat rag) to protect the face and neck during sandstorms against the sand and the sun. In extremely hot and dry conditions a wet sweat rag worn loosely around the neck will assist in body cooling.

Combat boots wear out quickly in desert terrain, especially if the terrain is rocky. The leather drys out and cracks unless a nongreasy mixture such as saddle soap is applied. Covering the ventilation holes on jungle boots with glue or epoxies prevents excessive sand from entering the boots. Although difficult to do, keep clothing relatively clean by washing in any surplus water that is available. When water is not available, air and sun clothing to help kill bacteria.

Change socks when they become wet. Prolonged wear of wet socks can lead to foot injury. Although dry desert air promotes evaporation of water from exposed clothing and may actually promote cooling, sweat tends to accumulate in boots.

Soldier/marines may tend to stay in thin clothing until too late in the desert day and become susceptible to chills--so respiratory infections may be common. Personnel should gradually add layers of clothing at night (such as sweaters), and gradually remove them in the morning. Where the danger of cold weather injury exists in the desert, commanders must guard against attempts by inexperienced troops to discard cold weather clothing during the heat of the day.

Compared to the desert battle dress uniform (DBDU) the relative impermeability of the battle dress overgarment (BDO) reduces evaporative cooling capacity. Wearing underwear and the complete DBDU, with sleeves rolled down and under the chemical protective garment, provides additional protection against chemical poisoning. However, this also increases the likelihood of heat stress casualties.

HYGIENE AND SANITATION

Personal hygiene is absolutely critical to sustaining physical fitness. Take every opportunity to wash. Poor personal hygiene and lack of attention to siting of latrines cause more casualties than actual combat. Field Marshal Rommel lost over 28,400 soldiers of his Afrika Corps to disease in 1942. During the desert campaigns of 1942, for every one combat injury, there were three hospitalized for disease. Hygiene and sanitation are covered in detail in FM 21-10. This section highlights some of the points that are of special importance to the commander in the desert.

Proper standards of personal hygiene must be maintained not only as a deterrent to disease but as a reinforcement to discipline and morale. If water is available, shave and bathe daily. Cleaning the areas of the body that sweat heavily is especially important; change underwear and socks frequently, and use foot powder often. Units deployed in remote desert areas must have a means of cutting hair therefore, barber kits should be maintained and inventoried prior to any deployment. If sufficient water is not available, troops should clean themselves with sponge baths, solution-impregnated pads, a damp rag, or even a dry, clean cloth. Ensure that waste water is disposed of in an approved area to prevent insect infestation. If sufficient water is not available for washing, a field expedient alternative is powder baths, that is, using talcum or baby powder to dry bathe.

Check troops for any sign of injury, no matter how slight, as dirt or insects can cause infection in minor cuts and scratches. Small quantities of disinfectant in washing water reduces the chance of infection. Minor sickness can have serious effects in the desert. Prickly heat for example, upsets the sweating mechanism and diarrhea increases water loss, making the soldier/marine more prone to heat illnesses. The buddy system helps to ensure that prompt attention is given to these problems before they incapacitate individuals.

Intestinal diseases can easily increase in the desert. Proper mess sanitation is essential. Site latrines well away and downwind of troop areas and lagers. Trench-type latrines should be used where the soil is suitable but must be dug deeply, as shallow latrines become exposed in areas of shifting sand. Funnels dug into a sump work well as urinals. Layer the bottom of slit trenches with lime and cover the top prior to being filled in. Ensure lime is available after each use of the latrine. Flies area perpetual source of irritation and carry infections. Only good sanitation can keep the fly problem to a minimum. Avoid all local tribe camps since they are frequently a source of disease and vermin.

DESERT SICKNESS

Diseases common to the desert include plague, typhus, malaria, dengue fever, dysentery, cholera, and typhoid. Diseases which adversely impact hydration, such as those which include nausea, vomiting, and diarrhea among their symptoms, can act to dramatically increase the risk of heat (and cold) illness or injury. Infectious diseases can result in a fever; this may make it difficult to diagnose heat illness. Occurrences of heat illness in troops suffering from other diseases complicate recovery from both ailments.

Many native desert animals and plants are hazardous. In addition to injuries as a result of bites, these natural inhabitants of the desert can be a source of infectious diseases.

Many desert plants and shrubs have a toxic resin that can cause blisters, or spines that can cause infection. Consider milky sap, all red beans, and smoke from burning oleander shrubs, poisonous. Poisonous snakes, scorpions, and spiders are common in all deserts. Coastal waters of the Persian Gulf contain hazardous marine animals including sea snakes, poisonous jellyfish, and sea urchins.

Skin diseases can result from polluted water so untreated water should not be used for washing clothes; although it can be used for vehicle cooling systems or vehicle decontamination.

The excessive sweating common in hot climates brings on prickly heat and some forms of fungus infections of the skin. The higher the humidity, the greater the possibility of their occurrence. Although many deserts are not humid, there are exceptions, and these ailments are common to humid conditions.

The following are additional health-related considerations when operating in a desert environment:

- The most common and significant diseases in deserts include diarrheal and insectborne febrile (i.e., fever causing) illnesses-both types of these diseases are preventable.
- Most diarrheal diseases result from ingestion of water or food contaminated with feces. Flies, mosquitoes, and other insects carry fever-causing illnesses such as malaria, sand fly fever, dengue (fever with severe pain in the joints), typhus, and tick fevers.
- There are no safe natural water sources in the desert. Standing water is usually infectious or too brackish to be safe for consumption. Units and troops must always know where and how to get safe drinking water.
- Avoid brackish water (i.e., salty). It, like sea water, increases thirst; it also dehydrates the soldier/marine faster than were no water consumed. Brackish water is common even in public water supplies, Iodine tablets only kill germs, they do not reduce brackishness.
- Water supplies with insufficient chlorine residuals, native food and drink, and ice from all sources are common sources of infective organisms.

PREVENTIVE MEASURES

Both diarrheal and insectborne diseases can be prevented through a strategy which breaks the chain of transmission from infected sources to susceptible soldiers/marines by effectively applying the preventive measures contained in FM 21-10. Additional preventive measures are described below:

- Careful storage, handling and purification/preparation of water and food are the keys to prevention of diarrheal disease. Procure all food, water, ice, and beverages from US military approved sources and inspect them routinely.
- Well-cooked foods that are "steaming hot" when eaten are generally safe, as are peeled fruits and vegetables.
- Local dairy products and raw leafy vegetables are generally unsafe.
- Consider the food in native markets hazardous. Avoid local food unless approved by medical personnel officials.

- Assume raw ice and native water to be contaminated-raw ice cannot be properly disinfected. Ice has been a major source of illness in all prior conflicts; therefore, use ice only from approved sources.
- If any uncertainty exists concerning the quality of drinking water, troops should disinfect their supplies using approved field-expedient methods (e.g., hypochlorite for lister bags, iodine tablets for canteens, boiling).
- Untreated water used for washing or bathing risks infection.
- Hand washing facilities should be established at both latrines and mess facilities. Particular attention should be given to the cleanliness of hands and fingernails. Dirty hands are the primary means of transmitting disease.
- Dispose of human waste and garbage as specified in FM 21-10. Additional considerations regarding human waste and garbage are—

  - Sanitary disposal is important in preventing the spread of disease from insects, animals, and infected individuals, to healthy soldiers/marines.

  - Construction and maintenance of sanitary latrines are essential.

  - Burning is the best solution for waste.

  - Trench latrines can be used if the ground is suitable, but they must be dug deeply enough so that they are not exposed to shifting sand, and they must have protection against flies and other insects that can use them as breeding places.

  - Food and garbage attract animals-do not sleep where you eat and keep refuse areas away from living areas.
- Survey the unit area for potential animal hazards.
- Shake out boots, clothing, and bedding before using them.

Section III. Environmental Effects on Equipment

Conditions in an arid environment can damage military equipment and facilities. Temperatures and dryness are major causes of equipment failure, and wind action lifts and spreads sand and dust, clogging andjamming anything that has moving parts. Vehicles, aircraft, sensors, and weapons are all affected. Rubber components such as gaskets and seals become brittle, and oil leaks are more frequent. Ten characteristics of the desert environment may adversely affect equipment used in the desert:

- Terrain.
- Heat.
- Winds.
- Dust and sand.
- Humidity.
- Temperature variations.
- Thermal bending.
- Optical path bending.

- Static electricity.
- Radiant light.

The relative importance of each characteristic varies from desert to desert. Humidity, for example, can be discounted in most deserts but is important in the Persian Gulf.

TERRAIN

Terrain varies from nearly flat, with high trafficability, to lava beds and salt marshes with little or no trafficability. Drivers must be well trained in judging terrain over which they are driving so they can select the best method of overcoming the varying conditions they will encounter. Techniques for driving and operating equipment in desert conditions are contained in Appendix C.

Track vehicles are well suited for desert operations. Wheel vehicles may be acceptable as they can go many places that track vehicles can go; however, their lower speed average in poor terrain maybe unacceptable during some operations. Vehicles should be equipped with extra fan belts, tires, (and other items apt to malfunction), tow ropes (if not equipped with a winch), extra water cans, and desert camouflage nets. Air-recognition panels, signal mirrors, and a tarpaulin for crew sun protection are very useful. Wheel vehicles should also carry mats, or channels as appropriate, to assist in freeing mired vehicles.

The harsh environment requires a very high standard of maintenance, which may have to be performed well away from specialized support personnel. Operators must be fully trained in operating and maintaining their equipment. Some types of terrain can have a severe effect on suspension and transmission systems, especially those of wheel vehicles. Tanks are prone to throw tracks when traveling over rocks.

Track components require special care in the desert. Grit, heat, and bad track tension accelerate track failure. Sprockets wear out quickly in sandy conditions. Track pins break more easily in high temperatures and high temperatures also increase rubber/metal separation on road wheels. Proper track tension is critical, as loose track is easily thrown and excessive tension causes undue stress on track components.

Increase the unit PLL of tires and tracks as sand temperatures of 165 degrees Fahrenheit are extremely detrimental to rubber, and weaken resistance to sharp rocks and plant spines, Items affected by mileage such as wheels, steering, track wedge bolts and sprocket nuts, and transmission shafts, must be checked for undue wear when conducting before-, during-, and after-operations maintenance.

HEAT

Vehicle coding and lubrication systems are interdependent. A malfunction by one will rapidly place the other system under severe strain. In temperature extremes, all types of engines are apt to operate above optimum temperatures, leading to excessive wear, or leaking oil seals in the power packs, and ultimately, engine failure. Commanders should be aware which types of vehicles are prone to excessive overheating, and ensure extra care is applied to their maintenance. The following are considerations for ensuring engines do not overheat:

- Check oil levels frequently to ensure proper levels are maintained (too high may be as bad as too low), that seals are not leaking, and oil consumption is not higher than normal.

- Keep radiators and air flow areas around engines clean and free of debris and other obstructions.
- Fit water-cooled engines with condensers to avoid steam escaping through the overflow pipe.
- Cooling hoses must be kept tight (a drip a second loses 7 gallons of fluid in 24 hours).
- Operators should not remove hood side panels from engine compartments while the engine is running as this causes air turbulence and leads to ineffective cooling.

Batteries do not hold their charge efficiently in intense heat. Check them twice daily. The following are additional considerations for maintaining batteries in intense heat:

- Change battery specific gravity to adjust to the desert environment (see vehicle TMs for details).
- Keep batteries full, but not overfilled, and carry a reserve of distilled water.
- Keep air vents clean, or vapors may build up pressure and cause the battery to explode.
- Set voltage regulators as low as practical.
- Increase dry battery supplies to offset high attrition rate caused by heat exposure.

Severe heat increases pressure in closed pressurized systems such as the M2 fire burner unit, and increases the volume of liquids. Ensure that the working pressure of all equipment is within safety limits and be careful when removing items such as filler caps.

Treat Halon fire extinguishers with care. High temperatures may cause them to discharge spontaneously. Put wet rags on them during the hottest part of the day to keep them cooler.

Some items of equipment are fitted with thermal cutouts that open circuit breakers whenever equipment begins to overheat. Overheating is often caused by high ambient temperatures, and can be partly avoided by keeping the item in the shade or wrapping it in a wet cloth to maintain a lower temperature by evaporation.

Flying time and performance of helicopters are degraded as the altitude and heat increases. Helicopter performance is also affected by humidity. Aircraft canopies have been known to bubble under direct heat and should be covered when not in use.

Ammunition must be out of direct heat and sunlight. Use camouflage nets and tarpaulins to provide cover. Ammunition cool enough to be held by bare hands is safe to fire.

Wood shrinks in a high-temperature, low-humidity environment. Equipment, such as axes carried on track vehicles, can become safety hazards as heads are likely to fly off shrunken handles.

Radiators require special attention. Proper cooling-system operation is critical in high-temperature environments. Check cooling systems for serviceability prior to deployment. Local water maybe high in mineral content which will calcify in cooling systems. Distilled water is better since tap water contains chemicals that will form a crusty coating inside the radiator and will ultimately clog it. A mixture of 40 percent antifreeze and 60 percent water is usually acceptable-check your appropriate technical manual to be certain.

During movement, and at operation sites where extremely hot temperatures exist, continuous protection from the heat is necessary for medical items and supplies, which deteriorate rapidly.

Air and all fluids expand and contract according to temperature. If tires are inflated to correct pressure during the cool of night, they may burst during the heat of day. If fuel tanks are filled to the brim at night, they will overflow at midday. Servicing these items during the heat of day can result in low tire pressure, overheating of tires, and a lack of endurance if the fuel tanks were not filled to their correct levels. Air pressure in tires must be checked when the equipment is operating at efficient working temperatures, and fuel tanks must be filled to their correct capacity as defined in the appropriate technical manual. These items should be checked several times a day and again at night.

The major problem with radios in a desert environment is overheating. The following steps can help prevent overheating of radios:

- Keep radios out of direct sunlight.
- Place a piece of wood on top of the radio. Leaving space between the wood and the top of the radio will help cool the equipment. Operating on low power whenever possible will also help.
- Place wet rags on top of radios to help keep them cool and operational. Do not cover the vents.

Any oil or fuel blown onto a cooler (heat exchanger) will gather and quickly degrade cooling. Fix even slight leaks promptly. Do not remove cooling ducts or shrouds. Check them for complete coverage-use tape to seal cracks. Do not remove serviceable thermostats if overheating occurs.

WINDS

Desert winds, by their velocity alone, can be very destructive to large and relatively light materiel such as aircraft, tentage, and antenna systems. To minimize the possibility of wind damage, materiel should be sited to benefit from wind protection and should be firmly picketed to the ground.

DUST AND SAND

Dust and sand are probably the greatest danger to the efficient functioning of equipment in the desert. It is almost impossible to avoid particles settling on moving parts and acting as an abrasive. Sand mixed with oil forms an abrasive paste.

Lubricants must be of the correct viscosity for the temperature and kept to the recommended absolute minimum in the case of exposed or semiexposed moving parts. Lubrication fittings are critical items and should be checked frequently. If they are missing, sand will enter the housing causing bearing failure. Teflon bearings require constant inspection to ensure that the coating is not being eroded.

Proper lubrication is crucial for success. Wipe off all grease fittings before you attach the grease gun and after use. Keep cans of grease covered to prevent sand contamination. Preserve opened grease containers by covering and sealing with plastic bags. Use of grease cartridges in lieu of bulk grease is preferred. All POL dispensing tools must be stored in protected areas to prevent contamination. Place a tarpaulin, or other material, under equipment being repaired to prevent tools and components from being lost in the sand. The automotive-artillery grease possesses a significantly high-temperature capability. If not available, an alternative is general purpose wide-temperature range (WTR) aircraft grease.

Oil should be changed about twice as often under desert conditions as under US or European conditions, not only because grit accumulates in the oil pan, but also because noncombusted low-octane fuel seeps down the cylinder walls and dilutes the reservoir. Diluted lubricants cool less effectively, and evaporate at the higher temperatures generated during engine operation. Oil changes and lubrication of undercarriage points at more frequent intervals will prolong engine and vehicle life under desert conditions. Units employed in desert environments should reevaluate their engine oil requirements and plan accordingly.

Keeping sand out of maintenance areas is critical due to the strong possibility of sand or dust entering the cylinders or other moving parts when the equipment is stripped. Baggies, cloth, or plastic can be used to protect open or disassembled components from blowing sand and dust. The same applies for disconnected water, oil, or other fluid lines. Be sure to cover both ends of the connection if stored. It is essential to have screens against blowing sand (which also provides shade for mechanics). The surrounding ground may be soaked in used oil or covered with rocks to bind it down. Mechanics must keep their tools clean.

Dust and sand can easily cause failure of such items as radio and signal distribution panels, and circuit breakers, and cause small electrical motors to burn out. Wheel and flight control bearings may require daily cleaning and repacking, and engines should be flushed of contaminants daily.

Rotor heads have reduced life, requiring more frequent inspections than in temperate climates. Pay particular attention to sand-caused wear on rotor heads, leading edges of rotor blades, and exposed flight controls. Over 200 pounds of dirt has been known to accumulate in the fuselage area of helicopters operating in desert conditions. These areas must be routinely checked and cleaned to prevent a pound-for-pound reduction in aircraft-lift capability.

Filters must be used when refueling any type of vehicle, and the gap between the nozzle and the fuel tank filler pipe must be kept covered. It takes comparatively little dirt to block a fuel line. Fuel filters will require more frequent cleaning and will need to be checked and replaced often. Engine oil should be changed more often and oil filters replaced more frequently than in temperate climates.

Compression-ignition engines depend on clean air; therefore, examine and clean air cleaners of every type of equipment at frequent intervals. The exact interval depends on the operating conditions, but as a minimum, should be checked at least daily.

Air compressors are valuable pieces of equipment in the desert. They are essential for cleaning air filters and removing dust and sand from components. Intake filters require cleaning daily.

Windblown sand and grit, in addition to heat, will damage electrical wire insulation over a period of time. All cables that are likely to be damaged should be protected with tape before insulation becomes worn. Sand will also find its way into parts of items such as "spaghetti cord" plugs, either preventing electrical contact or making it impossible to join the plugs together. Use a brush, such as an old toothbrush, to brush out such items before they are joined. Electrical tape placed over the ends of spaghetti cords also works.

Radio is the primary means of communications in the desert. It can be employed effectively in desert climates and terrain to provide the reliable communications demanded by widely dispersed forces. However, desert terrain provides poor electrical ground, and a counterpoise (an artificial ground) is needed to improve the range of certain antennas.

Some receiver-transmitters have ventilating ports and channels that can get clogged with dust. These must be checked regularly and kept clean to prevent overheating. Mobile subscriber equipment may require the deployment of additional radio access units (RAU) AN/VRC-191. These assemblages are the primary link for the mobile subscriber radio telephone terminal (MSRT) AN/VRC-97s which are located down to battalion level. The normal operating range of the receiver-transmitter used with these radios may only be 10 kilometers in the desert.

Dust and sand adversely affect the performance of weapons. Weapons may jam or missiles lock on launching rails due to sand and dust accumulation. Sand- or dust-clogged barrels lead to in-bore detonations. Daily supervised cleaning of weapons is essential. Particular attention should be given to magazines which are often clogged, interrupting the feeding of weapons. Cover missiles on launchers until required for use. To avoid jamming due to the accumulation of sand, the working parts of weapons must have the absolute minimum amount of lubrication. It may even be preferable to have them totally dry, as any damage caused during firing will be less than that produced by the sand/oil abrasive paste. Paintbrushes are among the most useful tools to bring to the desert; they are extremely effective in cleaning weapons and optics.

Take precautions to prevent exposure of floppy disks and computers to dust or sand. Covering them in plastic bags is a technique that has worked for several different units. A number of units have successfully operated PLL computers in inflatable medical NBC shelters (MIS 1). This technique has obvious drawbacks since the shelter was not designed for this; however, until a materiel fix is developed, this sort of innovation may be necessary. Compressed air cans, locally purchased from computer vendors, will facilitate the cleaning of keyboards and other components of computer systems.

All optics are affected by blown sand, which gradually degrade their performance due to small pitting and scratching. It is necessary to guard against buildup of dust on optics, which may not be apparent until the low light optical performance has severely deteriorated. It may be advisable to keep optics covered with some form of cellophane film until operations are about to start, especially if the unit is in a sandstorm. A cover that has no sand on the underside should also be used and must be secured so it cannot vibrate against the wind screen. Both of these measures are equally important to tactical security as sun reflected from these optics will reveal positions.

Sand and dirt can easily accumulate in hull bottoms of armored vehicles. This accumulation, combined with condensation or oil, can cause jamming of control linkages. Sand accumulation at the air-bleeder valve can inhibit heat from escaping from the transmission and result in damage to the transmission. Park tactical wheeled vehicles with the rear facing the wind to keep sand out of the radiator. Tracked vehicles should park to protect the engine compartment (grille doors away from wind) from the same sort of damage. The operator's checks and services increase in importance in this environment.

HUMIDITY

Some deserts are humid. Where this is the case, humidity plus heat encourages rust on bare metal and mold in enclosed spaces such as optics. Bare metal surfaces on equipment not required for immediate use must be kept clean and very lightly lubricated.

Items such as optics must be stored in dry conditions; those in use should be kept where air can circulate around them, and should be purged at frequent intervals. Aircraft must be washed daily, particularly if there is salt in the air, using low-pressure sprays.

TEMPERATURE VARIATIONS

In deserts with relatively high-dew levels and high humidity, overnight condensation can occur wherever surfaces (such as metal exposed to air) are cooler than the air temperaturc. Condensation can affect such items as optics, fuel lines, and air tanks. Drain fuel lines both at night and in the morning (whenever necessary). Clean optics and weapons frequently. Weapons, even if not lubricated, accumulate sand and dirt due to condensation.

THERMAL BENDING

Weapon systems such as the tank cannon are affected in several ways by the desert. One is thermal bending, which is the uneven heating and cooling of a gun tube due to ambient temperature changes. Modem tanks, like the MI, have been designed to compensate for these factors. The muzzle reference system (MRS) allows the crew to monitor any loss of gun sight relationship and to comet for any error using the MRS update at regular intervals. MI-series tanks are equipped with a thermal shroud, allowing for more even heating and cooling of the gun tube. Both factors can greatly reduce the accuracy of a tank weapon system. By boresighting at regular intervals and constant monitoring of the fire control system, the tank crew can maximize its readiness. "Gun tube droop" can be countered using the MRS update at least four times in a 24-hour period: at dawn as part of stand-to; at noon to compensate for gun tube temperature chang: before EENT, for TIS reticle confirmation; and at 0100 hours to compensate for gun tube temperature changes.

OPTICAL PATH BENDING

The apparent illusion of target displacement is commonly called refraction. Under certain light and environmental conditions, the path of light (line of sight) may not appear to travel in a straight line. Refraction may cause problems for tank crews attempting engagements at ranges beyond 1,500 meters. Figure 1-17 shows an example of optical path bending in the desert. Refraction may occur in the following conditions:

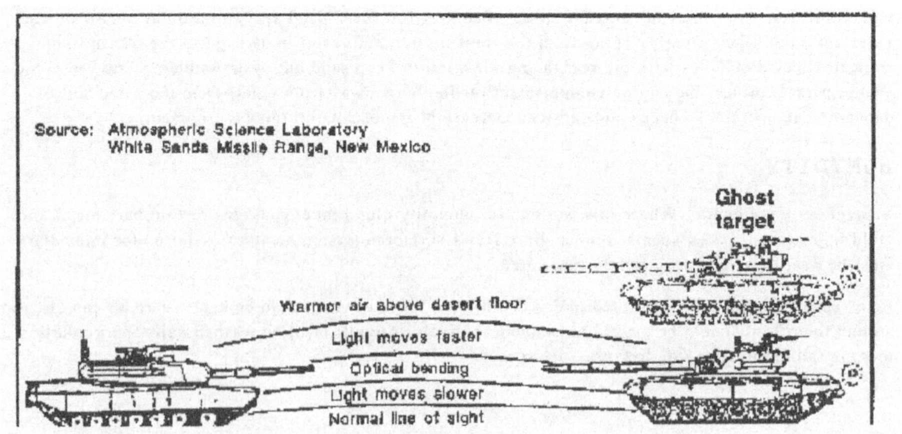

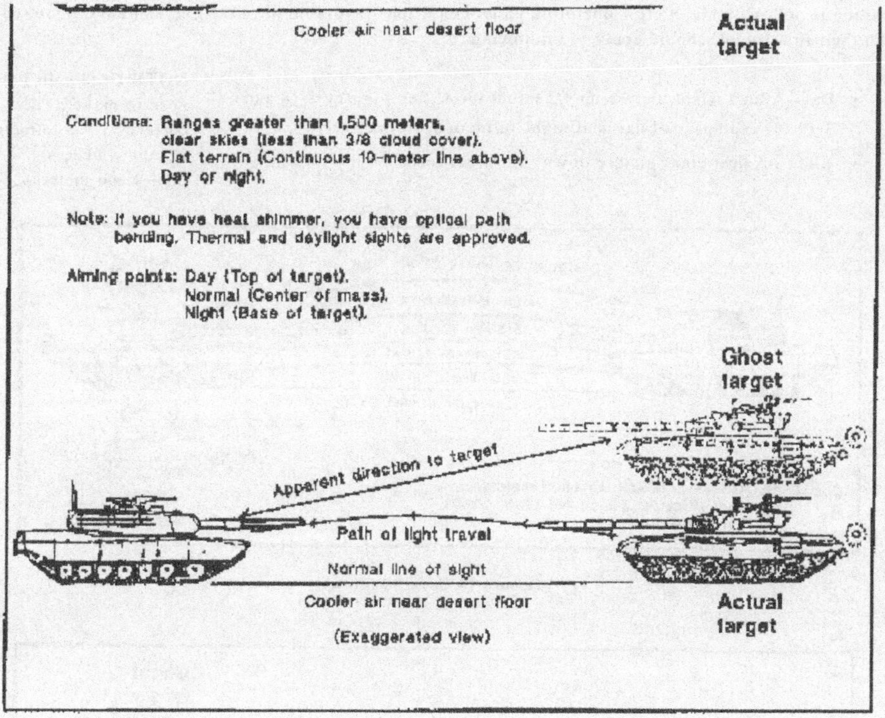

Figure: 1-17. Optical path bending in the desert.

- Day—Clear sky, flat terrain, winds less than 10 miles per hour.
- Night-Clear sky, flat terrain, winds under 4 miles per hour.

NOTE: Any time heat shimmer is present, refraction may also exist.

The effect of refraction is to make the target appear lower during the day; the sight picture, though it appears center of visible mass to the gunner, is actually below the target. This may result in a short round. At night, the effects are the opposite and may result in an over round. Crews must not be fooled by what appears to be a good range from their laser range finder (LRF); the laser beam will refract with other light rays and still hit the desired target.

The most effective measure available to the crew to minimize refraction is an elevated firing position. A position at least 10 meters above intervening terrain will generally negate any effects. If this type of position is not available, a crew operating under conditions favorable to refraction, and having missed with their first round, should apply the following:

- Day--Adjust sight picture up 1/2 target form. See Figures 1-18 and 1-19 for examples of day and night refraction, respectively.
- Night—Adjust sight picture down 1/2 target form.

NOTE:Crews do not need to make a correction for refraction at ranges of less than 1,500 meters.

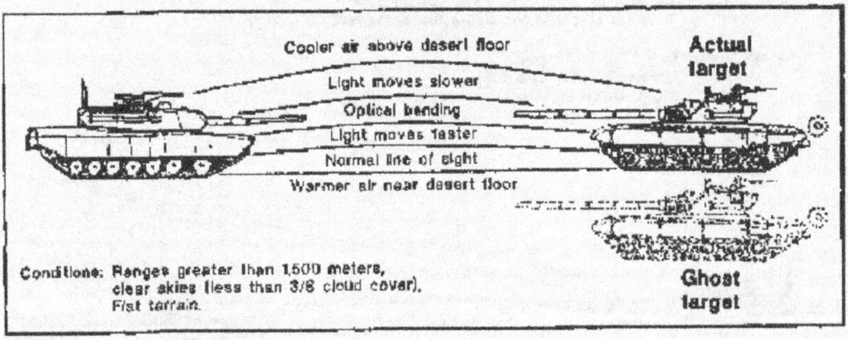

Figure 1-18. Day refraction.

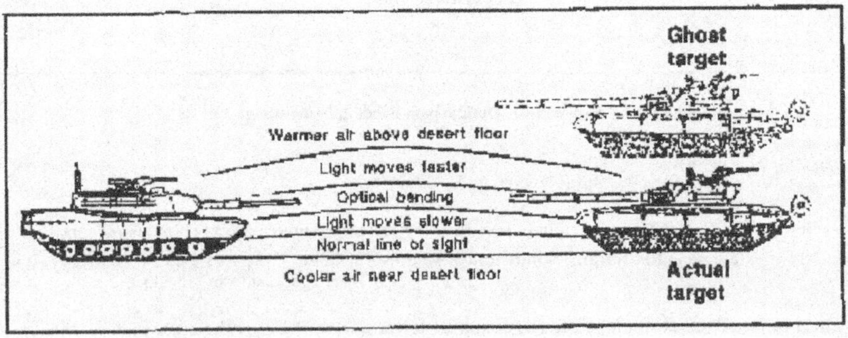

Figure 1-19. Night refraction

STATIC ELECTRICITY

Static electricity is prevalent and poses a danger in the desert. It is caused by atmospheric conditions coupled with an inability to ground out due to dryness of the terrain. It is particularly prevalent with aircraft or vehicles having no conductor contact with the soil. The difference of electrical potential between separate materials may cause an electrical discharge between them when contact is made, and if flammable gases are present, they may explode and cause a fire. Poor grounding conditions aggravate the problem. Be sure to tape all sharp edges (tips) of antennas to reduce wind-caused static electricity. If you are operating from a fixed position, ensure that equipment is properly grounded.

Establish a metal circuit between fuel tankers and vehicles before and during refueling. Ensure the fuel tankers and vehicles are grounded (for example, by a cable and picket or by a crowbar). Grounding of vehicles and equipment should be accomplished in accordance with appropriate operations manuals.

Static electricity is also a hazard with helicopter sling loads. Exercise care when handling and transporting unlike materials that might generate static electricity. Also turn off all switches, uncouple electrical connectors, and ground vehicle or aircraft electrically-operated weapon systems before reaming. Static electricity will also ruin circuit boards and other electronic equipment.

RADIANT LIGHT

Radiant light may be detrimental to plastics, lubricants, pressurized gases, some chemicals, and infrared tracking and guidance systems. Items like CO^2 fire extinguishers, M13 decontamination and reimpregnating kits, and Stinger missiles must be kept out of constant direct sunlight. Optics have been known to discolor under direct sunlight (although this is unusual), so it is wise to minimize their exposure to the sun's rays.

CHAPTER 2

PREPARATIONS FOR DESERT OPERATIONS

This chapter describes preparations for deployment and training for operations in a desert environment. A force sent to a desert theater will fight with the equipment it has in accordance with current doctrine. While equipment and doctrine can be modified to suit the new environment, much will depend on how well soldiers/marines and leaders have mastered their individual training. Units that have trained in Germany and the United States will have the basic technical and tactical skills that can be adapted for desert warfare. Well-trained troops and leaders can adapt quickly to the peculiar conditions of the new environment. If their individual and collective skills have been neglected, no amount of desert lore will remedy the situation.

Section I. Factors to be Considered When Preparing for Desert Operations

When a unit is alerted for operations in a desert environment the commander must first answer some or all of the following questions:

- To what country is the unit going?
- What are the climatic and terrain conditions of that country?
- Will the unit be taking its own equipment overseas?
- What is the tentative timeline for departure? Will there be a period of time where the unit has time for individual training while the vehicles are in transit?
- What unit equipment is being sent overseas and what items will it require for modification (including camouflage painting)?
- What special equipment does the unit require for desert operations?
- What special maintenance is required for weapons and equipment before deployment to or arrival in a desert environment?
- Are there personnel in the unit who—

    - Have desert experience as observers or controllers?

    - Have any experience in desert conditions?
- Are all personnel physically fit?
- How many soldiers/marines are nondeployable?
- What types of operations are expected?

Once these and other questions have been answered the commander must develop a program to bring his unit to a level where it is fully capable of successfully operating in harsh desert conditions. To do this, first set a list of priorities for both individual and unit training. The training priorities listed below are shown as a guide only. They can be modified as necessary depending on the state of readiness of the unit when it is first alerted for desert employment.

Section II. Individual Training

In order to fight and survive in desert operations troops must fully understand the desert environment. The objective of individual training is to prepare the individual for operations in a desert environment. This requires both mental and physical preparation.

To the extent practicable, troops should be acclimatized before arrival in the area of operations. The requirement for acclimatization will vary slightly between individuals, but physical conditioning (fit soldiers/marines acclimatize more easily) is a part of the acclimatization process. Acclimatization should take place in climatic conditions that are similar, or slightly more strenuous, than those of the prospective area of operations.

CAMOUFLAGE AND CONCEALMENT

Camouflage and concealment training may be divided into concealment from the ground and concealment from the air. Particular attention must be paid to movement, color, shadow, and deception. Units should practice erecting and disassembling camouflage netting in order to become more efficient. Well-trained crews can save time and headaches. Camouflage and concealment are equally important for combat service support troops. Appendix E contains information about desert camouflage and concealment techniques.

DRIVER TRAINING

Because of the absence of established roads in desert areas, driving requires experience, individual skill, and physical endurance on the part of the vehicle operator. Driver training exercises should be long and arduous to expose vehicle operators to the rigors of the desert as well as to the effects of fatigue. The need for dispersing and avoiding preceding vehicles is stressed when operating over crusted surfaces or when the trail deteriorates while operating over sand (except suspected mine areas). Training should be directed toward driver proficiency in dune areas, choice of the best ground, selection of proper gear ratios, and driver knowledge and appreciation of the exact capabilities of his vehicle. Driver skill should be developed in taking maximum advantage of momentum, gear shifting, estimating and utilizing proper speeds, and avoiding sudden driving or braking thrusts. Additional driving techniques are contained in Appendix C.

SURVIVAL, EVASION, AND ESCAPE

Convincing a soldier/marine that he is capable of surviving in the desert environment strengthens his Self-confidence, and thus his morale. FM 21-76 contains details on survival, evasion, and escape in the desert, but the following points concerning FM 21 -76 should be included in desert survival, evasion, and escape training due to their importance:

- It is unlikely that wells will be poisoned. However, some wells in the North African desert have such strong concentrations of mineral salts that water taken from them may lead to intestinal irritation and subsequent illness.

- Although water is undoubtedly the most important factor in survival, a soldier/marine should not discard his personal weapon or any navigational equipment except in extreme circumstances. Mirrors of any type should be retained for signaling aircraft or other ground forces.

FIELD TRAINING

Following minimum preliminary training in garrison, desert living can only be practiced in the field, often as part of unit training. Important aspects that should be covered include:

- The effects of heat, including possible dehydration and salt loss (the need to maintain the body fluid level).
- The effects of temperature variations.
- The effects of cold weather in the desert.
- First aid for heat illnesses. Each soldier/marine should be issued a memory aid card showing symptoms and immediate treatment.
- Maintenance of morale and the ability of the individual to accept the challenge of the desert. Self discipline and common sense.
- Environmental effects such as those of sand, wind, and light.
- Water discipline.
- Hygiene and sanitation.
- Comet clothing equipment, including how to wear and maintain and clothing.
- Precautions against snakes, scorpions, insects, and disease-bearing organisms.

To the extent possible, the commander should train his unit in terrain and environmental conditions similar to what he expects to find in the operational area. It would be both shortsighted and dangerous for example, to allow water for bathing if the expected operational area is totally waterless. To further accustom the troops to hardships, contact with garrison or other urban areas should be kept to the minimum except for medical or welfare reasons. Once field training has started, necessary supplies should be brought to field locations and items that are unlikely to be available in the operational area (commercial soft drinks and foods) should not be permitted. To gain the maximum value from this training, the unit's exposure to outside influences should be kept to a minimum.

ENEMY ORGANIZATIONS AND TACTICS

This can be taught in garrison on sand tables and map maneuvers, followed by tactical exercises without troops (TINT), and unit exercises in the field. If enemy equipment is available it should be brought to the unit so it can be studied firsthand.

DESERT NAVIGATION

Troops must be thoroughly briefed on the type of terrain and the general environment they will encounter, including-

- Water sources, if any.

- Landmarks or significant permanent terrain features.
- Friendly and enemy areas of operation.
- Prevailing winds.
- Whether or not the local populace is pro or con the US.

This information will assist navigation by reconnaissance units or individuals who become separated from their units. Although maps are the most obvious navigation aids, numerous types of equipment and techniques are available to assist soldiers/marines during desert operations.

Maps

Although maps used in field training will be those of the local area, sufficient maps of the operational area should be obtained to allow distribution for study and possible use during garrison training. This is particularly important if the operational maps use foreign words to describe terrain, such as sebhka, summan, hidiba, and dikaka.

In addition, the grid system on some maps differs from the universal transverse mercator grid system on US maps. In many Middle East countries that were previously under British influence, for example, the Palestine grid system is used on military maps. These maps, generally last surveyed during World War II or the following decade, are widely used, not only in the area of Palestine, but also in Egypt and much of Saudi Arabia. And since they are commonly produced in either 1:100,000 or 1:50,000 scale, they do not mesh with standard US maps. In some instances, accurate maps may not be available. An alternative is to draw the grid lines on attached blank sheets of paper. This method can be highly effective when used in conjunction with navigational aids such as the GPS and LORAN.

Latitude and Longitude Conversions

One of the oldest systematic methods of location is based upon the geographic coordinate system. By drawing a set of east-west rings around the globe (parallel to the equator), and a set of north-south rings crossing the equator at right angles and converging at the poles, a network of reference lines is formed from which any point on the earth's surface can be located. The distance of a point north or south of the equator is known as its latitude. The rings around the earth parallel to the equator are called parallels of latitude or simply parallels. Lines of latitude run east-west but north-south distances are measured between them. A second set of rings around the globe at right angles to the lines of latitude and passing through the poles are known as meridians of longitude or simply meridians. One meridian is designated as the prime meridian. The prime meridian of the system we use runs through Greenwich, England, and is known as the Greenwich meridian. The distance east or west of a prime meridian to a point is known as its longitude. Lines of longitude (meridians) run north-south but east-west distances are measured between them.

Geographic coordinates are expressed in angular measurement. Each circle is divided into 360 degrees, each degree into 60 minutes and each minute into 60 seconds. At any point on the earth, the ground distance covered by one degree of longitude is about 111 kilometers (69 miles); one second is equal to about 30 meters (100 feet). The ground distance covered by one degree of longitude at the equator is also about 111 kilometers, but decreases as one moves north or south, until it becomes zero at the poles. For example, one second of longitude represents about 30 meters at the equator but at the latitude of

Washington, D. C, one second of longitude is approximately 24 meters.

Geographic coordinates appear on all standard military maps, and on some they may be the only method of locating and referencing the location of a point. The four lines that enclose the body of the map (neatlines) are lines of latitude and longitude. Their values are given in degrees and minutes at each of the four comers. In addition to the latitude and longitude given for the four corners, there are, at regular intervals along the sides of the map, small tick marks extending into the the body of the map. Each of these tick marks is identified by its latitude and longitude value. Different methods exist for converting longitude/ latitudes to the military grid system. Special equipment such as the global positioning systems have the capability to convert longitude and latitudes to grid coordinates, or this may be accomplished through manual means outlined in FM 21-26. For more information on map reading and land navigation techniques consult FM 21-26.

Navigation Aids

Navigation aids vary in sophistication and complexity and may include the following:

Sun Compasses/Sextants

These systems can be used on moving vehicles and require accurate timekeeping.

Lensatic Compasses

Individual compass error and local deviation must be known before using the lensatic compass. The lensatic compass cannot be used with any accuracy on dense steel vehicles such as tanks. A crew member should dismount to obtain an azimuth. It is unreliable near large quantities of metal, and can also be affected by underground mineral deposits. Power lines also adversely affect the lensatic compass.

Gyro Compass

An efficient gun azimuth stabilizer (a gyroscope) used on fairly flat ground is useful for maintaining direction.

Fires

Planned tracer fire assists in maintaining bearings, and field artillery and mortar concentrations, preferably smoke (or illumination at night), are useful checks on estimated locations.

Distance Recorders

It is essential to record distance moved, which may be done by using a vehicle odometer.

Beacons

These are particularly useful for aircraft navigation, but can also permit the enemy to locate friendly forces. It may be necessary to place them in open desert with unit locations being marked at certain distances and bearings from them.

Radars

Provided the position of a radar is known, it can measure range and bearings and, therefore, the position of a vehicle.

Aerial Photographs

The advantage of aerial photographs, particularly to aviators, is their ability to show up-to-date views of the variations in color and texture of the desert soil.

Global Positioning Systems

The GPS is a space-based, radio-positioning navigation system that provides accurate passive position, speed, distance, and bearing of other locations to suitably equipped users. The system assists the user in performing such missions as siting, surveying, tactical reconnaissance, artillery forward observing, close air support, general navigation, maneuver, and ground-based forward air control. It can be operated in all types of weather, day or night, anywhere in the world; it may also be used during nuclear, biological, and chemical warfare. It is important to remember these types of devices are aids to navigation; therefore, users should continuously plot their positions. In the event of a GPS failure, you can revert to more traditional navigation and position determination methods.

Position and Azimuth Determining System

The PADS is a highly mobile, self-contained, passive, all-weather, survey-accurate position/navigation instrument used by field artillery and air defense artillery units for fire support missions. The system provides roil-time, three-dimensional coordinates in meters, and a grid azimuth in roils. It also gives direction and altitude. The PADS can be used by the land navigator to assist in giving accurate azimuth and distance between locations. A unit requiring accurate information as to its present location can use PADS to provide this information.

Position Location Reporting System

The position location reporting system (PLRS)/joint tactical information distribution system (JTIDS), hybrid (PJH), is a computer-based system. It provides near real time, secure data communications, identification, navigation, position location, and automatic reporting to support the need of commanders for information on the location, identification, and movement of friendly forces. The PLRS is based on synchronized radio transmissions in a network of users controlled by a master station. The major elements of a PLRS community include the airborne, surface vehicular, and man-pack users; the PLRS master station; and an alternate master station. The system can handle 370 user units in a division-size deployment per master station with a typical location accuracy at 15 meters. The man-pack unit weighs 23 pounds and includes the basic user unit, user readout, antenna, backpack, and two batteries.

Dead Reckoning

The simplest system of navigation is known as dead reckoning. This is a means of finding where an individual is located by a continuous plotting of where he has been. More exactly, dead reckoning consists of recording and plotting a series of courses, each measured as to the distance and direction horn a known point, to provide a plot from which the position can be determined at any time. In the desert, the direction traveled is determined with a compass and the distance is measured by counting paces or reading the odometer of a vehicle. Detailed information on navigation by dead reckoning is contained in FM 21-26.

Shadow-Tip Method

This method provides orientation by reading the way the sun casts shadows. To use the shadow-tip method, find a fairly straight stick about 1 meter long and follow these steps:

- Step 1. Push the stick into the ground at a fairly level, brush-free spot where a distinct shadow will be cast. The stick need not be vertical; inclining it to obtain a more convenient shadow, in size or direction, does not impair the accuracy of the shadow-tip method.
- Step 2. Mark the tip of the shadow with a small peg, stick, stone, twig, your finger, hole in the sand, or other means. Wait until the shadow's tip moves a few inches (if you use a 1-meter stick, 10 to 15 minutes should be enough time),
- Step 3. Mark the new position of the shadow's tip.
- Step 4. Draw a straight line from the first mark to the second mark and extend it about a foot past the second mark.
- Step 5. Stand with the toe of the left foot at the first mark and the toe of the right foot past the line you drew.

You are now facing true north. Find other directions by recalling their relation to north. To mark directions on the ground (to orient others), draw a line at right angles to the first line, forming a cross and mark the directions.

If you cannot remember which foot to place on the first rock (see step 5), remember this basic rule for telling east from west: the sun rises in the east andsets in the west (but rarely due east or due west). The shadow's tip moves just the opposite. Therefore, the first shadow-tip mark is always in the west direction and the second mark in the east direction, everywhere on earth. Figure 2-1 depicts finding your direction by using the shadow-tip method.

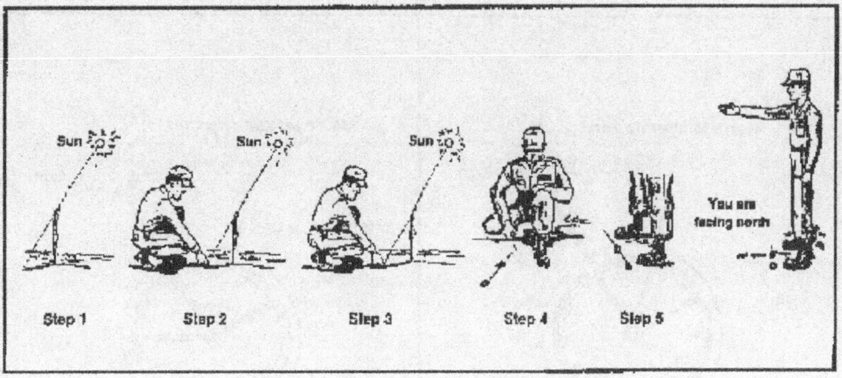

Figure 2-1.Finding direction Using shadow-up method.

Watch/Sun Method

An ordinary analog watch (with hands) can be used to determine the approximate true north in the North and South Temperate Zones. The North Temperate Zone is north of the equator and the South Temperate Zone is south of the equator. The temperate zones extend from latitude 23-1/2 degrees to 66-1/2 degrees in both hemispheres. In the North Temperate Zone only, the hour hand is pointed toward the sun (see Figure 2-2). A north-south line can be found midway between the hour hand and 12 o'clock. This applies to standard time. For daylight savings time, the north-south line is found midway between the hour hand and 1 o'clock. If there is any doubt as to which end of the line is north, remember that the sun is in the north, and remember that the sun is in the eastern part of the sky before noon and in the western part in the afternoon.

The watch may also be used to determine direction in the South Temperate Zone (see Figure 2-2). However, it is used a bit differently. Twelve o'clock is pointed toward the sun, and the north-south line will be halfway between 12 o'clock and the hour hand. If on daylight savings time, the north-south line lies midway between the hour hand and 1 o'clock.

On cloudy days, place a stick at the center of the watch and hold it so that the shadow of the stick falls along the hour hand in the North Temperate Zone. In the South Temperate Zone, the shadow falls along a line from the center of the watch through 12 o'clock. Direction is then determined using the appropriate technique.

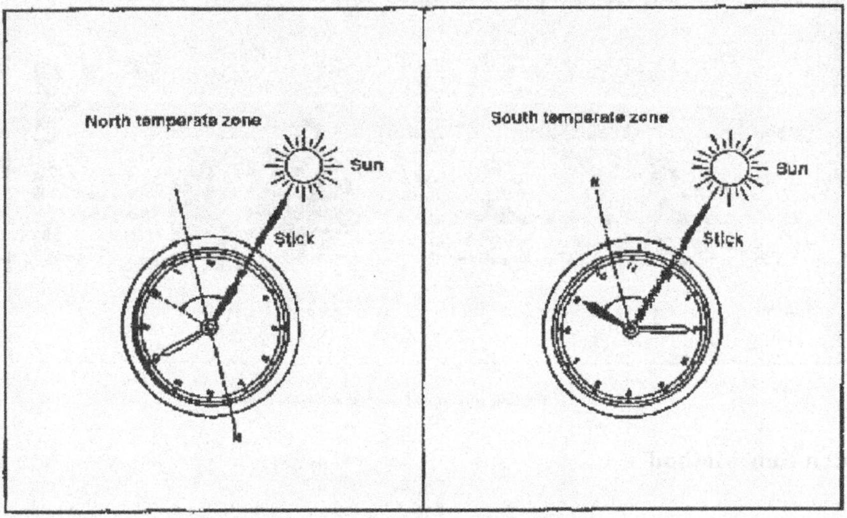

Figure 2-2. Using a watch to find north.

Celestial Navigation Method

North of the equator, locating your direction at night can be determined by locating the North Star. To find the North Star, look for the Big Dipper. The two stars at the end of the bowl are called pointers. In a straight line out from the pointers is the North Star (at about five times the distance between the pointer stars). The Big Dipper rotates slowly around the North Star and does not always appear in the same position (see Figure 2-3).

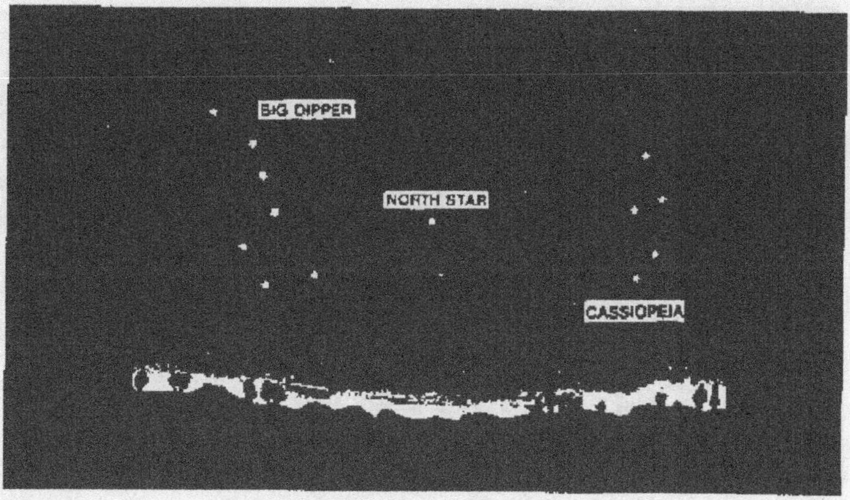

Figure 2-3. *Big* Dipper.

The constellation Cassiopeia can also be used. This group of five bright stars is shaped like a lopsided M (or W when it is low in the sky). The North Star is straight out from the center star, about the same distance as from the Big Dipper. Cassiopeia also rotates slowly around the North Star and is always almost directly opposite the Big Dipper. Its position, opposite the Big Dipper, makes it a valuable aid when the Big Dipper is low in the sky, possibly out of sight because of vegetation or high terrain features.

South of the equator, the constellation Southern Cross will help you locate the general direction of the south and, from this base, any other direction. This group of four bright stars is shaped like a cross that is tilted to one side. The two stars forming the long axis, or stern, of the cross are called the pointers. From the foot of the cross, extend the stem five times its length to an imaginary point (see Figure 2-4). This point is the general direction of south. From this point, look straight down to the horizon and select a landmark.

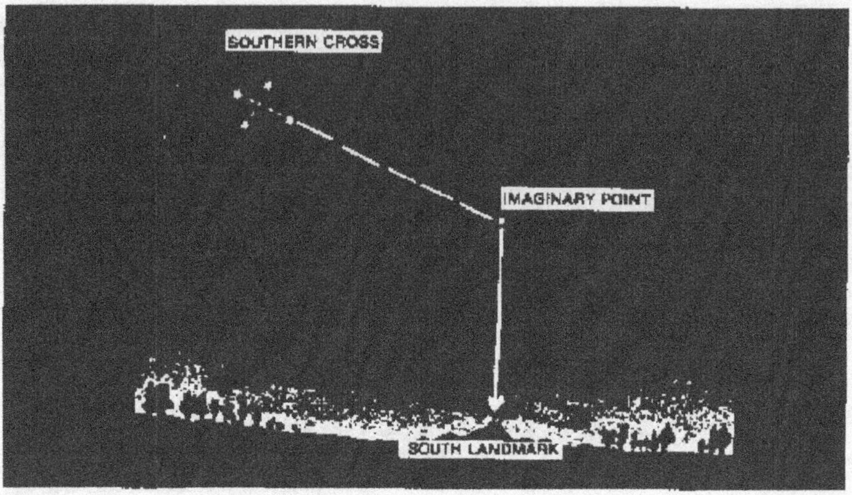

Figure 2-4. Southern Cross.

OPERATIONAL AREA (HOST COUNTRY)

A description of the host country should cover only those facts that apply to forthcoming operations, for example:

- Geographic description.
- Climate (throughout the year).
- Religious issues and constraints.
- Cultural differences and special considerations, important customs, and the behavior expected of US military personnel. (Such as speaking to a woman in some Arab countries, which can be offensive to the local inhabitants.)
- Population density.
- Industry and agriculture.
- Language(s) (phrase books may be issued).
- Communications and transportation network.
- The armed forces (and possibly police), including organization, equipment, and rank structure.
- The situation that has led to the introduction of US forces and reasons why US forces are being introduced. No soldier/marine should have to question why he is fighting for a country other than his own, if this is the case.

Treatment of these subjects will vary in degree according to category. Personnel who may require additional information such as the country's history, can find it in the appropriate DA 550-series pamphlet.

DESERT MANEUVER

Chapter 3 describes the influence of the desert environment on tactical operations. This subject should first be taught to a limited number of leaders and commanders as a theoretical subject, down to platoon level. Leaders should then train their units during unit training. The emphasis should be on small unit tactics, including combined arms operations. Additional subject matter that should be covered includes -

- Terrain in the operational area, emphasizing differences and similarities with the training areas the units will use.
- Application of concealment, using terrain and artificial means such as smoke, and the application of maneuver techniques.
- Mobility in the desert.
- Command and control techniques for desert operations.
- Navigation.
- Conduct of fire in desert operations.
- Resupply during desert operations.
- Special equipment techniques.

GENERAL TRAINING

Nearly all equipment will be affected in one way or another by the environment as described in Chapter L The purpose of this training is to train operators. Training should include-

- Likely effects on the equipment they operate.
- Efficient operations of the equipment within the limits imposed by the environment, including tactical limitations of the equipment, for example, helicopters may have difficulties flying NOE; and radios will normally be operated on reduced output due to the environment and enemy ECM.
- Preventive maintenance-employing any special techniques required by the desert environment. The appropriate equipment technical manual or lubrication order provides specific information concerning hot climate operations and maintenance.
- Basic desert recovery and repair techniques, including defensive measures, and camouflage required during recovery and repair operations.

Instruction must be oriented toward the expected operational area. For example, it is possible to keep radios cool by using ice packs, but if ice packs are not going to be available in the area of operations, then do not teach troops this technique as it will not be practiced.

SPECIAL MAINTENANCE AND SUPPLY TECHNIQUES FOR STAFF AND LEADERS

Special maintenance techniques that need to be addressed are the same as those taught to specialists, however, they only need to emphasize aspects that ordinarily require control or supervision, or affect the employment of equipment in desert terrain. This training should include any special handling techniques required in the operational area. The importance and difficulties of supply in desert operations are described in Chapter 4. Training should be modified according to-

- Modified tables of organization and equipment (MTOE) and mission of the unit.
- Supply situation expected in the area of operations.
- Capabilities of logistic units likely to support unit operations with special attention given to units not normally found in conventional operations, for example, well-drilling teams, refrigeration assets available in the theater, and transportation cargo carrier companies.

NUCLEAR, BIOLOGICAL, AND CHEMICAL (NBC) TRAINING

Wearing protective clothing, flak vests and masks in the desert environment will make a person extremely uncomfortable. Troops should not participate in strenuous activity while wearing protective clothing until they are acclimatized. Training in MOPP gear should become progressively more strenuous. Physical training sessions in field protective masks progressing to foot marches in MOPP 4 is a recommendation. Use of protective clothing in severe desert heat is described in Chapter 1 and Appendix D. Points that should be emphasized during training are-

- The value of being uncomfortable rather than dead.
- The need to avoid heat illness by¬

  - Reducing the labor rate to the minimum, and delaying work until cooler hours.

  - Maintaining proper body water and salt levels, particularly during a time of chemical threat.

  - Detecting the first symptoms of heat illness in others by constant vigilance.

  - Increasing the time factor of an operation as troops will move slowly when wearing protective clothing.

DESERT TERRAIN APPRECIATION

When training soldiers/marines to appreciate desert terrain, leaders should focus on the effects of the different types of desert terrain on the capabilities and limitations of unit equipment. Highlight the impact of the terrain on vehicular trafficability, fields of fire, and observation in the likely operational area. When possible, crews and small unit leaders should learn to appreciate desert terrain from practical experience in terrain as nearly similar as possible to that in the likely combat zone.

MEDICAL TRAINING CONSIDERATIONS

The unit surgeon can provide valuable information on the medical implications of operations in the desert environment. He can advise the unit commanders on measures to take to ensure training includes preventive medicine concepts essential to keeping nonbattle injuries to a minimum. Bum casualties should also be a medical training consideration as these will be the most likely casualties in a mechanized environment. Hydration and mouth-to-mouth resuscitation of injured personnel in a field protective mask should also be a training consideration. Nonbattle casualties, due to a lack of consideration of preventive medicine concepts, can far outnumber combat casualties.

Section III. Unit Training

When determining unit training requirements, the commander must first consider the training level of his unit when alerted for deployment. When time does not permit a comprehensive training program, the commander must concentrate on those areas where his unit is least proficient, considering the priorities previously described. In order to operate in the desert environment, the unit must, above all, be physically fit, so physical conditioning is of paramount importance.

COMBINED ARMS AND SERVICES TEAM

The greatest combat power results when leaders synchronize combat, combat support, and combat service support systems to complement and reinforce one another. The slice concept refers to CS and CSS units task organized to support a particular maneuver or combined arms unit. Leaders should routinely practice habitual relationship and cross attachment of units. Training as combined arms teams is critical to successful desert operations.

PHYSICAL CONDITIONING AND ACCLIMATIZATION

To the extent possible, physical conditioning and acclimatization should take place simultaneously. When a unit is training in a hot environment, begin physical training at night or during the cooler part of the day and work up to rigorous training, such as foot marches in open sand terrain at midday. Emphasis on mounted operations in desert warfare does not imply that foot marching can be totally disregarded. Physical conditioning must be continued after arrival in the area of operations.

Medical advice should always be available during periods of physical training in hot weather. Training several days prior to deployment to combat the effects ofjet lag works well, if performed as a command directive. Avoiding coffee and alcohol, and drinking plenty of water will also assist in overcoming the effects ofjet lag. Planning for the change in time zones and developing a sleep plan are recommended and will facilitate readiness.

WEAPONS TRAINING

Soldiers/marines must train to proficiency at all ranges but accuracy at maximum effective range both in daytime and nighttime must be emphasized. Firing should also be practiced during the heat of day to condition troops to heat haze and mirages. Emphasis must also be placed on maintenance of individual weapons in view of sandy desert conditions.

NBC TRAINING

The purpose of unit NBC training is to train individual soldiers/ marines to become proficient as a team while wearing protective clothing and masks, and when in combat vehicles-while buttoned up. This training should be conducted both in daytime and nighttime, until the unit can operate effectively under these conditions. Additional information and guidance on the effects of the environment on NBC weapons are contained in Appendix D.

MARCH DISCIPLINE

Although of particular importance to combat support and combat service support units, all units should be trained in tactical road marches. Training should emphasize-

- Off-road movement over open terrain. Irregular spacing when moving in convoy.
- The need to maintain sufficient distances between vehicles to preclude "dust blindness."
- Actions to be taken when stuck in sand, and when a vehicle breaks down.
- Vehicle camouflage.
- Removal of tracks which would reveal friendly locations.
- The need for dispersion when halted.
- Air defense drills.
- Practicing incoming artillery drills.

OBSTACLES AND BARRIERS

In some desert areas, natural obstacles such as wadis or other terrain features can be found. Often, however, it will be necessary to use artificial obstacles if enemy movement must be slowed. A minefield, to be of any tactical value in the desert, must usually cover a relatively large area, so mechanical means and engineer support are required. Since there are often too many avenues of approach to be covered with mines, it is usually best to employ mines to cover any gaps between units, especially at night. Minefields are most effective when they can be covered by observation or fire. During unit training, soldiers/marines should be trained to lay mines wearing gloves, since human scent attracts desert animals who may attempt to dig them up. Emphasis should be on antitank minefield since combat vehicles are the most dangerous threat.

SCOUTING, SURVEILLANCE, AND PATROLLING

Effects of the environment on scouting, observation, and surveillance techniques are described in Chapter 3. Effects of the environment on surveillance, target acquisition, and night observation devices are described in Chapter 1.

ADJUSTMENT AND CONDUCT OF FIRE

The principles for adjustment and conduct of fire in the desert are the same as for operations in more temperate climates. However, the following considerations, somewhat peculiar to desert operations, should be kept in mind:

- Obscuration from sand, dust, smoke, or a combination of these, can affect direct-fire adjustment.
- There may be major inaccuracies of initial rounds from indirect-fire weapons due to misjudgment of target location.
- The target may be concealed by sand or dust if rounds land short, on, or near the observer target line.
- Heat haze and mirages can mislead gunners and observers as to target location. This condition can particularly affect antitank guided missile gunners.

Direct-fire gunners may have to depend on flank observers, who may be any individual on the battlefield equipped with a radio. If observation is lost, subsequent corrections are very unlikely to cause a second round hit. The following considerations can help to overcome the obscuration or sending problem:

- An observer requesting indirect fires needs to ensure that initial rounds land beyond the target to preclude short rounds obscuring the target, and then adjust accordingly.
- By remembering the greatest impact of heat haze (which varies throughout the day) is on ATGM gunners-when both gunner and target are within 2-3 feet of the desert surface.

AIR DEFENSE

In desert operations any type of unit, be it tank, infantry, trains, tactical operations centers, or supply points, can expect to be a target for air attack. Air attacks may be from fighter bomber aircraft using cannons, missiles, bombs, napalm, and machine guns, or from attack helicopters using machine guns, rockets, or missiles.

Enemy air superiority should be assumed during all field training and simulated fighter bomber attacks and attack helicopter missions should be flown against the unit whenever possible. When practical, aerial photographs of positions should be taken, and pilots interviewed to assist in the critique of air defense, both passive and active. Points that should be emphasized during training are mentioned in the following paragraphs.

Passive air defense measures should be taken routinely. When stopped for any period of time, take every advantage of whatever cover and concealment are available. As previously described, natural cover and concealment will be difficult to find in many desert areas. Nevertheless, vehicles, particularly unarmored vehicles, should be irregularly dispersed and dug-in, or revetments provided. When appropriate, air guards, trained in aircraft recognition, should be posted, with clear instructions on actions to be taken when aircraft are sighted. Units not being attacked by aircraft, but in close proximity to the attack, may desire to remain stationary in order to avoid detection. Reducing infrared signatures is also a passive anti-air consideration. Artificial camouflage can be used as described in Appendix E.

Active air defense techniques used in desert operations are the same as those described in other doctrinal manuals, appropriate to the level of command. However, at small unit level, additional emphasis should be given to air defense using small arms. When combat vehicles on the move are engaged by enemy aircraft, their immediate action will depend on whether or not they are maneuvering in contact with the enemy. If they are in contact they should continue to maneuver, relying on overwatch elements and air defense artillery to engage attacking aircraft.

Vehicles about to be engaged by enemy aircraft in open desert where cover is not available, should move perpendicular to the attacking aircraft to evade rocket or machine gun fire. Engage the aircraft with small arms fires, if possible. Meanwhile, the remainder of the unit should mass small arms fire to the aircraft's front. Sudden variations in course may also distract the pilot.

COMMUNICATIONS

Good communications in desert terrain will often depend on the state of mind of the operators. They must be enthusiastic, persistent, and determined to make and maintain contact. Unit training should concentrate on ECCM techniques. When conducting field training, higher headquarters can provide assistance in the form of small teams to jam unit nets. Practice actions to be taken when radio contact is lost due to heat (described in Appendix C).

FRATRICIDE CONSIDERATIONS

The following are fratricide considerations when operating in a desert environment and may be topics for fratricide awareness training:

- Because of the absence of easily identifiable terrain features in the desert, knowing your exact location can be especially important. Be in the right place at the right time. Use position location/navigation (GPS) devices; know your location and locations of adjacent units (left, right, leading and follow on). Synchronize tactical movement.
- Ensure positive target identification. Review vehicle/weapons ID cards; know at what ranges and under what conditions positive ID of friendly vehicles/weapons is possible. This is especially important in the desert due to the likelihood of weapons being able to fire at their maximum ranges.
- Maintain situational awareness-be aware of current intelligence, unit locations/dispositions, denial areas (minefields/FASCAM), contaminated areas (e.g., ICM and NBC), SITREPs, and METT-T. This can be more difficult in desert environments because navigation is more difficult.
- Conduct individual and collective (unit) fratricide awareness training. This includes target identification/recognition; fire discipline; and leader training.
- Use common language/vocabulary. Use doctrinally correct, standard terminology and control measures (e.g., fire support coordination line (FSCL), zone of engagement, and restrictive fire line (RFL)).
- Consider the effects of key elements of terrain analysis on fratricide. These include observation and fields of fires, cover and concealment, obstacles and movement, key terrain, and avenues of approach.
- Gun tube orientation can also assist in avoiding/preventing fratricide.

CHAPTER 3

OPERATIONS IN DESERT CONDITIONS

In the two previous chapters the environment and its effects on personnel and equipment and preparations for desert operations were described. This chapter describes desert operations and is divided into three sections.

Section I. How the Desert Environment Affects Tactical Operations

The key to success in desert operations is mobility, and this was clearly evident in ground operations in Desert Storm. The tactics employed to achieve victory over Iraq were wide and rapid flanking movements similar to those Rommel and Montgomery executed in North Africa.

Trafficability and cross-country movement become critical to desert operations when using these tactics. Trafficability is generally good in the desert and cross-country movement is a lesser problem, but not always. Salt marshes can create NO-GO conditions during the rainy season. Sand can also bog down traffic and make foot movement slow and exhausting. The steep slopes of dunes and rocky mountains can make vehicular movement a NO-GO. The wadis create cross-compartmented terrain. The banks of these stream beds can be steep and unconsolidated. Then, when it rains, it becomes a torrent of dangerously rushing water, leading to flat lake beds that can create NO-GO mud conditions. Rock quarries and mining areas can also adversely affect mobility and trafficability. Often these areas are not reflected on maps. Satellite imagery can be helpful in identifying these areas, as was the case in Kuwait during Operation Desert Storm. In rocky terrain, tires can easily be punctured by sharp angular debris; however, overall movement is mostly uninhibited. Given ample fuel, water, and other resources, units can go around natural and man-made obstacles.

Movement can easily be detected because of sand and dust signatures left due to the loose surface material. In an actual engagement, this may not be all that bad because a unit is obscured from direct fire while advancing, but the element of surprise may be lost. Moving at night becomes the logical choice. The dust is still there, and vehicles (which should be widely spaced) can get separated. But at night, reflection of the sun's rays from glass, mirrors, or metal, which can give away movement and positions up to 20 kilometers away, is not a concern.

Using the ability to make fast and wide flanking movements, a unit can encircle and cut off enemy forces. The Israeli forces under General Ariel Sharon did just that to the Egyptian Third Army in the 1973 War, and the British did the same to the Italians in North Africa in January 1941. In Desert Storm, the night-fighting AH-64 helicopters, combined with field artillery fires, made for an unbeatable team in this regard.

Land navigation is a challenge during movement in the wide expanses of many arid regions. There are few landmarks to key on, and maps and even photos can become dated quickly, especially in the sandy deserts where dunes migrate. The global positioning system (GPS) with the small lightweight GPS receivers (SLGRs) is a major aid for desert operations.

Refuel and resupply operations require periods in which forces assume the defense, but only temporarily. The flat sandy desert topography that is characteristic of Saudi Arabia is not conducive to defense, compared to rocky plateau topography. In mountains and canyons, a defensive posture can be favorable. Controlling the passes, as mentioned earlier, can essentially close off vast areas to an attacker and make it extremely costly for him.

While a unit is in the defense, it needs both ground and air reconnaissance to detect enemy movement at long range. Obstacles must be placed in all types of topography, primarily to slow advances and channel columns. Neglecting these security measures in the flat sandy regions can lead to disaster.

MILITARY ASPECTS OF THE TERRAIN

The following paragraphs describe how terrain affects tactical operations in the desert. This discussion follows the outline of the terrain analysis process summarized by the factors of OCOKA.

Observation and Fields of Fire

Observation and fields of fire are generally excellent in most desert areas. The atmosphere is stable and dry, allowing unrestricted view over vast distances, but this can also be a problem. Range estimation by "gut feeling" is subject to error. The effective ranges of weapons can easily be reached, and a correct estimation of maximum ranges is critical for all weapons, especially for wire-guided munitions.

Flat desert terrain permits direct-fire weapons to be used to their maximum range. Open terrain and a predominantly clear atmosphere generally offer excellent long-range visibility; but at certain times of the day visibility may be limited or distorted by heat.

Two primary considerations in the desert environment are longer range observation, and fields of fire at the maximum effective ranges for weapons. However, rapid heating and cooling of the atmosphere hinder these factors and cause distortion of ranges to the aided and unaided eye. Mechanical and electronic means must be used to verify estimated ranges such as GSR and laser range finders. Boresight and zero more frequently at standard ranges.

The desert is not absolutely flat, so weapons are sited to provide mutual support. Dead space is a problem. Even though the landscape appears flat, upon closer inspection it can be undulating with relatively deep wadis and depressions. These areas must be covered by indirect fire.

When on the offense, attacks should be initiated with the sun at or near one's back whenever possible. This eliminates most shadows that degrade optical weapon guidance and makes visual target acquisition difficult.

When there is no usable dominant terrain available, the only means of observation may be from an aeroscout, or limited to short-range observation by the vehicle commander. Other visibility problems are caused by heat distortion. Heat waves at the desert surface cause images to shimmer making positive identification difficult and degrade depth perception. Ranges to targets may be distorted from heat rising from the desert surface. Use range finders to verify correct distances. Be prepared to use bracketing techniques with large adjustments to hit an enemy target with artillery.

Radars are unlikely to be affected by heat haze so they could be valuable on flat terrain during midday heat if optical vision is hopelessly distorted; however, they arc almost useless in sandstorms. Image intensification is of limited value in sandstorms, and depends on the phase of the moon at night. If there is no moon, use artificial illumination outside the field of view of the system.

Since thermal imagery devices depend on the difference between ambient temperature and equipment temperature, they are more useful at night than in the day. Because of the distinct advantages of thermal sights, they should be used as the primary sighting systems for vehicles so equipped.

Correction of field artillery fires, especially those of larger pieces, may be complicated by dust hanging in the air following the impact of ranging rounds. Forward observers should consider placing initial rounds beyond a target rather than short of the target. Observation of fires, especially direct fires by tanks, may be difficult due to dust clouds, so wingmen may have to observe direct fires.

Cover and Concealment

Cover and concealment are generally scarce in the desert. The flat sandy deserts provide little if any natural cover or concealment, especially from aerial attack or reconnaissance. Ground concealment and protection from fire can be found behind dunes or in wadis. Troops must be aware of the potential for flash floods when using wadis for ground concealment.

Some arid regions have vegetation that can provide limited concealment from ground observation. In rocky, mountainous deserts, cover and concealment are best found behind boulders and in crevices. Daytime vehicular movement eliminates nearly any possibility of surprise, as dust trails created by the traffic can be spotted for miles. At night noise and light discipline is critical, as both sound and light travel great distances because of the unobstructed flatness of the terrain and atmospheric stability. Camouflage can be effectively employed to improve on natural cover and concealment. See Appendix E for additional information on concealment and camouflage.

Obstacles

Natural obstacles do exist in the desert, and arid regions are well suited for man-made obstacles. The wadis and steep slopes of escarpments, mountains, hills, and dunes hinder cross-country movement. Sand dunes may stretch for miles and prevent direct movement across their length. These sand dunes are often more than 100 feet in elevation and consist of loose sand with high, steep downwind faces that make vehicular traversing next to impossible. Aerial reconnaissance immediately before any large movement is advisable because sand dunes migrate with shifting winds and they may not be where maps or even photographs show them.

In the Desert Storm area, the salt marshes have a crust on the top that can deceive a vehicle driver. These dry lake beds can become obstacles, especially in the wetter seasons when the water table is higher. A top crust forms on the surface, but below the crust the soil is moist, similar to marsh conditions. The surface may look like it has good trafficability, but the crust will collapse with the weight of a vehicle, and the vehicle becomes mired. The high premium on fuel and time makes it costly to go around these natural obstacles.

Sandy deserts are ideal for employing minefield. Although windstorms can reveal previously buried mines, these mines can still channel movement and deny access to certain areas. The battles of the Bi'R Hacheim Line and El Alamein were influenced by minefield. Other obstacles include ditches, revetments, and barriers, such as the Bar Lev Line along the Suez Canal, made by bulldozing sand mounds or by blasting in rocky mountainous areas to close passes.

Key Terrain

Key terrain in the desert can be any man-made feature, mountain pass, or source of water, and of course, high ground. Because there are few man-made features in the desert, those that do exist can become important, perhaps even key.

Passes through steep topography are also likely to be key, again because they are so few. The North African campaigns of World War II focused on the control of passes, specifically the Solium and Halfaya. In the Sinai Wars between Egypt and Israel, the Mitla, Giddi, and Sudar passes were key. In Afghanistan, control of the mountain passes provided the Mujahideen safe haven from the Soviets. Oases, where wells exist, become important for water resupply. The high ground in desert terrain is usually key terrain. The relative flatness and great distances of some deserts, such as in Iraq, make even large sand dunes dominant features.

Avenues of Approach

Avenues of approach are not clearly defined in arid regions. The vast, relatively flat areas permit maneuver from virtually any direction. This point became obvious to units establishing defensive positions in Desert Storm. Wide envelopments are possible, as demonstrated in the Desert Storm ground campaign. Modem sensor technology, limited natural concealment, and improved observation make the element of surprise a challenge. Yet, surprise was achieved during Desert Storm-Iraqi commanders were shocked when they discovered US tanks in their perimeters.

The major limitation with respect to avenues of approach may be fuel. The great distances a unit must travel to outflank enemy positions require significant amounts of fuel and complicate resupply. In mountainous and canyon topography avenues are much more limited, and the wadis and valleys are likely to be the only possible access routes. Any roads that do exist are probably in the valleys. Nevertheless, none of the considerations outlined above are reasons to disregard flanking movements.

MANEUVER

Army operations are ideally suited to desert environments. Its thrust of securing and retaining the initiative can be optimized in the open terrain associated with the desert environments of the world. In that environment, the terrain aspect of METT-T offers the potential to capitalize on the four basic tenets of the doctrine initiative, agility, depth, and synchronization.

Initiative

Israeli efforts in 1967 and initial Egyptian assaults in 1973 clearly illustrate the effects of initiative in the desert environment.

Agility

The Egyptian success in 1973 was negated by their failure to ensure agility. Conversely, the Israeli actions on the flanks of the Egyptian force demonstrated the effects of a force capable of rapid and bold maneuver.

Depth

Depth does not necessarily relate to distance. In the nonlinear battlefield offered by the desert, depth often equates to an agile reserve force of sufficient size to counter enemy efforts into flanks and rear areas. Depth is also a concept of all-around defense for forces-the ability to fight in any direction.

Synchronization

To a large measure, the German successes against the British in the Western Desert were due to their ability to synchronize their operating systems. More recent events illustrate this tenet between and internal to, operating systems. Heavy/light operations have demonstrated that light forces can be key to achieving tactical and operational momentum. The Israeli airmobile assault against supporting artillery in the 1967 battle of Abu Ageila is a good example of the effective use of light forces in this type of environment.

Maneuver must be at the maximum tactical speed permitted by the terrain, dust conditions, and rate of march of the slowest vehicle, using whatever cover is available. Even a 10-foot sand dune will cover and conceal a combat vehicle. Air defense coverage is always necessary as aircraft can spot movement very easily due to accompanying dust clouds. In some situations movement may be slowed to reduce dust signatures. Rapid movement causes dramatic dust signatures and can reveal tactical movements.

Another consideration during maneuver is dust from NOE flight, which can be seen as far as 30 kilometers. This is especially true when the enemy is stationary. Aeroscouts must use caution to avoid blundering into enemy air defense weapons.

To achieve surprise, maneuver in conditions that preclude observation, such as at night, behind smoke, or during sandstorms. In certain circumstances, there may be no alternative to maneuvering in terrain where the enemy has long-range observation. Then it is necessary to move at the best speed possible while indirect fires are placed on suspected enemy positions. Speed, suppressive fires, close air support, and every other available combat multiplier must be brought to bear on the enemy.

Tactical mobility is the key to successful desert operations. Most deserts permit good to excellent movement by ground troops similar to that of a naval task force at sea. Use of natural obstacles may permit a force to establish a defensive position that theoretically cannot be turned from either flank; however, these are rare. Desert terrain facilitates bypassing enemy positions and obstacles, but detailed reconnaissance must be performed first to determine if bypassing is feasible and will provide an advantage to friendly forces.

Dismounted infantry may be used to clear passes and defiles to eliminate enemy ATGM positions prior to the mounted elements moving through.

Avenues of approach of large forces may be constrained due to limited cross-country capability of supply vehicles coupled with longer lines of communications. The limited hard-surface routes that do exist are necessary for resupply.

RECONNAISSANCE

Reconnaissance is especially important in desert environments. Reconnaissance is a mission undertaken to obtain information by visual observation, or other detection methods, about the activities and resources of an enemy, or about the meteorologic, hydrographic, or geographic characteristics of a particular area. The desert environment may influence any or all of these techniques. The environmental effects on troops and their equipment may also influence observation techniques, or the frequency of vehicle and equipment maintenance that is required. Reconnaissance produces combat information. Combat information is a by-product of all operations, acquired as they are in progress. Reconnaissance, however, is a focused collection effort. It is performed prior to or in advance of other combat operations, as well as during that operation, to provide information used by the commander to confirm or modify his concept. Cavalry is the Army corps or division commander's principal reconnaissance organization.

Surveillance is a primary task of Army cavalry during reconnaissance operations. Surveillance is the systematic observation of airspace or surface areas by visual, aural, electronic. photographic, or other means. Scouts, ground and air, are the principal collectors of information. Scouts and their associated equipment are particularly affected by the environmental aspects of deserts. They require equipment that enhances their senses allowing them to conduct mounted and dismounted surveillance with stealth, at long-range, and in limited visibility, all of which can be adversely influenced by the desert environment.

SECURITY

Security operations obtain information about the enemy and provide reaction time, maneuver space, and protection to the main body. Security operations are characterized by aggressive reconnaissance to reduce terrain and enemy unknowns, gaining and maintaining contact with the enemy to ensure continuous information, and providing early and accurate reporting of information to the protected force. Security operations may be affected by various aspects of the desert environment including the sun, wind, sand, vegetation, sandstorms, terrain, and heat. Security operations include-

- Screen.
- Guard.
- Cover.

Counterreconnaissance is an inherent task in all security operations. Counterreconnaissance is the sum of actions taken at all echelons to counter enemy reconnaissance and surveillance efforts through the depth of the area of operation. It is active and passive and includes combat action to destroy or repel enemy reconnaissance elements. It also denies the enemy information about friendly units.

COMMAND, CONTROL, AND COMMUNICATIONS

The following paragraphs describe command, control, and communications considerations when operating in a desert environment.

Command

The effort to synchronize battlefield operating systems during the planning process can be negated by the failure to continue the synchronization effort during the preparation phase of a mission. This is especially true in the construction of engagement areas for defensive operations. Direct fire, indirect fire, and obstacles are linked, and the adjustment of one requires the adjustment of all. The commander must know and have a feel for what his unit can do, how long his unit takes to accomplish a mission, and what he really wants his unit to accomplish.

Adjustment of the elements of the battlefield operating systems can unravel the focus of a commander's intent. This is especially true in open terrain. Tactical commanders should personally direct the synchronization of engagement areas. Obstacles should be positioned, indirect fires adjusted, and direct fires rehearsed under the personal supervision of the commander. The commander controls operations using a highly mobile command group located well forward. He personally directs the battle, but must not be drawn into personally commanding an isolated segment of the force to the detriment of the remainder of the command. As previously mentioned, dry desert conditions can sometimes reduce radio signal strength and create unforeseen blind spots, even in aircraft operating nap of the earth.

Units may employ either a jump TOC or retransmission stations to facilitate communications with rear areas, as maneuver units are unlikely to be in one place very long. (If wire is used it should be buried to a minimum depth of 12 inches to avoid damage from track vehicles or shell fire.) There must be plenty of slack in the line to allow for sand shift and accurate map plots of buried wire should be kept. If overhead wire must be used, it should be mounted on posts erected in the form of tripods to avoid falling during severe weather.

Air or vehicle mounted liaison officers can be used if units are stationary or under listening silence. They should be proficient in navigation and sufficiently equipped to facilitate parallel planning. Liaison officers are highly effective and should be employed at every opportunity.

Continuous Operations

Continuous operations are affected by a number of factors in a desert environment. Fatigue is probably the foremost degrader of performance. Performance and efficiency begin to deteriorate after 14 to 18 hours of continuous work and reach a low point after 22 to 24 hours. Most tasks involving perceptual skills begin to show a performance degradation after 36 to 48 hours without sleep. Soldiers/marines cease to be effective after 72 hours without sleep. Performance decreases dramatically in an NBC environment and sleep becomes more difficult in MOPP gear. Sleep deprivation coupled with the environmental factors of the desert and the stresses of combat can significantly affect mission accomplishment.

The two categories of personnel who can be expected to show signs of fatigue first are young immature soldiers/marines who are not sure of themselves and seasoned old soldiers/marines upon whom others have relied and who have sustained them at cost to themselves. Commanders and leaders often regard themselves as being the least vulnerable to fatigue. Tasks requiring quick reaction, complex reasoning, and detailed planning make leaders the most vulnerable to sleep deprivation. Leaders denying themselves sleep as an example of self-control is extremely counterproductive. These factors are complicated by the environmental aspects of desert operations and should be considerations for operational planning.

Control

Clear identification of engagement areas is necessary to facilitate the massing and distribution of fires. In the absence of identifiable terrain, target reference points (TRPs) can be created with damaged/destroyed vehicles that are moved into required locations at the direction of commanders invested with the responsibility for specific engagement areas. Other types of TRPs could be used. For example, marker panels, visible and infrared chemical lights, flags, and white phosphorus/illumination rounds could be used. The construction or fabrication of TRPs must be resourced and well planned in order to be effective. For example, how will TRPs be replaced for subsequent defensive operations? Another common problem is TRP proliferation, which makes TRPs difficult to identify when each echelon of command has allocated too many TRPs.

Pyrotechnics are usually more effective in desert climates than in temperate climates; however, heat mirages and duststorms may impair or restrict their use. Even heliographs (signal mirrors) may be useful as they are directional and therefore can aid security. Sound communications are usually impractical due to distance, vehicular noise, and storms, but can be used for local alarm systems.

Colored flags with prearranged meanings can be used as a means of communication in the flat open terrain of the desert. Colored flags tied to antennas may also assist in vehicle/unit recognition during limited visibility operations and offensive operations.

As previously described, the desert offers excellent fields of fire. Tanks and heavy antitank weapons should be sited to take advantage of their long range and accuracy. Firing first and accurately are the most important considerations in desert operations.

Target identification is the recognition of a potential military target as being a particular object (such as a specific vehicle, by type). At a minimum, this identification must determine the potential target as friendly or threat (identify friend, foe, or neutral [IFFN]). Because it is easy to become disoriented, it is often necessary to mark sectors of fire on the ground with poles or rocks, if available.

Communications

Communications support is also adversely affected by high temperatures. The heat causes anomalies in radio and other electrical transmissions, and radio battery life is reduced. Radio range is shorter during the day than at night. At night, range improves but static electricity may cause interference. FM communications range can be reduced by as much as 50 percent because of high temperatures. HF ground wave propagation over the dry sandy soil is reduced.

Night communications make communications security a concern, as it always should be. Experience in Desert Shield and Desert Storm indicates vastly expanded ranges of FM radios. Communications between units 40 to 50 kilometers apart was not unusual. Communications obviously affect command and control as well as intelligence collection and dissemination, and their importance must not be underestimated.

COMBAT SUPPORT

A force operating in the desert must be a balanced force with combat support and combat service support-it must be a combined arms team. While principles of combat support operations are found in doctrinal manuals dealing with a specific arm of service, there are some techniques that must be modified or emphasized in the desert.

INTELLIGENCE

The relative importance of intelligence sources may vary from that expected in more conventional areas. Enemy prisoners of war require immediate interrogation as the flexibility of operations will rapidly make their information outdated. Information given by civilians encountered in desert operations should be treated with caution unless corroborated. Military intelligence teams located in the area of operations can determine if these EPWs and civilians are in fact what they say they are, or infiltrators sent to harass the rear area and commit acts of sabotage. Electronic support measures are a major source of intelligence in desert warfare. Enemy activity, or the lack of it, is a good source of information; so punctual, accurate reports by all sources, both positive and negative, are necessary.

FIRE SUPPORT

The Allies in North Africa in 1942 found that placing small field artillery units in support of small maneuver units gave the units a sense of security, but produced limited results. Field artillery was effective only when massed (battalion or higher) and only when continued for some time because of the protective posture and mobility of the target. Typically, the control of massed fires was the responsibility of the division artillery.

The Allies in North Africa in 1942 experienced heavy casualties from Axis units overrunning the artillery positions after penetrating the armor and infantry positions. Often, the Axis units would attack from the east at one time, from the west later, and from several directions simultaneously. At first, the Allies simply emphasized direct fire. Later, the Allies attached antitank gun units to the artillery battalions to increase the artillery's antitank ability.

When armor and infantry units move, the artillery must move with them. The most useful technique is for the artillery to move in a formation with a lead vehicle so that, immediately upon stopping, the artillery is in a position or formation to deliver fire in any direction and simultaneously defend the position from any direction. The Allies in North Africa in 1942 and units in Desert Shield/Storm found that the armor and infantry units would outdistance the artillery unless the artillery moved with them. The artillery moved within 2-3 kilometers of the leading troops to provide responsive fires. The armor and infantry provided protection for the artillery. The whole group moved in one cohesive formation, sometimes in a large box or diamond formation.

Due to the fluid nature of desert operations and the possibilities for excellent enemy observation, close and continuous field-artillery support for all levels of the force is necessary. Field-artillery pieces should be at least as mobile as the force they are supporting. Crews must be proficient in direct fire and prepared to defend against a ground attack.

Due to the threat of immediate counterbattery fire, field artillery units must be prepared to move into position, fire, and rapidly displace to another position. A battery should be prepared to displace several times a day.

Field artillery units employed in desert operations should be equipped with the most sophisticated survey devices available. Manual systems are slower and not necessarily as accurate, thus affecting tactical employment and reducing response time.

Aerial observation may often be extremely difficult due to enemy air defense, so most adjustment is by ground observers. How the environment affects observation of fires was described previously in this chapter in the paragraph, "Observation and Fields of Fire." Recompute weather conditions frequently as weather conditions can change rapidly from the morning to the evening, and thus affect the accuracy of fires.

Fires are planned as in temperate climates. When there are no significant terrain features along a route of advance, targets are planned using coordinates.

A moving force in a desert is at a disadvantage in comparison with a stationary unit due to lack of concealment and the presence of dust clouds. The defender may engage with missiles from an unexpected direction or from terrain features of no apparent significance. The attacker must be prepared to rapidly shift fires to suppress unforeseen targets. Tactical aircraft may be used to suppress or destroy targets. Targets for aircraft can be marked with indirect- or direct-fire smoke. White phosphorus or illuminating rounds set for low-air burst are also effective.

Indirect fires are used to slow the enemy advance, to suppress enemy weapons and observers, and to conceal movement between positions using smoke. Defensive operations in deserts are characterized by long-range engagement with tanks and ATGMs.

AIR DEFENSE

Identification of friend or foe is difficult. Throughout the entire theater of operations there will be numerous weapon systems that are common to both sides of the conflict. The individual soldier/marine is going to be faced with the monumental task of separating friend from foe by more than just from the recognition of the manufacturer or silhouette of a piece of equipment. This will be true of both air and ground systems. This identification problem will be compounded by the nonlinear battlefield where the focus of operations will not be separated by a line.

The desert is an outstanding environment for employing aircraft. Every unit must be extremely proficient at passive and active air defense. The Allies in North Africa and the Israelis in the Middle East found that dispersion limited the effects of air attacks, and small arms air-defense techniques were effective. Almost every weapon in North Africa had a secondary antiaircraft or antitank mission.

Emphasize to each unit that, when in position, units must disperse very widely making a less lucrative target. When moving in column and under air attack, units must move at least 40 to 50 meters off the road because aircraft normally have nose guns trained on troth sides of the road. A vehicle on the road or on both sides of the road will die.

Because of the wide open spaces characteristic of many deserts and the relatively large areas associated with desert operations, forces fighting in the desert should be reinforced with additional air-defense weapons. Still, there may not be sufficient dedicated air-defense systems to fully cover the force. When this is the case, commanders must be especially careful when establishing air-defense priorities in view of relatively long lines of communication and the tendency to maneuver over relatively large areas. In any event, all units must include a scheme for countering air attacks in their battle plans using both active and passive measures.

Although Army armored and mechanized infantry division air-defense weapons are tracked, this does not necessarily apply to corps medium-altitude air defense units. However, Army corps surface-to-air missile (SAM) units have considerably greater ranges and are equipped with more sophisticated early warning and control systems. Some corps units should be employed well forward. These weapons will have to displace by section to ensure continuous coverage.

Air-defense units should be located close to elements of supported units to provide for ground defense. When the supported unit moves, the air defense unit must also move, which requires careful coordination to ensure that movement of the supported unit is not delayed. Airspace management difficulties are compounded in the multinational environment. SOPS should be exchanged among multinational forces to lessen the confusion of airspace management.

ENGINEERS

Engineer operations in the desert are similar to those in temperate climates although there are fewer natural terrain obstacles to be crossed. Depending on the terrain anticipated in the operations area, a dry-gap crossing capability may have to be obtained from corps support units. Important tasks for engineers in desert operations include-

- Mobility/countermobility/survivability support, including construction of obstacles, logistics facilities and routes, field fortifications, airfields, and helicopter landing pads.
- Water supply.
- Topographic support (map-making).

Mobility

The vastness of the desert makes mobility a prime concern. Roads are usually scarce and primitive. Cross-country mobility is possible in some areas, but trafficability is poor in soft sand, rocky areas, and salt flats. Engineers assist maneuver by reducing slopes, smoothing rock steps, and bridging dry gaps.

Expanded engineer reconnaissance capability will be needed to identify routes, existing obstacles, and minefield locations. Flat, open areas provide good sites for aircraft landing strips; however, in most cases the soil must be stabilized. Normally, desert soil produces extensive dust and has limited weight-bearing capacity.

Engineers use various agents to alleviate severe dust conditions (diesel, JP4, or oil mixtures for example). This is particularly critical in reducing engine wear in areas supporting rotary wing aircraft. It is also important along heavily traveled roads and in cantonment areas. Engineers also use soil-stabilization techniques to increase soil-bearing capacity for airstrips and MSRs.

The application of the fundamentals of breaching-suppress, obscure, secure, and reduce-and the organization of the force in terms of supporting, breaching, and assaulting elements, are even more important in the desert due to the enhanced observation and fields of fire. However, the desert does offer greater opportunities to bypass enemy obstacles because of the greater range of mobility afforded by desert terrain. Caution must be exercised when choosing to bypass enemy obstacles since the bypass may lead the force to the enemy's kill sack.

The increased mobility in the desert makes it easier for the enemy to counterattack exposed flanks of attacking forces. Plan obstacles to protect flanks during offensive operations. Beyond conventionally emplaced minefield, FASCAM, which includes artillery-delivered mines, GEMSS, and air-delivered Gator munitions, are all systems that lend themselves to situational development. FASCAM and conventional minefield may be appropriate, but consider the time required to employ FASCAM when selecting this option. Artillery-delivered FASCAM does not deploy well in soft sand and removes a majority of your indirect-fire assets from the fight.

Countermobility

Due to the mobility inherent in desert operations, obstacles must be extensive and used in conjunction with each other and with any natural obstacles, and covered by direct and indirect fires. Isolated obstacles are easily bypassed.

Mines are easily emplaced in a sand desert, and blowing sand will effectively conceal evidence of emplacement. However, the following potential problem areas must be considered when emplacing mines:

- Large quantities of mines are required for effectiveness.
- Sand can cause malfunctioning.
- Shifting sand can cause mine drift.
- An excessive accumulation of sand over the mines can degrade performance.
- Sand may be blown away and expose the mines.

In suitable terrain, antitank ditches that exceed the vertical step of enemy main battle tanks may be used. Because antitank ditches cannot be conceded, they must be dug so they do not outline a defensive front or flank. They have the advantage of not requiring as much logistic support as minefield. They must be covered by observation and fire to prohibit enemy infantry using them as ready-made trenches.

Because of limited off-road mobility of most combat service support vehicles, considerable engineer efforts may be necessary to construct and maintain routes forward to maneuver units. Local resources, such as salt-marsh mud laid on sand can be used. Track vehicles should not use these routes since they could easily ruin them.

Most desert regions have a natural terrain structure that restricts maneuver such as sandy dunes, rocky plateaus, mountains, and wadis. These structures must be interpreted rapidly and correctly, and then reinforced with obstacles to fix, turn, or disrupt enemy movement, according to the commander's plan.

Minefield and antitank ditches are the primary means of creating obstacles in the desert. Antitank ditches require extensive preparations, but they are effective when adequate preparation time is available. Many desert villages have irrigation ditches that can be used tactically. Other countermobility methods are generally not effective. Road craters, for example, are usually easy to bypass. In sandy areas, ditches can easily be filled in, so they are not good obstacles. Opportunities for bridge destruction are rare, and local materials for expedient obstacles are scarce.

Engineers and combat forces should coordinate the siting of planned obstacles to support the defensive concept. In defensive operations the effectiveness of obstacles requires synchronization.

Survivability

Desert terrain varies from region to region. Generally, however, observation is excellent and concealment is difficult. Deserts provide little cover and concealment from ground-based observers and even less from aircraft. These conditions make modem weapon systems more lethal in deserts than in any other environment.

In the desert, hull and turret defilades for tactical vehicles are essential. This allows the defending force to take advantage of their long-range weapon systems in the face of enemy fires. Dispersion and frequent moves are other survivability techniques that can be used.

The preparation of fortifications in the desert is difficult. Fortifications in sandy soil often require revetments. In rocky plains or plateaus it may be impossible to dig. To counter this problem, build up emplacements with rocks and use depressions.

Camouflage is very effective when properly employed; however, patterns and techniques must be carefully selected to match the local desert environment. Camouflage nets should be provided for all equipment. See Appendix E for additional comments on desert concealment and camouflage.

Desert Survivability Positions

Defensive positions are very vulnerable to offensive fire due to long-range observation and fields of fire in the desert. This, coupled with a lack of natural obstacles, may lead the commander to invest the bulk of his engineer effort into survivability positions. Survivability positions enhance the ability of all direct-fire elements to survive indirect-fire and to return fire on the enemy. Survivability positions are normally more important than antitank ditches, especially in open terrain. See Figures 3-1 through 3-6, for examples of survivability positions. The following are some things you should or should not do when preparing survivability positions:

DO-

- Ensure adequate material is available.
- Dig down as much as possible.
- Maintain, repair, and improve positions continuously.
- Inspect and test position safety daily, after heavy rain, and after receiving direct and indirect fires.
- Revet excavations in sandy soil.
- Interlock sandbags for double-walled constructions and corners.

DO NOT-

- Fail to supervise.
- Use sand for structural support.
- Forget to camouflage.
- Drive vehicles within 6 feet of a position.
- Overfill sandbags.
- Put troops in marginally safe bunkers.
- Take shortcuts.
- Build above ground unless absolutely necessary.
- Forget lateral bracing on stringers.

- Check stabilization of wall bases.
- Fill sandbags approximately 75 percent.
- Construct to standard.
- Use common sense.

The commander's responsibilities during construction of survivability positions are to—

- Protect troops.
- Continuously improve and maintain unit survivability.
- Provide materials.
- Periodically inspect.
- Plan and select fighting position sites.
- Get technical advice from engineers, as required.

In a combat situation, it may be necessary to improvise construction of a survivability position by using materials not normally associated with the construction. Some examples of field-expedient materiel are-

Wall Revetment

- Sheet metal.
- Corrugated sheet metal.
- Plastic sheeting.
- Plywood.
- Air mat panels.
- Air Force air-load pallets.

Overhead Cover Stringers

- Single pickets.
- Double pickets.
- Railroad rails.
- "I" beams.
- Two-inch diameter pipe or larger.
- Timbers 2" x 4", 4 " x 4 ", and larger.
- Reinforced concrete beams.
- 55-gallon drums cut in half longitudinally.
- Large diameter pipe/culvert, cut in half.
- Precast concrete panels, 6-8 inches thick.
- Airfield panels.
- Air Force air-load pallets.

Wall Construction (Building up)

- Sand-grid material.
- 55-gallon drums filled with sand.
- Expended artillery shells filled with sand.
- Shipping boxes/packing material.
- Prefabricated concrete panels.
- Prefabricated concrete traffic barriers.

Stand-Alone Positions

- Prefab concrete catch basins, valve pits, and utility boxes.
- Military vans.
- Connexes or shipping containers.
- Large diameter pipe/culvert.
- Steel water tanks.
- Other storage tanks (cleaned and ventilated).
- Vehicle hulks.

- Shipping pallets.

The following is a suggested inspection checklist to follow when preparing survivability positions:

- Location is sited tactically sound.
- Low profile is maintained.
- Materials are of structural quality (standard construction material).
- Excavation-walls are sloped.
- The setback for overhead is a minimum of 1 foot or 1/4 the depth of cut.
- Stringers¬

    - Are firmly on a structural support.

    - Have lateral bracing emplaced along supports.
- 2" x 4" or 2" x 6" stringers are used on the edge; the strength is on the depth of the lumber.
- Supports¬

    - Stringers are firmly on supports.

    - Supports extend past the excavation by 1/2 the depth of cut.
- Revetments—

    - Quality of construction is checked.

    - Sheeting is supported by pickets.

    - Pickets are tied back.
- Overhead cover¬

    - Quality of structural layer is inspected.

    - Quality of dust layer-plywood or panels-is inspected.

    - Layer is cushioned at least 18 inches deep.

The one-man fighting position is the individual's basic defensive position. The one-man fighting position with overhead cover (see Figure 3-1) provides protection from airburst weapon fragments. A good position has overhead cover that allows the soldier/marine to fire from beneath it. Stringers extend at least 1 foot on each side of the position to provide a good load-bearing surface for overhead cover.

Figure 3-1. One-man Fighting position with overhead cover.

Generally, the two-man fighting position is preferred over a one-man position since one soldier/marine can provide security while the other is digging or resting. The position can be effectively manned for longer periods of time; if one soldier/marine becomes a casualty, the position is still occupied. Further, the psychological effect of two men working together permits occupation of the position for longer periods. Overhead cover also improves the position's effectiveness; it is made as described for the one-man position (see Figure 3-2).

Figure 3-2. Two-man fighting position wiih overhead cover.

Fighting positions for machine guns are constructed so the fires are to the front or oblique; the primary sector of fire is usually oblique so the gun can fire across the unit's front. The position is shaped so the gunner and assistant gunner can get to the gun and fire it to either side of the frontal direction. Overhead cover is built over the middle of the position (see Figure 3-3). It is constructed as described for the one-man position.

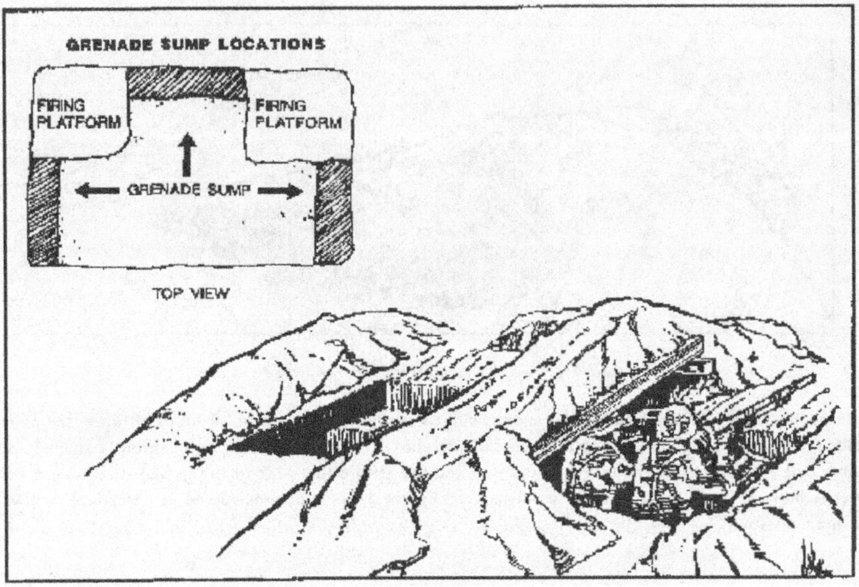

Figure 5-3. Machine-gun position with overhead cover.

Protective shelters and fighting bunkers are usually constructed using a combination of the components of positions mentioned thus far. Protective shelters are primarily used as command posts, observation posts, medical aid stations, supply and ammunition shelters, and sleeping or resting shelters. Figure 3-4 shows an example of a command bunker.

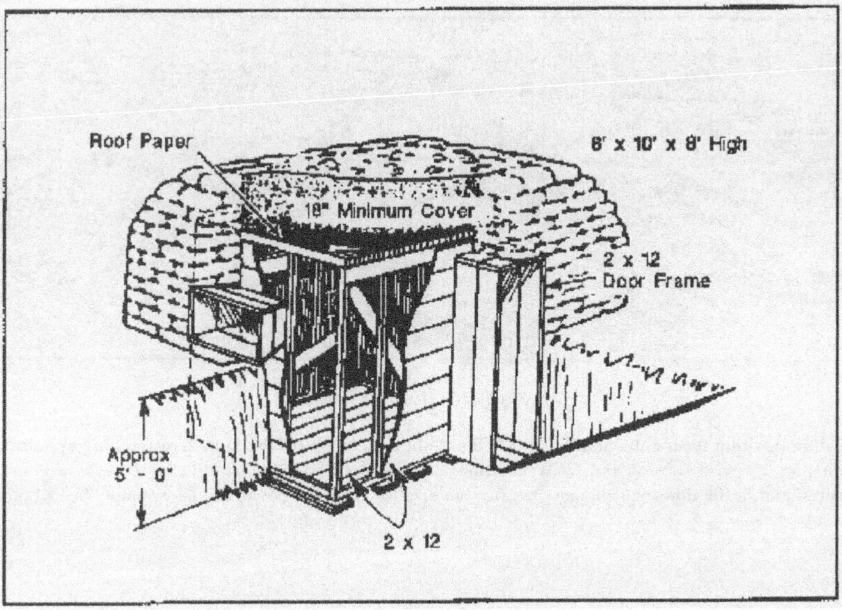

Figure 3-4. Command Bunker .

The Dragon position requires some unique considerations. The soldier/marine must consider the Dragon's extensive backblast and muzzle blast, as well as cleared fields of fire. When a Dragon is fired, the muzzle extends 6 inches beyond the front of the position, and the rear of the launcher extends out over the rear of the position. As the missile leaves the launcher, stabilizing fins unfold. Therefore, the soldier keeps the weapon at least 6 inches above the ground when firing to leave room for the fins. A waist-deep position will allow the gunner to move while tracking a target. Because of the Dragon's above ground height, soldiers/marines should construct frontal cover high enough to hide the soldier's/marine's head and, if possible, the Dragon's backblast. The soldier/marine must dig a hole in the front of the position for the biped legs. If cover is built on the flanks of a Dragon position, it must cover the tracker, missiles, and the gunner. Overhead cover that would allow firing from beneath it is usually built if the backblast area is clear (see Figure 3-5).

Figure 3-5, Dragon Position.

A fighting position for the dismounted TOW must not interfere with the launch or tracking operations of the weapon. As with Dragon and LAW positions allowances for backblast effects are necessary. Backblast and deflection requirements restrict the size of overhead cover for the weapon. See Figure 3-6.

Figure 3-6.Dismounted TOW position.

Designers of fighting positions and protective positions in desert areas must consider the lack of available cover and concealment. Fighting positions should have the lowest profile possible, but mountain and plateau deserts have rocky soil or "surface chalk" soil which makes digging difficult. In these areas, rocks and boulders are used for cover. Because target acquisition and observation are relatively easy in desert terrain, camouflage and concealment, as well as light and noise discipline, are important considerations during position construction.

Indigenous materials are usually used in desert position construction. However, prefabricated structures and revetments, if available, are ideal for excavations. Metal culvert revetments can be quickly emplaced in easily excavated sand. Sandbags and sand-filled ammunition boxes are also used to prevent side walk of positions from collapsing.

FM 5-103 discusses vehicle fighting positions in detail and should be consulted for more information. Figure 3-7 provides specifications for vehicle survivability defilade positions that can be dug by the D-7 dozer.

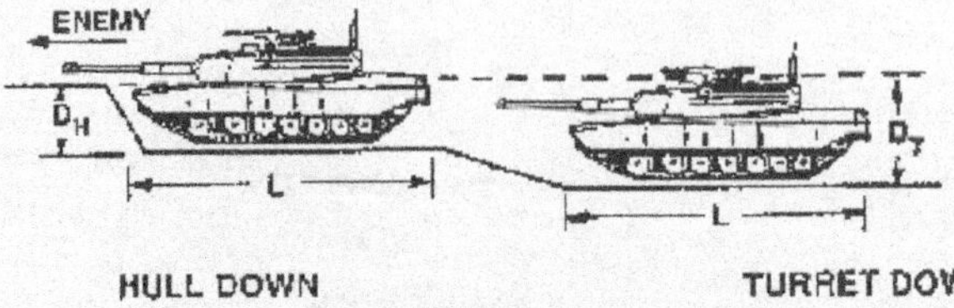

VEHICLE	L	D_H	D_T	WIDTH
M1 Tank	2 Dozer lengths	Top of battery box	Top of your head	1 1/2Blade widths
M60 Tank	2 Dozer lengths	One fist above battery box	Top of smoke stack	1 1/2 Blade widths
M2 BFV M3 CFV	1 1/2 Dozer lengths	Top of engine housing	Top of smoke stack	1 1/2 Blade widths
M113 APC	1 1/2 Dozer lengths	One fist above battery box	Top of air intake cover	1 1/4 Blade widths
M901 ITV /FISTV	1 1/2 Dozer lengths	Top of engine housing	Top of your head	1 1/4 Blade widths
M109 155 Howitzer	6 1/2 Dozer lengths	One fist above battery box	Not required	1 1/2 Blade widths
M110 8" Howitzer	6 Dozer lengths	Top of blade tip crown	Not required	1 1/2 Blade widths
M577 Command Post Vehicle	1 1/2 Dozer lengths	Top of your head	Not required	1 1/4 Blade widths

These dimensions are described with respect to you and your dozer. Techniques for construction of survivability positions are described in FM5-34, ENGINEER FIELD DATA.

Figure 3-7. Survivability defilade positions.

Logistics areas (BSA/DSA) require additional survivability support. Desert operations require that logistics concentrations such as BSAs and DSAs be given additional considerations for survivability support. These sites are large, relatively static, and difficult to camouflage. As a result, these support areas are vulnerable to enemy interdiction. Military vans or connexes should be covered with sandbags to improve protection. Additionally, if they are covered with heavy plastic, with plastic drapes over the entrances, protection against NBC effects can be improved.

Engineer digging assets, such as bulldozers, should be tasked to provide survivability support to these sites with particular emphasis placed on hardening ammunition and fuel storage locations. Caution should be used when digging foxholes and tank hide positions since some areas have a tendency to cave in.

Water Supply

Water supply is the most important mission of engineers in the desert. The search for water sources requires continuous, intensive reconnaissance. Water may be obtained by drilling beds of dry water courses, or by deepening dry wells. Once found, water must be made potable and stored or transported. Since water purification trucks may be high-priority targets and barely sufficient for the task, any force operating in the desert must be augmented with water supply units (including well drilling), water purification and water distillation teams, and transportation teams. Another possible water source is the reverse osmosis water purification unit (ROWPU). This unit is an ISO frame-mounted, portable water purification system capable of purifying water from almost any shallow well, deep well, and surface water or raw water source. The ROWPU is capable of removing NBC contaminants, minerals, and biological impurities. The single greatest benefit of the reverse osmosis process is the ability to desalinate sea water. The ROWPU is capable of producing potable water at a rate of 600 gph. The ROWPU is powered by a 30-kilowatt generator set.

Topographic Support (Map-making)

Large areas of the world's deserts are not covered by maps of any useful tactical scale. Existing maps are frequently inaccurate and increase the difficulties of navigation. Therefore, engineer topographic companies must augment the force by preparing, printing, and distributing up-to-date maps of the operational area. USAF, Army, and Marine aviation support can be used to produce gridded maps from aerial photography of the area forward of the line of contact.

MILITARY POLICE

Combat support provided well forward by military police will continue in desert operations, although over increasingly extended distances. MP tactical and physical security will be of special importance over extended lines of communication, such as petroleum pipelines and viaducts transporting water over long distances. Protection of these critical items demands both active and passive measures, including overflight by returning aircraft or overwatch by convoy movements. The storage sites for water, food, POL, and ammunition have historically been principal targets for enemy action, and consequently must receive augmented security.

The indefinite conditions and number of roadways will require increased circulation control points to direct traffic, redirect stragglers, and provide information so that throughput forward to the fighting elements will be expedited. Military police are especially valuable when the combat commander must employ concentration or economy of force in the face of the enemy to gain a favorable combat ratio. MPs can secure the roadways, enforce priority movement, and prevent any delay of the elements undertaking passage of lines to blocking or defensive positions. MPs can also assist in the handling of EPWs.

US AIR FORCE SUPPORT

US Army and Marine forces fighting in the desert can expect to be supported by USAF tactical fighter-bomber and airlift aircraft. Close air support by USAF tactical fighter bombers is more important in desert warfare in view of lack of concealment, relatively large areas of operations, and mobility of forces employed by each side. Air support in a desert environment has advantages over more temperate areas of operations. For example, it is easier to locate targets; visual observation is normally far superior to that in temperate climates; and ground movement is more readily apparent.

Air attacks can be handicapped by lack of covered approaches, but increased visibility permits engagement from standoff ranges. When flying close air support missions it is important for pilots to be able to differentiate between enemy and friendly forces. Use panels or other visual or electronic identification means to assist in identification.

Because of the extended lines of communication likely in desert operations, USAF tactical airlift should be used whenever possible. This is particularly true of resupply operations conducted from a lodgement area to forward trains areas when considerable distances are involved.

Planning for air support must be as detailed as time permits to determine mission and armament requirements, time over target, and method of control. The joint air-ground operations system (AGOS) used to request and coordinate the use of US Air Force tactical air support is described in FM 100-26.

US NAVY SUPPORT

When the force is being supported by US Navy gunfire, or Navy or Marine aircraft, elements of a Marine air and naval gunfire liaison company (ANGLICO) are attached to Army ground forces. The mission of the company is to support an Army division by providing control and liaison agencies for employment of this support.

ANGLICO platoons and teams can advise commanders on the capabilities, limitations, and employment of naval gunfire, and USN or USMC air support. Platoons are normally placed with brigades or higher headquarters, and air and gunfire support teams placed with battalion task forces. Although the company has organic vehicles and some combat services support capability, its elements generally require additional administrative and logistical assistance from the supported unit. In order to communicate with Army/Marine units, additional communications equipment may also have to be provided. Additional information on ANGLICO employment can be found in FM 31-12.

COMBAT SERVICE SUPPORT

Combat service support for desert operations is described in detail in Chapter 4. When planning a desert operation it is necessary to consider the following factors:

- The speed of supply may be slowed and lines of communication can be vulnerable due to the distances between units. Except for Class V and sometimes Class III, resupply should beat night for reasons of security.
- A great demand for water can tie down large quantities of transport and may involve laying pipelines. Water is vital, so consider the water situation during every operation estimate.
- Increased maintenance is required due to heat, sand, and dust damage to equipment. This not only increases the repair work load, but also increases demand for replacement items due to increased wear.

Supply

Offensive operations in this environment may involve considerable expenditure of ammunition and high POL consumption. Units must carry maximum combat supplies, and plans for resupply must be widely disseminated and clearly understood. Use every opportunity for resupply.

Due to the importance of combat service support, attacking the rear will be more immediate y effective at lower levels of command in the desert than in temperate climates. In the case of water, for example, the enemy must be able to obtain resupply. Degradation of the enemy's trains places him in a situation where his troops must maneuver against the attacker regardless of the planned scheme of defense.

Combat service support must be reliable and timely, using vehicles that can travel over difficult terrain to reach combat units. In the desert, more than anywhere else, the commander must ensure that he has support that is capable of maintaining his unit for a specified period of time, even if the logistic line of communication is temporarily broken.

The mobility and freedom of tactical maneuver are tied to the ability of the logistic chain to supply maneuver units. Two alternatives are available: increase the rate of supply, probably requiring more vehicles, or prestock, which ties units to the stocked area. Some important supply considerations are outlined below:

- Class I. It is often impractical to supply hot rations from mess trucks, especially when the unit is subject to enemy air reconnaissance or target acquisition devices. T rations and B rations are the usual method of troop feeding.
- Class III. Daily requirements for POL in desert operations can be expected to be high. Estimates for POL requirements should take into consideration large-scale maneuver inherent in desert operations.
- Class V. Estimates of ammunition requirements should reflect the high level of commitment that can be anticipated in desert operations.

Maintenance

Disabled vehicles are vulnerable targets. Both disabled vehicles and the maintenance vehicles used in working on them must be concealed during the day, and strict light and sound discipline imposed at night. Maintenance contact teams should carry Class IX supplies that have a quick turnover.

Section II. Offensive Operations

This section discusses offensive operations as they are modified by desert terrain.

GENERAL

The main purpose of offensive operations in desert terrain is to destroy the enemy, Operations may be undertaken to secure key or decisive terrain, to deprive the enemy of resources or decisive terrain, to deceive and divert the enemy, to develop intelligence, and to hold the enemy in position. Destruction of the enemy can be accomplished by concentrating friendly forces at a weak point in the enemy's defense and destroying enemy combat units, or by driving deep into the enemy's rear to destroy his combat service support and cut his lines of communication. No force can survive in the desert for long without combat service support.

An imaginative commander is not bound by terrain constraints in seeking and destroying the enemy. Due to the scarcity of key terrain in the desert, normally the only constraints placed upon a maneuvering force is its ability to maintain the only constraints placed upon a maneuvering force is its ability to maintain responsive combat service support and to protect its combat service support from enemy attack. The longer the lines of communication become, the more susceptible they are to being cut.

In most deserts, the scarcity of large areas of defensible terrain means that an enemy force has at least one flank open to attack, The attacking force must seek this flank and attempt to maneuver around it into the enemy's rear before the enemy can react and block the envelopment with mobile reserves.

Successful offensive operations depend on rapid, responsive, and violent maneuver, seeking a vulnerable enemy flank while exposing none to the enemy. The enemy, realizing the danger of remaining stationary in this terrain, may choose to conduct spoiling attacks or to counterattack. The resulting meeting engagement between the two attacking forces will often be a series of flanking actions and reactions with success going to the one who can find the other's unguarded flank first.

Attacking forces may conduct or participate in movement to contacts or hasty or deliberate attacks. Within a division, lead elements of forward units may be conducting a deliberate attack on the enemy's weak point or flank to open a gap for following units to move through and exploit success. Lead units of the exploiting force will be conducting a movement to contact and hasty attacks to overcome pockets of enemy resistance. Regardless of the type of operation being conducted, attacking units use the fundamentals for offensive operations described in FM 100-5/FMFM 6-1 and other tactical manuals, modified to suit the terrain.

FUNDAMENTALS OF THE OFFENSE

The attacker must conduct active and aggressive reconnaissance to the front, flanks, and rear, not only to locate and identify enemy obstacles, units, weak points, and flanks, but also to give early warning of threats to his flanks and combat service support elements. A moving force is at a disadvantage in the desert due to a lack of concealment. Therefore, it is necessary to push reconnaissance units as far out from the main body as possible to allow early warning and to deny the enemy close-in observation.

Information gathered by this reconnaissance must be passed promptly to all units. In the desert, a negative report may be as important as an enemy sighting. Commanders and staffs must avoid the two extremes of either passing too little information or overwhelming their subordinates with useless trivia. Similarly, reconnaissance units must also avoid extremes. There is a very real possibility that extensive reconnaissance in one area will alert the enemy of intended operations in that area. Therefore, the need for reconnaissance must be tempered with the need for deception. In fact, reconnaissance may even serve as a deceptive measure to draw the enemy's attention away from the real objective or area of operations.

Concentrate on overwhelming combat power. Mass is achieved in both time and space. Units must be able to rapidly concentrate at a given time and place, and then disperse just as rapidly to avoid offering a lucrative target to the enemy. Concentration does not necessarily mean that vehicles and men are massed in a small area, but that units have the ability to place an overwhelming concentration of fires on the enemy.

Mutual support is as important in the desert as in temperate climates. Due to the large distances covered by maneuver in the desert, mutual support does not mean that any one unit is always in position to fire against an enemy threatening another unit. However, units must be capable of maneuvering in support of one another without disrupting the scheme of maneuver.

Concentration requires movement, and possibly weakening of forces facing the enemy in another part of the zone. Due to the enemy's observation capabilities, movement should take place at night or in conditions of limited visibility whenever possible. Deception measures play an important part in concentration, either to mislead the enemy as to the strength or true intentions of the opposing forces, or their avenues of approach. In this environment of negligible concealment, deception cannot be overemphasized.

The enemy's objective is to stop and destroy the attacking force by direct and indirect fires, obstacles, and counterattacks. The attacker must in turn suppress enemy weapon and surveillance systems to degrade their effects and their intelligence-gathering capability.

Attack helicopters and high-performance aircraft are extremely useful due to their ability to maneuver and apply firepower over a large battlefield in a short time. So, suppression of enemy air defense has a high priority during offensive operations. The destruction of enemy antitank capabilities must also have a high priority due to the shock potential of armor in the desert. No target that has a long-range antitank capability should be disregarded. Good gunnery and well-planned fire distribution are preeminent.

In featureless desert terrain, the requirement to shock, overwhelm, and destroy the enemy demands accurate reconnaissance to identify actual positions from false positions, and excellent navigation so that a commander may be certain of the deployment of his forces. Reconnoiter to find a gap or assailable flank (without alerting the enemy that the area is being reconnoitered) and concentrate the main body to go through or around it with suppressive fires on the flank(s). A gap must be wide enough to allow one unit to bypass another unit that could be stalled. Obstacles are likely to be placed so that attempts to go around them will often lead the attacker into a tire sack. Equipment capable of breaching obstacles must be located well forward.

ENVIRONMENTAL CONSIDERATIONS

As a general rule, a force attacking in daylight should try to wait until the sun is comparatively low and position behind it. This enables enemy targets to be plainly seen without their shadows, while the defenders are handicapped by glare, mirages, and haze. It is not always possible (nor essential) for the sun to be directly behind the attackers. To rely on this leads to a stereotyped method of attack which could become evident to the defenders. The commander of a maneuver force should attempt to keep the sun somewhere on a 3,200-mil arc to his flanks or rear, giving a wide choice of angle of attack.

Dust is an observational hazard to a maneuvering force, especially where there is little or no wind. Teams should move in echelon with overmatching elements on the upwind side, and observers and attack helicopters should operate well to the flank. Since it is impossible to disguise movement during daylight, the assault should be as rapid as possible to minimize enemy reaction time.

The decision to move through a sandstorm will depend on the unit's distance from the enemy, trafficability, the presence of minefield, and the direction and density of the storm. If the advancing unit is caught in a storm blowing from the enemy's direction, the safest alternative is to halt until it abates, although this may not always be possible. In some situations it may be possible for platoons to form close column, using taillights only, and continue movement. When the storm is blowing toward the enemy it is possible (and extremely effective) to conduct an attack immediately behind the storm.

In certain circumstances equipment or positions that are camouflaged and are less than 1 meter from the ground are invisible to an observer at the same height out to approximately 2,000 meters. At the same time, mirages allow observation of objects below the horizon, although these may be distorted, enlarged, or fuzzy to the point of being unrecognizable. These effects often depend entirely on the angle of the sun to the observer and are best combated by-

- Maintaining observers as high above the desert floor as possible, even if only in hull-down positions behind sand dunes.
- Allowing a vehicle's crew on one side of a position to warn a crew on the other side of a possible threat to his front by crews observing over wide areas.

Many offensive operations take place at night. Observation in these conditions varies according to the amount of ambient light. During nights when the moon is full or almost full, the clear desert sky and ample ambient light allow good observation, both with the naked eye and with night observation devices. Maneuvering units using night-vision devices must continually scan the surrounding terrain to pickup enemy activity that normally would be acquired by peripheral vision in the daylight.

The desert night is extremely dark when there is little or no moon. Under these conditions passive-vision devices, with the exception of thermal imagery, are of little value unless artificial light is used. Active light sources will have to be relied upon. Employment of artificial light must be strictly controlled by the headquarters directing the operation to maintain surprise. As a general rule, direct-fire weapons should not illuminate their target themselves, as their vision will be obscured by debris kicked up due to muzzle blast. Following contact, when some targets should be on fire, passive devices can be used.

MANEUVER

If the terrain permits masking of maneuvering units, and trafficability is good, normal fundamentals of fire and maneuver are used. Trafficability may be restricted by rocky terrain as in the Golan Heights, or the ground may be so flat that the defender has total observation of the area. Movement in these circumstances requires speed of maneuver, deception, and considerable suppression to degrade enemy observation and fires. Frontal attacks should be avoided, especially in conditions of restricted trafficability. It is preferable to maintain pressure on enemy units in unfavorable terrain, while other forces find enemy weaknesses in terrain that is more favorable for an attack.

Lack of clearly defined terrain features complicates navigation and phased operations. Units conducting an enveloping maneuver are apt to lose direction unless routes have been carefully reconnoitered by the maximum number of leaders.

Section 111. Defensive Operations

This section discusses defensive operations as they are modified by desert terrain.

GENERAL

It is unlikely that a US force will be fully deployed in a desert country before an enemy attack. The more probable situation, assuming a secure lodgement area, will be that part of the force will be in position supporting an allied force, while the remainder is moving in by air and sea. Tactically, the allied force will be outnumbered, so the initial mission will be to gain time until the entire force is present in the operational area. This will require a defensive posture initially, but a defense undertaken so aggressively as to convince the enemy that his offensive action will be too costly in personnel and equipment to be worth maintaining. The enemy will be well aware that US forces are arriving in the area, and will make every effort to conclude his operation successfully before the force is fully prepared for combat operations.

The force may conduct defensive operations during subsequent stages of the operation for any of the reasons described in FM 100-5/FMFM 6-1. Portions of the force may be required to defend the important types of terrain described below:

- Man-made features such as ports, key logistic installations, roads, railroads, water pumping stations, airfields, and wells.
- Natural features, such as mountain passes, or dominating ground, such as Mount Hermon on the border of Syria and Israel, or the Solium escarpment near the sea between Libya and Egypt.
- Key or decisive terrain that need not necessarily be a major feature, but one whose loss will inhibit the force in some manner. For example, the loss of terrain relatively close to a lodgement area may hinder the planned rate of buildup.

With the exception of the above cases, the retention of desert terrain normally makes little difference to the final outcome of battle. This does not mean that a commander has complete discretion to move his force wherever and whenever he wishes, as this movement will affect the dispositions of other US forces or allies. It means that possession of terrain is less important than the destruction of enemy forces. Although it will be necessary to dominate certain terrain or retain freedom to maneuver in large areas of the desert, there is no more sense in permanently occupying such areas than occupying a patch of sea. Assuming equal equipment capabilities for both opposing forces, the critical factor in defense will be the force ratios involved and the state of morale and training of the opposing forces.

A defense using aggressive maneuver at all levels is the best way to destroy large numbers of enemy without being destroyed in the process. If the defending force fails to remain mobile and active, the enemy will easily outflank it and strike directly at vital targets, such as the lodgement area. It is almost certain that one flank or the other will be open as were the south flanks of the British and German forces in Egypt and Libya in 1940-43. Since it will not be possible to maintain an unbroken line between strategic obstacles, air and ground security forces must be positioned in width and depth to guard against an enemy trying to outflank the defender.

Obstacles, both natural and artificial, are used to slow, contain, or isolate enemy units in order to defeat and destroy his units one at a time. Forward units block the enemy and canalize him into one or two avenues where he can be engaged from the flank. A reserve can then counterattack to destroy any remaining enemy.

Mutual support is normally a factor of time rather than weapon range due to the large areas to be covered. Gaps in initial positions may have to be accepted between and within task forces; although the ideal is to site units in such a manner that forces in at least two positions can engage an enemy maneuvering on any one of them. This greatly reduces any possibility of defeat in detail. When gaps exist they must be kept under surveillance. The defensive plan must include provisions for maneuvering to fire on any part of a gap before the enemy can move through it. A unit's area of responsibility must be defined by higher headquarters and should be clearly identifiable on the ground, which, due to the absence of significant terrain features, may require marking by artificial means.

FUNDAMENTALS OF THE DEFENSE

The fundamentals of defense are described in FM 100-5/FMFM 6-1 and in doctrinal manuals appropriate to each level of command. The following paragraphs discuss some points to remember in desert operations as they apply to the fundamentals of defense.

Reconnaissance and security units and force surveillance systems must focus on-

- What is the enemy's short-term objective?
- What are the enemy's avenues of approach, and what force is employed on each of them?
- Are the apparent movements real or feints?

As soon as these questions have been answered the commander will be able to maneuver to destroy the enemy. Until they are confirmed he can do nothing more than react to enemy initiatives, This is dangerous in any circumstance and doubly so in the desert as the side with the greatest potential for maneuver is more likely to win.

Direct-fire weapons must be used to their maximum effective range both by day and night. Limitations in night-vision equipment cannot be allowed to reduce depth or frontages; so plans for field artillery or mortar illumination are made for defense during limited visibility.

It is essential that all elements of a force retain their tactical mobility and efficient communications so that they can immediately react to changes in the commander's plans. Each individual weapon must be sited in a number of firing positions, even though vehicular movement may be exposed to air attack. Infantry fighting vehicles must remain in positions where they are concealed, capable of giving fire support to the dismounted squad, and available for immediate remounting.

Combined arms teams are essential to give the commander the capability he requires to fight the defensive battle. Defending forces orient on primary enemy approaches but units must also be prepared for attack from any other direction. Itis neither possible nor necessary to have maximum firepower in all directions, provided weapons can be moved to threatened areas before the enemy reaches them. Air cover or an air defense umbrella is necessary for a successful defense.

It is rare to find positions where any substantial part of the unit area of operations can be protected by natural obstacles. This require extensive use of artificial obstacles, depending on time, personnel, and combat service support available. Obstacles are used to divide the enemy force to improve local force ratios, and to slow the enemy's advance, thus permitting a flank attack. Conventional minefield must be clearly marked on the friendly side and recorded to avoid unnecessary losses if friendly forces later maneuver over the area.

STRONGPOINTS

Strongpoints are rare in desert warfare; however, they may be necessary to defend an oasis, mountain pass, or other key terrain essential to the defender's scheme of maneuver. When it is necessary to deny terrain to an enemy force, it is far better to initiate the defense well forward of the terrain feature, conduct the defense in depth, and destroy the enemy or force him to break off his attack before he reaches the critical feature.

In some cases the level of fortification and the deployment of the enemy may be a function of time, or the enemy's intention and his understanding of what our forces are intending to do. The effectiveness of these strongpoints depends on the range of fires, the level of fortifications, and the decision of the opponent to attack them.

Deeply dug and well-prepared strongpoints surrounded by a minefield and having underground accommodations are usually used in the desert. Although these strongpoints may be neutralized by air or artillery fire and bypassed, eventually they will have to be assaulted. If they have been carefully sited and are well defended they can be quite effective. Variations of the strongpoint defense are used in rear operations. Combat service support units will use this method in perimeter defenses or base-cluster defenses. See Figure 3-8 for an example of a strongpoint and Figure 3-9 for an example of a strongpoint holding key terrain.

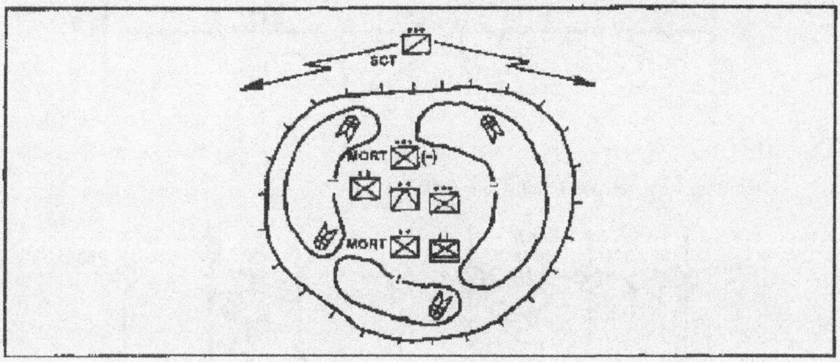

Figure 3-8, Strongpoint

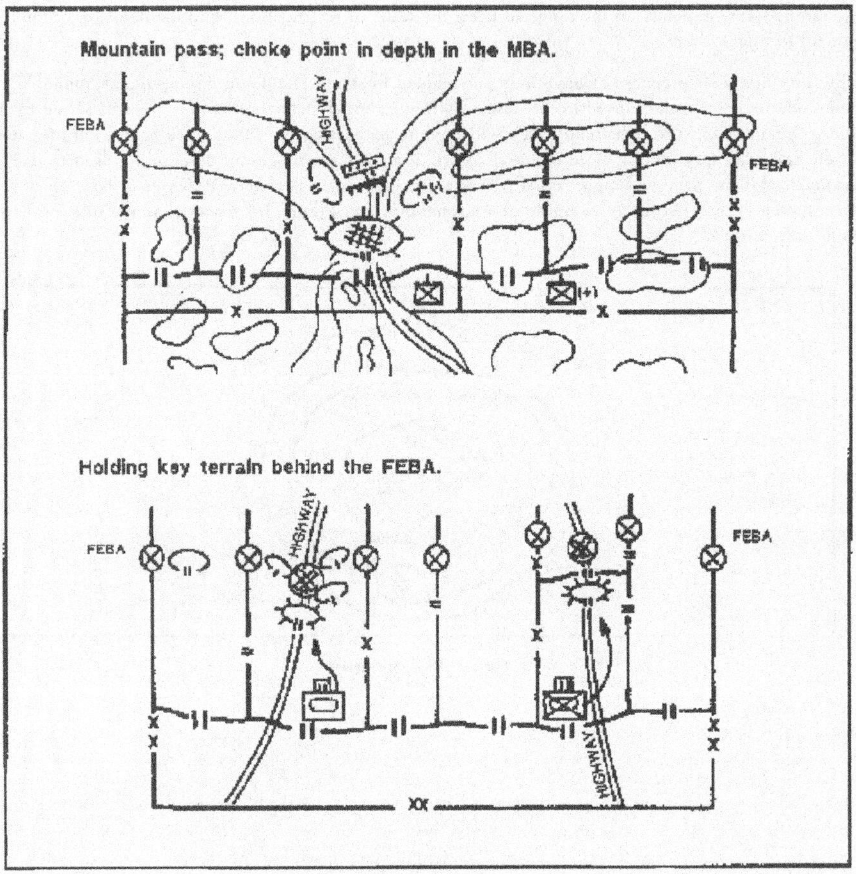

Figure 3-9, Strongpoints holding key terrain.

REVERSE SLOPE DEFENSE

The use of the reverse slope defense takes on added importance in the desert. Concealment is hard to achieve in the open desert. Detection of a unit's location invites both direct and indirect fires in abundance. The use of reverse slope positions will deny the enemy direct observation of positions until he is within the range of direct-fire weapons. Reverse slopes can even be found on seemingly flat desert floors; an intervisibility line will provide the reference for the establishment of engagement areas to support a reverse slope defense. A common misconception is that the desert is flat, when in fact, deserts are normally very uneven, with large breaks in the terrain.

Desert environments give special significance to the terrain aspect of METT-T. Commanders at all levels should place emphasis on the impact of desert terrain as it relates to the other factors of METT-T. The reverse slope defense in desert terrain warrants special considerations.

Direct-fire positions should be placed at the maximum effective ranges from the intervisibility line. This is where the enemy cannot see or engage a force with direct fire until he is within its engagement area. He can only deploy limited forces at a time. This allows the defender to mass fires on one portion of the enemy force at a time. The attacking force will have difficulty in observing and adjusting indirect fires. Obstacles may not be seen by the enemy until he is upon them and force him to breach under massed frees. Observation posts (OPs) positioned forward to see the advance of the enemy can influence the fight through indirect fires. The OPs can direct indirect fires on enemy forces that are slowed or stopped outside direct-fire ranges.

This defensive technique may be used in all defensive missions. Light infantry units use the reverse slope for protection against enemy long-range fires and to reduce the effects of massive indirect fires (artillery and close air support). The reverse slope defense brings the battle into the defender's weapons' ranges. Use of the reverse slope provides an opportunity to gain surprise.

The goal is to cause the enemy to commit his forces against the forward slope of the defense, resulting in his force attacking in an uncoordinated fashion across the crest. A reverse slope defense is organized on the portion of a terrain feature or slope that is masked from enemy direct fires and observation by the topographical crest, and extends rearward from the crest to maximize the range of the defender's weapon systems. See Figure 3-10 for an example of a reverse slope defense and Figure 3-11 for the organization of the reverse slope defense.

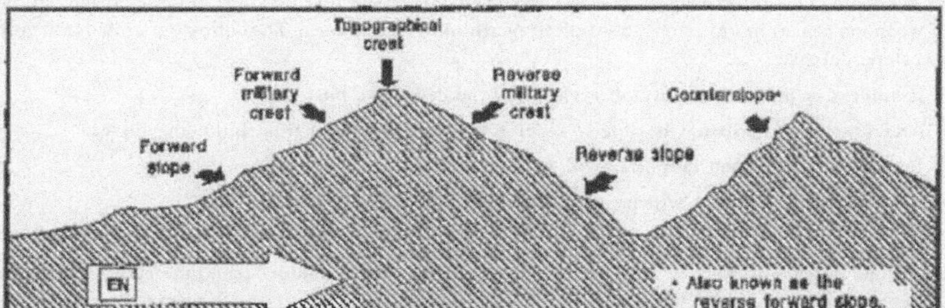

Figure 3-10. Reverse slope defense.

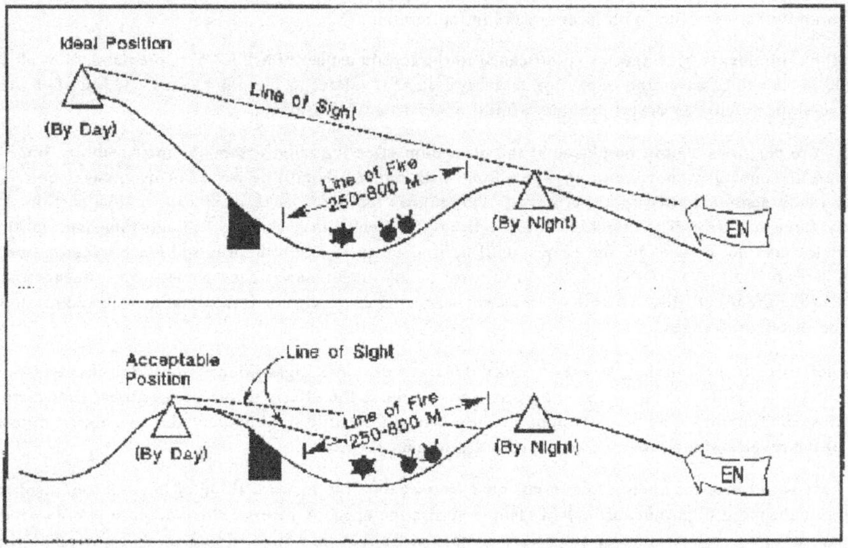

Figure 3-11. Organization of the reverse slope defence.

A disadvantage may be that the maximum ranges of weapon systems may not be employed due to the terrain available. The desert may be the best environment for the reverse slope defense. It may allow the use of weapons at maximum ranges as well as facilitating advantages. The following are advantages of a reverse slope defense:

- It hinders or prevents enemy observation of the defensive position.
- Attacking forces will not be able to receive direct-fire support from following forces.
- Enemy long-range antitank fires will be degraded.
- Attacking enemy forces will be silhouetted on the crest of the hill.
- Engineer work can be conducted away from direct-fire and observation from the enemy.

Reverse slope defense is not one concept, but a series of concepts that produce the potential for success. The concepts are-

- Pursue offensive opportunities through surprise and deceptive actions, with the intent of stealing the initiative, imposing the commander's will on the enemy, and breaking the enemy's morale.
- Afford the defender a variety of options in positioning his troops, with each option designed to draw the enemy into unfamiliar terrain.
- Enhance light infantry effectiveness and survivability.

A hasty or deliberate reverse slope defense may be considered when any of the following conditions exist:

- When the forward slope lacks cover and concealment, and effective enemy fire makes that position untenable.
- When the terrain on the reverse slope affords appreciably better fields of fire than those available on the forward slope.
- When it is desirable to avoid creating a distortion or dangerous salient in friendly lines by relying on forward slope positions.
- When it is essential to surprise and deceive the enemy as to the unit's true defensive positions or main effort.
- When seeking to gain protection from the enemy as he is massing fires.

DELAY OR WITHDRAWAL

When it is necessary to delay or withdraw, a desert offers many advantages to the defender. Long-range fields of fire allow engagements at maximum effective range of direct-fire weapon systems, and disengagement before the defender's position. However, dust clouds created by a moving force make it necessary to disengage under cover of smoke or darkness. Even a sandstorm can be used to the advantage. Field artillery, US Air Force fighter bombers, and attack helicopters can also be used to allow a ground maneuver unit to disengage and move rapidly to the next position.

When it is necessary to trade space for time, often a counterattack to destroy enemy advance units will do more good than trying to defend longer from an intermediate position.

Commanders at all levels should clearly understand the scheme of maneuver concept of the operation, and what it is they are expected to do, especially if communications should fail. Plans must include provisions for alternate means of communication. Routes should be clearly marked and reconnoitered to the maximum extent practical.

Due to the distances involved and constantly changing task organization, passage of lines is more difficult to coordinate and control. Pay extra attention to the identification of vehicles, routes of passage, signals, and coordination of movements.

Deception should be a part of all desert retrograde operations. The object of deception is to conceal the fact that a retrograde operation is taking place and that units are thinning out. Smoke and dummy positions can be used, false radio messages transmitted, and even dust clouds used to deceive the enemy.

ENVIRONMENTAL CONSIDERATIONS

In the desert it is necessary to modify the techniques of defense as described in doctrinal manuals applicable to each level of command and according to the mission, the fundamentals described in the preceding paragraph, and to the environmental considerations that are described in the following paragraphs.

Observation

The enemy will try to attack when the sun is low and behind him so as to dazzle the defender. The defender's observers must be as high as possible above the desert floor to see the advancing enemy as soon as possible.

Active light sources can be detected from great distances, especially during nights with low ambient light. Positive control of active light sources must be maintained until the battle is joined. Even then, the force equipped with passive devices will have the advantage over the force that is not equipped with these devices.

Heat from combat vehicles can give an enemy using thermal imagery devices a complete picture of the defensive scheme. So, combat vehicles should not prematurely occupy battle positions at night.

Sandstorms

Sandstorms may be used by the enemy to hide an offensive operation especially if the storm is blowing from the enemy's direction. When this is the case, units should immediately occupy their battle positions before the storm arrives. The unit should remain there until it ends, ready to fire and maneuver against the attacker after the storm abates. If vehicle patrolling is possible, a scout platoon or similar unit should cover all gaps, preferably moving in pairs, and on straight lines in view of navigational difficulties.

Terrain

From the point of view of a defending brigade or battalion task force commander, avenues of approach will often seem unlimited. Long-range observation must be maximized and scouts employed well forward to offset this problem. Radars should also be used extensively to provide early warning. It is necessary to identify the enemy's main effort early in order to move to concentrate.

Lack of concealment, especially from the aerial detection, prohibits units from occupying firing positions until just before engaging the enemy. Combat vehicles must displace immediately after engagement or risk destruction. Because of frequent displacement, mutes between battle positions should be reconnoitered and marked when possible, without revealing the scheme of defense. Smoke must be used frequently to conceal movement.

TACTICAL DECEPTION OPERATIONS

Analysis of desert operations from World War II to the present day indicates that tactical deception and surprise are clearly linked to the ability to move and mass forces during periods of limited visibility.

Operational planning should emphasize night movement of units. To minimize the problems of dust and to enhance deception, movement should be accomplished using multiple routes. Place priority on training to support this requirement. Associated with night movement is the requirement for night passage through lanes in minefield and forward passage through friendly forces.

In every modem desert war, deception has played a major role. The lack of concealment leads commanders to believe that with a reasonable reconnaissance effort they can gain an accurate picture of the enemy's dispositions. Reconnaissance by German, British, Israeli, Egyptian, and Syrian forces in modern desert warfare has been sufficient to detect the presence of combat forces in the desert. Deception has been successfully used in each of the modem desert conflicts to mislead commanders.

Since the desert environment makes it difficult to hide forces, the alternative is to make them look like something else-trucks and plywood made to look like tanks, and tanks made to look like trucks.

The movement of personnel and equipment and the placement of logistic support installations are normally indicators of a force's intent. The movement of empty boxes or pallets of ammunition and the establishment of fuel storage areas with real or dummy assets can deceive the enemy as to planned offensive actions. Use minimal actual transportation assets and make numerous, visible trips to simulate a large effort.

There are many examples of successful deception efforts by US forces from World War II In September 1944, the 43rd Cavalry Reconnaissance Squadron (Reinforced) occupied a 23-mile fronton the left flank of XX (US) Corps on the Metz Front. This squadron portrayed an armored division for several weeks and was so successful that the German Order of Battle Maps showed the 14th (US) Armored Division (AD) to be in the area. The 14th AD was not even in Europe at the time. Expertise in deception operations is critical to success.

Deception plays a key part in offensive operations and has two objectives: the first objective is to weaken the local defense by drawing reserves to another part of the battlefield. This may be done by making a small force seem larger than it is. The second objective is to conceal the avenue of approach and timing of the main attack. Some deception methods that can be used in offensive operations are-

- Using dummy units and installations.
- Using phony radio traffic.
- Using movement and suppressive fires in other areas timed to coincide with the real attack.
- Using small convoys to generate dust clouds.
- Filling ration boxes with sand and stacking them at landfills.
- Moving trucks into and out of the area giving it the appearance of being a storage facility or logistic base.
- Emulating damage to induce the enemy to leave important targets alone. For example, ragged patterns can be painted on the walls and roof of a building with tar and coal dust, and covers placed over them.
- Stacking debris nearby and wiring any unused portions for demolition. During an attack, covers are removed under cover of smoke generators, debris scattered, and demolitions blown. Subsequent enemy air photography will disclose a building that is too badly damaged to be used. Troops using the building after an attack must guard against heat emissions after dark and care must be taken to control electromagnetic emissions.
- Using phony minefield to simulate live minefields. For example, disturb the ground so that it appears that mines have been emplaced and mark boundaries with appropriate warnings.
- Making a real minefield to appear as phony or camouflaging it. For example, once a real minefield is settled, a wheel or a specially made circular wooden tank track marker can be run through the field, leaving track or tire marks to lure the enemy onto live mines. Antipersonnel mines should not be sown in such a field until the track marks have been laid. Another method is to leave gaps in the mechanically laid field, run vehicles through the gaps, and then close them with hand-laid mines without disturbing the track marks.
- Using decoys to confuse the enemy as to the strength of friendly forces and the unit's identity, or to conceal unit movement by being sited in a position after the real unit has moved. For more information on deception operations, see FM 90-2.

LONG-RANGE SURVEILLANCE OPERATIONS

Desert characteristics affecting LRS operations are: lack of water (a major problem), scarcity of vegetation, extensive sand areas, extreme temperature ranges, brilliant sunlight, and usually excellent observation. Movement using animals, vehicles, or by foot may be considered and is generally restricted to darkness. More training in land or air navigation and terrain orientation procedures may be necessary.

AIR ASSAULT OPERATIONS

An air assault task force provides commanders with truly unique capabilities. They can extend the battlefield, move, and rapidly concentrate combat power like no other forces.

An air assault task force uses the helicopter to move to and close with the enemy. Initial assault elements must be light and mobile. They are often separated from weapon systems, equipment, and materiel that provide protection and survivability on the battlefield. Thus, an air assault task force may be particularly vulnerable in a desert environment to enemy-

- Attack by aircraft and air defense weapon systems during the movement phase due to differences in desert effects on observation and fields of fire.
- Attacks (ground, air, artillery) during the loading and unloading phases and at other times when the infantry is not dug in.
- Small arms fire that presents a significant threat to helicopters.
- Artillery or other fires that may destroy helicopters and air assault forces during PZ (pickup zone) or LZ (landing zone) operations.

Air assault operations are discussed in greater detail in FM 90-4. Marine Corps assault support ensures the rapid buildup of combat power and facilitates the quick maneuver of ground forces. See FMFM 5-35 for more information.

AIRBORNE OPERATIONS

The airborne division is organized to rapidly deploy anywhere in the world. It is the only US Army division with a rapid, strategic, combined arms, forced-entry capability. It will most likely be the initial force deployed for contingency operations. It is ideally suited and primarily designed to seize, secure, and repair airfields in order to provide an airhead for follow-on forces and to delay, disrupt, and reduce enemy forces.

Airborne operations can be adversely affected by various environmental considerations of the desert. High winds, sandstorms, and heavy rainfall or thunderstorms may impact on mission accomplishment. Planning considerations for combat service support are complicated by the fact that airborne forces will be the first Army forces in an immature and austere desert theater. For more information on airborne operations, see FM 90-26.

MARINE AIR GROUND TASK FORCE OPERATIONS

Marine operating forces are organized for combat as Marine air ground task forces (MAGTFs) composed of command, ground combat, aviation combat, and combat service support elements. The MAGTFs are closely integrated combat forces capable of rapid response to any crisis or contingency. Their Naval/Marine Corps expeditionary nature makes them ideal for immature and austere environments as was seen in Operations Desert Shield/Storm (Southwest Asia) and Restore Hope (Somalia).

A variety of types of MAGTFs may be formed in support of national strategy and rapid crisis response. The Marine expeditionary force (MEF) is the Corps' principal organization for combat and peacetime readiness, and is formed from the legislated division and aircraft wing teams. These MEFs provide a reservoir of integrated combined arms combat power that can be task organized to simultaneously execute a wide range of global missions. The MAGTFs are mission tailored and range in size from very powerful MEFs, capable of prosecuting operational campaigns against the most capable potential threat through rapidly deployable and employable Marine expeditionary units (MEUs), to small special purpose forces (SPMAGTFs) formed for specific missions or crises. In the early moments of Operation Desert Shield, a MEF provided the nation a powerful combined arms combat force to stand against aggression, while US forces and equipment and supplies were being assembled in Southwest Asia. For more information on MAGTF operations, see FMFRP 2-12.

MARITIME PRE-POSITIONING FORCE OPERATIONS

The maritime pre-positioning force (MPF) gives the nation an added dimension in mobility, readiness, and global responsiveness. The MPF program involves 13 ships, organized in three squadrons. Maritime pre-positioning squadron one (MPSRON-1) operates in the eastern Atlantic Ocean, MPSRON-2 in the Indian Ocean, and MPSRON-3 in the western Pacific. The MPF, when called upon, provides equipment and 30 (days of supplies for a 16,000-man Marine expeditionary brigade (MEB). The MEB's personnel and selected equipment can be airlifted quickly using roughly 250 airlift sorties to an objective area to join with its equipment at a secure site. Equipment and supplies can also be selectively off-loaded to support smaller MAGTFs. During Operation Desert Shield/Storm, all three MPFs were off-loaded in Southwest Asia providing immediate support to deploying forces. During Restore Hope, one MPF supported operations. For more information on MPF operations, see FMFM 1-5.

ARMY SPECIAL OPERATIONS FORCES

To meet our nation's global commitments, the Army maintains a balanced force of armored, light, special operations, and support forces for use across the operational continuum. Army special operation forces (ARSOF) are an integral part of the total Army force. ARSOF have five elements: special forces, rangers, Army special operations aviation (ARSOA), PSYOP, and civil affairs. These forces offer significant capabilities to the desert theater of operations. Details of ARSOF capabilities are discussed in FM 100-25.

MILITARY OPERATIONS ON URBANIZED TERRAIN (MOUT)

Because there are few man-made features throughout the expanse of the desert, those that do exist can become important, perhaps even key. Key terrain in the desert can be any man-made feature. Settlements (where a logistics base may be established), road junctions, shelters, and airfields, all become important, simply because they are so few in number. Growing villages and settlements straddle these lines of communication, and small villages may exist near water sources and other key terrain. It may be necessary to conduct MOUT operations to control these areas. In areas involved with Desert Storm, paved roads and even dirt roads were considered key terrain for both high-speed movement and for providing clearly defined directions and locations. Commanders must be prepared to fight on terrain that is constantly being modified by man. More information on conducting MOUT can be found in FM 90-10.

CHAPTER 4

COMBAT SERVICE SUPPORT

There is a lack of every kind of resource in the desert, especially of the sophisticated infrastructure of ports and railways for their high capacity for moving combat supplies. Logistical support is always a challenge, and an arid environment burdens all types-supply, aviation, communications, and maintenance. Commanders must be sensitive to the constraints, and those providing support must work to overcome them. A unit's tactical effectiveness in the desert depends to a large degree on the combat service support available. Equally, its vulnerability lies in its exposed lines of communications and the immobility of its bases of supply and support.

Inherent to the success of any tactical operation is continuous, sound, logistical planning for adequate supply, medical, and maintenance support. This is especially important in the desert because the greater distances used in maneuver and deployment complicate supply procedures. Other reasons are the shortage of locally available water and the increased maintenance requirements due to sand and dust damage. The effects of the environment on equipment are severe, requiring increased levels of support to maintain a standard level of efficiency. The extended supply lines required for expanded frontages call for special considerations and procedures to ensure adequate and timely supplies arrive to sustain combat in the desert.

US forces in the desert operate at the end of a long, perhaps tenuous line of communication. Cargo space must not be wasted to provide all the comforts of home. A significant difference in living standards between rear area support personnel and those in forward combat areas must be avoided since this can affect morale and weaken the ability of combat units to resist psychological warfare. Transportation priority must be given to minimum essential materials and the support base should be austere.

Section I. Base Development Plan

US forces deploying for operations in a desert environment should expect to begin operations from a lodgement area. When this is the case, it is necessary for the headquarters deploying the force to prepare a detailed base development plan. How the plan is developed will depend on a number of factors that are described below:

- The mission and size of the force. The size of the force depends on its mission and the operations it is expected to conduct. The size of US forces deployed for desert operations could vary from a small force conducting a show of force, to a joint task force capable of full-scale operations.
- Security of the lodgement area. A lodgement area will probably be secured by allied forces or US Marines before deployment of US Army forces into the operational area. However, it may be necessary to use US Army forces, either air-dropped or air-landed, to secure a lodgement area.

- Transportation of US forces into the lodgement area. It is probable that initial forces will be transported by air and follow-on forces by sea. Another possibility is that initial forces will be transported by sea with follow-on personnel being transported by air.
- Strategic lines of communication (LOC). The initial strategic LOC will probably be an air LOC. However, at some point in the operation a sea LOC will be established to convey the bulk of the supplies, supplemented by an air LOC to haul time-sensitive items.
- Theater lines of communication. Lines of communication within the operations area should be analyzed before selecting the lodgement area. The analysis should include ports available, airfields throughout the operational area, road nets, and railroads. It may be necessary to stage engineer construction units into the operational area early to improve existing facilities and LOCs or to construct new ones. In a single or multi-corps theater, a theater army headquarters provides overall management of CSS operations. It establishes priorities, assigns missions, and allocates resources in accordance with the theater army commander's concept of operations.
- Local resources. These are extremely important as they will affect logistics planning. Typical information about resources in the operational area that should be obtained before base development planning includes-

  -Airfields.

  -Water sources.

  -Fresh rations.

  -Labor supply.

  -Construction materials and available equipment.

  -Material-handling equipment at ports and airfields.

  -Local hospitals, maintenance capability, and storage sites.

  -Local power supply to include types and equipment.

  -Railroad rolling stock and gauge of tracks in local areas.

After consideration of the factors listed above, the lodgement area is selected Ideally, a lodgement area should have a deep water port and airfield suitable for heavy strategic airlift, located at the end of an adequate road or rail system suitable for an intratheater LOC. Once the lodgement area has been selected, then LOC-port units can be specially tailored for early deployment to the operational area.

Section II. Theater Support

THEATER HEADQUARTERS COMMAND

Should a friendly nation ask for military assistance either following an invasion or in anticipation of one, the first combative units to arrive in the theater will probably be a force designed to secure entry points. The assembly and movement of armor and mechanized forces will take time, time that the logistics staffs can put to good use preparing for the reception of these forces. Elements of the theater staff can assess the assets and facilities available in the following areas and make an estimate of the work, labor, equipment, and other resources required to support the buildup of the force.

Logistics Reconnaissance

Port capacity, cranes and off-loading equipment, storage facilities, the availability of stevedores, and local shipping assets at the main port facilities should be determined. Contract and requisition of host nation support assets can be arranged by initial theater staff personnel. The theater staff should also determine airfield facilities and capacities that are available. Railway assets that can be made available for logistics support and the movement of troops should also be determined.

Local Resources

The host nation in a Middle Eastern theater will probably be an oil producer and may be able to provide the bulk of our fuel requirements. However, if the host nation exports its oil in bulk crude, it may have only a limited refinery capacity to meet a local domestic demand for diesel, kerosene, high-octane fuel, and lubricants.

Fresh Rations

The host nation may be able to produce a limited supply of fresh rations but the bulk may have to be imported from neighboring countries. If there is arable land within the host nation, it may be possible to start farms to relieve the burden on the local economy and on that of neighboring friendly states.

Hospitals

The extent to which hospitals in the host nation can accept long-term patients until our own base hospitals can be established is also an issue of concern.

Power Supply

The share of the host nation's electric power supplies which can be offered to our forces must be determined. Initially, there is unlikely to be any shortage in power supply, but as a large force (including other allies) builds up there maybe a generating capacity problem. At the beginning of the campaign the difficulty may be in distributing electricity where it is required throughout the theater. It may be necessary to construct overhead or underground cables together with transformers.

Transient Camps

Sites should be located for transient camps for troops arriving in the theater and convalescent camps for the recovery of the sick and wounded.

Local Currency

Local currency may also be required for extended operations, both to pay troops and to locally purchase supplies.

Maps

Maps are a critical item that maybe more readily available through local survey teams or oil companies.

Interpreters

Interpreters will be required to communicate with the host-nation troops, contractors, and labor forces.

Stockpiles

Higher stock levels are required in the desert due to the following factors: use of the limited life of many perishable items in a harsh environment; the enormous distances stockpiles must be lifted and over inadequate transport systems; the loss of supplies due to sudden changes in the fortunes of war, and the time it takes to replace items from the US. The levels of each commodity to be held in theater and the proportion of the totals will be decided during the staff planning process. The distribution will depend on the tactical situation and the vulnerability of the lines of communication to enemy action.

Expenditure Rates

While the requirement for rations and water remain relatively constant, the expenditure of fuel and ammunition will vary far more, not just because of the fluctuation between quiet periods and intense operations, but because of the environment. The amount of driving in soft sand and the longer distances to be traversed combine to increase consumption beyond central European rates. Similarly, the expenditure of tank and artillery ammunition may be increased because of the open terrain.

Host-Nation Support

In a desert environment where resources, CSS personnel, and equipment are limited, the use of host-nation support assets can be vital to the success of an operation. Host-nation support assists in the accomplishment of missions and functions in support of US forces and enhances their capability to perform their wartime role.

All forms of peacetime transition to wartime, and wartime host-nation support should be included in the planning process.

Host-nation support includes-

- Government agency support such as police, fire companies, and border patrols, may be available to support US forces.
- Contractor support such as supplies and services, including laundry, bath, bakery, transportation, labor, and construction.
- Host-nation civilians may be able to provide needed skills for laborers, stevedores, truck drivers, managers, and technicians.
- Host-nation military units may provide traffic control, convoy escort, installation security, cargo and troop transport, POL storage and distribution, and rear operations.
- Host-nation facilities may be contracted and used for hospitals, headquarters, billets, maintenance shops, or other activities.

- Functional or area support maybe provided in the form of rail operations, convoy scheduling, air traffic control, and harbor pilot services.
- Services may be provided by the host nation for gymnasiums, recreation facilities, and other morale and welfare demands.
- Supplies and equipment needed for missions may be acquired locally, precluding or reducing materiel shipments from CONUS.

Section III. Corps Support Command

An Army corps support command (COSCOM) deploying to support desert operations must be carefully tailored to meet the needs of combat forces operating in a harsh environment. Requirements for long-haul truck companies, engineer construction battalions, water production units, and LOC-port units previously described, must be carefully weighed. A shortfall of these units could significantly impair combat operations. Organization of the COSCOM should be planned based on the factors described in the previous paragraphs, with particular attention given to-

- Number of troops to be supported.
- Quantity and types of equipment to be maintained.
- Tonnage to be handled.
- Available local resources and labor force.
- **Types** of units to be deployed to the theater of operations.

The organization of the COSCOM and a description of its tasks are provided in FM 63-3J. Initial corps forces entering the theater can be supported by a forward support battalion (FSB) of a division support command and a corps support battalion (CSB) of a corps. Once initial forces have arrived in the theater, additional tailored elements from the COSCOM must immediately follow, or even arrive first to minimize the requirement for the DISCOM cooperate such activities as ports or airheads.

Section IV. Division Support Command

As previously mentioned, combat service support units are high-priority targets for any desert enemy. In most cases, Army division support command (DISCOM) units will not be able to provide for their own security, considering the many ways in which they could be attacked. Air defense protection must be provided. It may even be necessary to provide a maneuver unit or additional MP units to secure DISCOM elements. Nearby maneuver units can also be designated to move to their defense-attack helicopters are especially suited for this purpose-and on-call fires should be planned by nearby field artillery units. Any pipelines in the division area must also be secured by any means at hand. Observation helicopters can be used to patrol pipelines.

Stocks should be kept as mobile as possible in the event rapid displacement is necessary. Stockpiling off vehicles must be held to a minimum, as should stockage levels. To the extent practical, supplies located forward of the division support area should be stored aboard vehicles to minimize the Possibility of having to leave them behind. For this purpose, a force operating in the desert should be augmented with additional transportation assets.

DISCOM organizations of the lead divisions in austere and immature theaters may be called upon to establish forward logistic bases. In these situations division assets may have to assume other support or transportation responsibilities temporarily until area support groups can establish support operations and transportation.

Section V. Combat Service Support Element

The Marine Corps' combat service support element (CSSE) is a task organized service support element of the MAGTF. Its composition is based on many factors, to include-

- MAGTF size (MEF, MEB, MEU, or SPMAGTF).
- MAGTF mission.
- Type of operation.
- Area of operation.

The considerations listed under Sections III and IV are also true for a CSSE. For more information on CSSE operations, see FMFM 4-1.

Section VI. Support Operations

Listed below are some of the factors that make support operations complicated.

SUPPLY

Consumption rates must often be developed after the force has operated for some time in the area. Water has to be found, purified, stored, and transported.

MAINTENANCE

There will be a greater demand for such items as filters, oils, and lubricants. More Class IX stores are required than normal, and the work load on maintenance units is much greater. Supply items and spare parts should be packed or wrapped as if to be air and water tight to prevent blowing sand from contaminating or damaging them. All echelons that request supplies and repair parts should be using the same or compatible equipment for requisitioning, with alternative means in place as a backup so there is little or no slow down in the reorder process.

CONCEALMENT

It is difficult to conceal trains areas. However, trains areas must be concealed to the best extent possible. These are soft targets in any environment and are high-priority enemy targets as their destruction (especially water, HETTS, and fuel supplies) effectively cripples the force.

DISTRIBUTION

Maneuver units may be farther apart, both in width and depth, than in temperate environments. They move more frequently and faster. Lines of communication are longer. Terrain away from the main supply mute (MSR) maybe such that it is only trafficked by cross-country vehicles, and then only with reduced payloads. Lack of significant terrain features may increase navigational problems, requiring local guides.

COMBAT SERVICE SUPPORT PLANNING

The commander's intent and METT-T analysis must dictate the CSS plan to support the tactical mission. However, CSS planners must not become locked into rigid CSS plans. The situation will dictate how trains are configured, echeloned, and controlled. Commanders and their staffs must use a logical and fast means to evaluate the battlefield and reach decisions. The military decision-making process described in FM 101-5 provides the framework within which the commander and staff interact to arrive at and execute a decision. Battlefield support must be planned to satisfy requirements during the following operational phases:

- Prior to D-day (before).
- Commitment to battle (during).
- Future mission (after).

All areas of CSS (man, arm, fuel, fix, move, and protect) must be considered during each operational phase to ensure an integrated, responsive plan of support. Support requirements must be projected and plans developed to satisfy these projected requirements. Supporting CSS plans should be as detailed as planning time permits.

CSS commanders and planners must thoroughly understand the tactical mission and plans and the commander's intent. They must know-

- What each of the supported elements will be doing.
- When they will do it.
- How they will do it.
- Where they will do it.
- What the priority of support is.
- Density of personnel and equipment being supported.

After analyzing the concept of the operation, CSS commanders and planners must be able to accurately predict support requirements. They must determine-

- What type of support is required.
- What quantities of support are required.
- The operational commander's priorities, by type and unit.

Using the support requirement of the tactical plan as a base, the support capabilities of the CSS structure are assessed. The staff must determine-

- What CSS resources are available (organic, lateral, and higher headquarters).
- Where the CSS resources are.
- When CSS resources will be available to maneuver units.
- How the FSB will make these resources available.

Based on this information, the staff must then develop support plans that apply resources against requirements in a manner that results in the most responsive support possible. Communications links must be established and maintained.

Orders that clearly describe tasks to be accomplished must be issued. Continuous follow-up must ensure tasks are being accomplished as planned.

CSS functions should be performed as far forward as the tactical situation and available resources will permit. They should be performed at or close to the site where the weapon system is located to lessen evacuation requirements. Support must be continuous, using immediately available assets. This will involve bringing ammunition, fuels, parts, end items, maintenance personnel, and occasionally replacement crews or individuals, to the forward elements such as battalion field trains, combat trains, and equipment downsites. Planning and execution emphasize the concept of providing support to forces in the forward areas.

CSS planners must know priorities for support. This is necessary to ensure that units with the highest tactical priority receive required support first. The commander and his staff provide mission directives, determine CSS requirements, and establish priorities within the unit.

Section VII. Security of Supply Routes

Long lines of communications require convoys from the support base to the combat forces, and convoys are subject to air attacks (as learned during World War II when convoys from Casablanca to Al Guettar Tunisia, were frequently targeted by Luftwaffe raids), Enemy ambushes on main supply routes (MSR) are always a threat in desert operations. Enemy patrols may also place nuisance mines on routes, especially at critical points such as defiles. Actions must be taken to minimize the threat to supply routes.

The MSR will be considerably longer in the desert. The logistical assets utilizing the MSR are extremely vulnerable and must be protected. It maybe necessary to allocate maneuver forces to maintain an open and relatively safe MSR. This will allow supplies to be pushed forward and casualties to be moved to the rear. Coordination must be accomplished between maneuver elements as to where responsibility will end and begin on the MSR during each phase of an operation. The MSR requires constant patrolling to ensure safest operations and continuance of supplies to the maneuver forces. Marking of the MSR facilitates security. Different techniques for marking MSRs range from chemical lights, to spray paint, to signs. MPs can also position themselves along the MSR to help guide units.

Section VIII. Supply

Time and distance factors developed on different terrain by experience are of little value in the desert. The absence of roads in forward areas, navigation problems, vulnerability of trains and supply installations to attack by ground forces or aircraft, sandstorms, and wide dispersion, all require a different appreciation of time for resupply operations.

CLASSES OF SUPPLY

Requirements for supplies vary from that of temperate climates according to the classification of supply. Differences that may be expected in any desert are described in the following paragraphs.

Class I

Until the theater is fully developed and ration requests can be implemented, ensure enough MRE rations for three to five days are stored on combat vehicles. Meals from this combat load are eaten only when daily Class I resupply cannot be accomplished. Frequency of unit feeding and use of A or B rations depends on the tactical situations, If possible, troops should receive at least one hot meal per day. Hot rations should be packed in platoon-size portions rather than consolidating company-size packages.

It is critical to plan for the cooling of water supplies. Troops will drink any potable water available to them; however, they would prefer chilled water. Commanders and staffs must plan for water coding systems, ice, and individual soldier/marine field-expedient devices. Troops fighting in the desert will likely be wearing the battle dress overgarment and body armor. This fact will impact on the planning for water consumption.

The key to having enough water in a task force conducting desert operations is the capacity of that force to store and transport it. Current water trailers are inadequate. The potential for water consumption is high when you consider personal use and consumption, decontamination, medical needs, messing operations, and maintenance uses. Possible solutions include converting fuel tankers to water tankers, the use of blivets, and local purchase of civilian water-holding tanks through host-nation support. Water is vital, yet local supplies may be scarce or nonexistent in a desert combat zone. If water is plentiful, as it is in mess around Tripoli and Benghasi in Libya, water supply should not be a problem, provided that normal water supply procedures are followed.

This paragraph deals with situations where local water is difficult to obtain. All units must maintain a continuous watch for possible sites such as oases, dry wells, dry water courses, open water (even marshes), or captured enemy dumps. These should be reported to the next higher headquarters, giving the location and quantity and flow, if possible. It is not the responsibility of these units to test the water for potability, which could be dangerous for untrained personnel. This task should be left to specialists.

Since distances between water points may be long, it may be desirable to augment the division with additional 5,000- and 2,500-gallon bulk water tankers, processed to haul water.

Priorities for water use should be established. See Appendix G for a suggested list of water priorities and additional information concerning drinking water needs, water requirements, and water heating rates. If vehicle decontamination is necessary, it will take a high priority. NonPotable water should be used for this task.

Class II

There is little change in Class II consumption. However, clothing variations, from tropical clothing to sweaters and sleeping bags, must be anticipated Requirements for items such as neck scarves and canteens will be increased as well as those for hand tools, since tools tend to get lost in the sand.

Class ill

There is a marked increase in oils and lubricants used in preventive maintenance however, the actual quantities depend on operating conditions. Some types of desert terrain can lead to greatly increased fuel consumption per mile moved or hours that equipment is used. Use of cans or fuel bladders in certain circumstances should also be considered as they allow fuel to be spread more evenly among cargo vehicles since a loaded fuel tanker's cross-country capability may be degraded in desert sand. HEMTTS may be a suitable replacement vehicle to solve the cross-country mobility problem.

Antifreeze requirements remain roughly the same as in temperate climates as antifreeze increases the boiling point of coolant and decreases wear on liquid-cooled engines. Various oils and lubricants are required in smaller user containers. This assists in preventing sand from contaminating larger containers since they would have to be moved from site to site, and opened and closed numerous times.

Class IV

The requirement for Class IV stores can be significantly more than in other theaters, and consumption of some items such as sandbags is greatly increased. Maximum use must be made of local materials. An engineer reconnaissance unit should be present in the theater from the initial buildup to establish what resources are available. All possible Class IV should be carried and incorporated into vehicle load plans when deploying. The construction of airstrips, minor port facilities, and rehabilitation of major port facilities and railways, are all engineer missions of particular importance in desert warfare.

Class V

Due to excellent firing conditions, and the need for extensive suppressive fires, ammunition consumption may be high. It may be necessary to restrict firing of certain types of ammunition once they have reached predesignated levels unless command approval is obtained. Battalion task force trains should contain a one-day supply of ammunition and missiles for all vehicles in the task force. Ammunition should be divided between combat and field trains when trains are echeloned. Units should keep ammunition as packaged until it can be uploaded on combat vehicles. This will protect the ammunition from sand that could cause weapon malfunction.

If artificial obstacles are to be employed, considerable quantities of mines will be required as minefield must be long and deep to be effective. Since extensive minefield will be preplanned, relatively few antitank mines need to be held in ammunition supply points forward of the division support area. When required, the quantities needed should be moved as close to minefield locations as possible. Only mines necessary to replenish unit basic loads used for local defense need be stocked forward of the division.

Class VI

The demand for Class VI supplies, especially beverages, is high. They are not, however, essential and if transportation is limited they are given a low priority, especially if refrigeration space is certain to be in short supply. Sundries packs can also be used.

Class VII

The demand for Class VII supplies depends greatly on maneuver and the intensity of the battle. The only variation that can be forecast is for refrigeration equipment, especially if it is necessary to move deceased personnel to the United States for burial.

Class VIII

Class VIII supplies may vary in type, but is unlikely that the overall quantity will vary significantly from that required in temperate climates.

Class IX

There is a large increase in demand for Class IX supplies due to environmental effects on equipment and the extra maintenance effort required. Small items with high usage rates should be held as far forward as team trains and may also be kept on fighting vehicles. Typical high consumption items are-

- Tires for wheel vehicles.
- Water pumps, gaskets, fan belts, water hoses, and clamps.
- All parts for ignition systems.
- Wheel and sprocket nuts, and wedge bolts.
- Spare caps for all liquid containers.
- Speedometers and cables (due to dead-reckoning navigation these are critical items).
- Filter elements.

FORECASTING

Due to extended lines of communication, consumption forecasts are very important in desert operations. Forecasts should be provided once a day and should include.

- A POL forecast for the next 24 hours.
- Status of the unit's basic ammunition loads.
- Equipment losses in the past 24 hours not previously reported.
- Status of reserve water and rations.
- Special supply shortages or maintenance problems not previously reported.

Section IX . Maintenance

In order to return equipment to battle as quickly as possible, repair of disabled equipment must be accomplished as close to the site of damage as possible. Evacuation should be limited whenever possible.

Due to unit dispersion, organizational maintenance personnel and direct-support contact teams will be thinly spread, so vehicle crews must be trained to make as many adjustments and repairs as they can. General guidelines for desert repair are-

- Repair only what is necessary to make the equipment combat effective.
- Recover and then evacuate to the nearest reasonably secure site, followed by on-the-spot repair.

An SOP for recovery and repair must be established either before or immediately upon arrival in the theater. The SOP should include-

- Guidelines for crew-level recovery and expedient repair.
- Recovery by organizational maintenance.

- Recovery by direct-support maintenance.
- Priorities for recovery by vehicle type.
- Limitations on field expedients. For example, the distance or time over which one tank is allowed to tow another tank considering the heat buildup in the transmission in this environment.
- Recovery of classified equipment such as crypto.
- Security and guides for recovery teams.

Section X. Personnel Support

POSTAL SERVICES

Mail is the soldier's/marine's link to family and friends. Inefficient distribution of mail can quickly undermine morale, regardless of the theater. Mail may be particularly affected by longer lines of communications in a desert theater of operations. Mail is important to the soldier/marine in the desert as it assists in defeating the sense of isolation caused by the environment and the necessary dispersion of units, It is especially important in the first few weeks to counter the shock of entering totally new terrain. Transportation of mail should be given a high priority on arrival in the theater of operations.

FINANCE SERVICES

The mission of finance support organizations during conflict is to provide high-priority support to the soldier/marine on an area basis. Mobile pay teams from corps-level finance organizations provide support to brigade-size units. Generally, finance support will not change in the desert environment.

LEGAL SERVICES

Legal service support will be provided to the commander and to troops by personnel of the division staff judge advocate section. This support will be on an as-required basis coordinated through personnel support channels. Legal advice will be available for the following areas:

- International law.
- Operational law.
- Foreign law.
- Status of forces agreements (SOFA).
- Rules of engagement.
- Claims and compensation payments.
- Public affairs.

RELIGIOUS SUPPORT

The chaplain is the staff officer responsible for implementation of the unit religious program. Included in this program are worship opportunities, administration of sacraments, rites and ordinances, pastoral care and counseling, development and mangement of the unit ministry team (UMT), advice to the commander and staff on matters of morals, morale as affected by religion, and ministry in support of combat shock casualty treatment. Many of the above elements may be affected by the religion of the host nation. With many of the deserts being in predominantly Muslim cultures, religious support may be affected and should be a consideration prior to deployment.

Section XI. Health Services

Medical unit requirements for desert operations are essentially the same as for temperate climates. It is essential that each brigade has an environmental sanitation team attached. When planning for medical support the following factors should be considered:

Increased dispersion and large areas over which battles are fought increases vehicle evacuation time. This problem can be further complicated if the enemy does not recognize the protection of the Red Cross, thereby inhibiting air evacuation within the range of enemy air defense weapons. The importance of units having trained combat lifesavers is critical to overcoming this. The reduction of the number of deaths due to slow evacuation time can be directly affected by the combat lifesavers available. One combat lifesaver per combat vehicle is an adequate number.

The comparatively long distances between units may limit the availability of medical aidmen to adequately support combat troops. Reinforcements may be required from the division medical battalion or from supporting corps level medical units. Augmentation should include vehicles as well as personnel.

The incidence of illness from heat injuries and diseases are higher than in temperate climates. Fevers, diarrhea, and vomiting, for example, cause loss of water and salt, which can culminate in heat illnesses. Cold weather injuries can also occur during a desert winter.

The mobility required of maneuver units will be inhibited if movement of any part of these units, including trains, is restricted by having to hold a number of casualties; therefore, the wounded and sick must be evacuated immediately.

In order to properly treat patients, all medical treatment facilities should be provided additional supplies of water. Medical personnel at all levels must assist tactical commanders in preventing or reducing heat casualties within their units.

Divisional medical units should be augmented with extra field ambulances from corps units. In an emergency, empty cargo trucks moving to the rear can be used for medical evacuation.

Evacuation from the battalion combat trains back to the brigade ambulance exchange point (AXP) or clearing station, will be performed by ground or air transportation. METT-T, availability of equipment, and the patient's condition, will be the determining factors on what method of transportation will be utilized.

The effects of nuclear weapons can be expected to be greater in desert terrain. Introduction of nuclear weapons by the enemy will greatly increase casualties and severely strain available medical resources. The same effects can be expected if the enemy introduces chemical weapons against unprepared troops.

Section XII. Naval and Air Force Assistance

NAVAL

During the initial stages of an operation it may be necessary to request logistic support from the US Navy. Ships, with the exception of a few special types, are neither designed nor equipped to give logistic support to ground forces. Limited support may be available if it is adequately coordinated in advance. A cruiser, for example, may have more than 20,000 gallons of water per day available beyond the requirements of the crew. Limited supplies of items such as bread may be available. Limited surgical and medical assistance may also be available.

AIR FORCE

The military air command (MAC) provides tactical airlift in support of the force. Air Force assistance must be coordinated with MAC to deliver personnel, supplies, and equipment forward to brigades and farther forward when necessary. Delivery is made by the most suitable means available, air landing, extraction, or airdrop. MAC also makes aircraft available for rearward movement of wounded persons or prisoners of war.

Section XIII. Other Combat Service Support Issues

MORTUARY AFFAIRS

The mortuary affairs program provides peacetime and wartime support to search, recover, identify, evacuate, and, when required, temporarily inter, disinter and re-inter deceased US military or civilian personnel, and allied and enemy dead. In addition, the program provides support to collect, inventory, store, and process personal effects (PE) of deceased, missing, captured, and medically evacuated US personnel and deceased allied and enemy personnel.

The goal of the mortuary affairs program is to search, recover, identify, and evacuate the remains of US military and certain civilian personnel from the theater of operations as long as feasible using available US Air Force aircraft.

The longer lines of communication required in desert environments may affect the evacuation of the deceased and therefore must be a planning consideration. Transport remains in palletized transfer cases when the tactical and logistical situation permits.

When the situation prohibits immediate evacuation, remains may have to be temporarily interred within the theater. When possible, the use of temporary cemeteries will be confined to echelons above corps. However, emergency war burial (mass burial) sites, as authorized by the theater commander, may be required as far forward as the brigade area. Desert environmental factors should be considered when establishing temporary interment sites, as the desert environment can significantly affect burial sites (strong winds, flash flooding and so forth. Remains will be buried with their personal effects in these burial sites to assist in the identification of remains when they are disinterred.

This concept will permit the use of host nation support (HNS) to dig and fill mass burial sites. Host nation support laborers will not actually handle or process remains or personal effects of US personnel, but will generally provide labor for interment operations. The mortuary affairs company commander has overall technical responsibility for the layout and survey accuracy of the cemetery. During World War II HNS/EPW labor was used successfully by mortuary affairs units.

CAPTURED MATERIEL AND PERSONNEL

In a desert theater of operations, where resources are scarce to begin with, the innovative use of captured materiel can be critical. This materiel can contribute to the retention of momentum by maneuver forces and decreases the need to consume our own supply stocks and to transport them to using units. Obvious sources are captured or overrun enemy fuel supply points, and materiel which may be used for barrier and fortification construction. Food and medical supplies may be used to feed and treat EPWs and civilians. Commanders and staffs must have a workable plan for handling EPWs. These potentially overwhelming requirements include health services, transportation, security forces, and so forth.

CLOTHING EXCHANGE AND BATH SERVICES

Clothing exchange and bath (CEB) services are provided by the supply and service company, when augmented. CEB services are requested through the brigade S4. The request must specify the location of the unit making the request, desired time for service, and range of clothing sizes for unit members. The requesting unit must be prepared to assist troops in setting up the CEB point.

NOTE: These activities may not be possible in the early stages of an operation, or not at all in forward areas. Additional effort will be required to provide these services in the desert.

APPENDIX A

DESERT COUNTRIES OF THE WORLD

This appendix provides brief descriptions of notable deserts of the world. Detailed information can be obtained from a terrain analysis from the Defense Intelligence Agency and the relevant volume of the DA Pamphlet 550-series. See Figure A-1 for desert locations of the world.

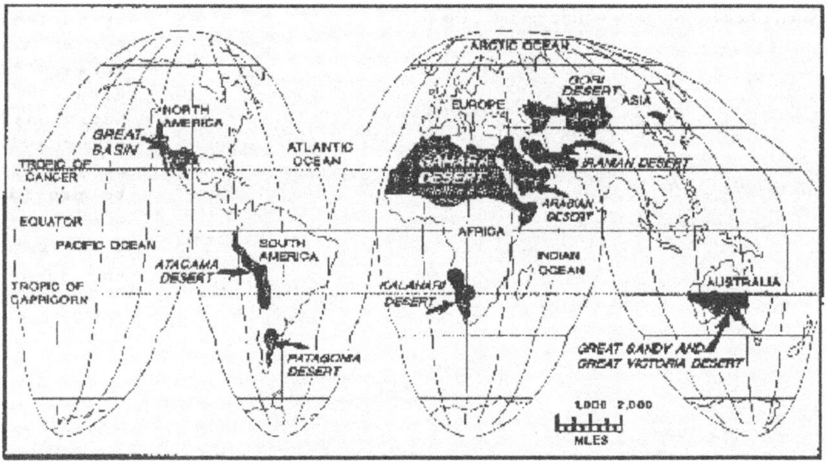

Figure A-1. Desens of the world

NOTABLE DESERTS OF THE WORLD

ARABIAN (EASTERN)	70,000 square miles	In Egypt between the Nile River and Red Sea, extending southward into the Sudan
ATACAMA	600 miles long	In northern Chile
CHIHUAHUA	140,000 square miles	In New Mexico, Texas, Arizona and Mexico

COLORADO	7,500 square miles	South of the Mojave desert, along the Colorado River in southern California
DEATH VALLEY	3,300 square miles	In eastern California and southwest Nevada
GIBSON	120,000 square miles	In the interior of western Australia
GOBI	500,000 square miles	In Mongolia and China
GREAT SANDY	150,000 square miles	In western Australia
GREAT VICTORIA	150,000 square miles	In western and southern Australia
KALAHARI	225,000 square miles	In southern Africa
KARA-KUM	120,000 square miles	In Turkmenia (former Soviet Union)
KAVIR	400 miles long	In central Iran
KYZLKUM	100,000 square miles	Kazakhstan and Uzbekistan (former Soviet Union)
LUT	20,000 square miles	In eastern Iran
MOJAVE	15,000 square miles	In southern California
NAFUD	40,000 square miles	Near Jawf in Saudi Arabia
NAMIB	800 miles long	Along the southwest coast of Africa
NUBIAN	100,000 square miles	In the Sahara in northeast Sudan
PAINTED DESERT	150 miles long In northern Arizona	In northern Arizona
SAHARA	3,500,000 square miles	In Northern Africa extending westward to the Atlantic (largest desert in the world)
SIMPSON	40,000 square miles	In central Australia
SONORAN	70,000 square miles	In southwest Arizona and southeast California
SYRIAN	100,000 square miles	In northern Saudi Arabia eastern Jordan, southern Syria,and western Iraq

Table 1-**1.** Desert counties of the world.

COUNTRY	AREA SQ MILES	POPULATION (MILLIONS)	LANGUAGE(S)	ADJACENT COUNTRIES
Algeria	919,487	24.2	Arabic French Berber	Morocco Western Sahara Mauritania Mali Niger Libya Tunisia
Bahrain	255	.480	Arabic English Farsi Urdu	Saudi Arabia Qatar
Egypt	386,660	53.3	Arabic English French Arabic	Libya Sudan Israel
Iran	636,293	51.9	Farsi Turkish Kurdish Arabic English	Iraq Turkey Former Soviet Union Afghanistan Pakistan
Iraq	167,924	17.6	Arabic Kurdish Assyrian Armenian	Syria Jordan Saudi Arabia Kuwait Iran Turkey
Israel	8,019	4.3	Hebrew Arabic English French	Egypt Jordan Syria Lebanon
Jordan	35,475	2.9	Arabic English	Israel Syria Iraq Saudi Arabia
Kuwait	6,880	1.9	Arabic English	Iraq Saudi Arabia
Lebanon	4,015	2.7	Arabic French Armenian English	Syria Israel

Table 1-1- Desert countries of the world (cont).

COUNTRY	AREA SQ MILES	POPULATION (MILLIONS)	LANGUAGES	ADJACENT COUNTRIES
Libya	679,359	4.0	Arabic Italian English	Egypt Sudan Chad Niger Algeria Tunisia
Oman	82,031	1.3	Arabic English Baluchi Urdu	UAE Saudi Arabia Yemen
Qatar	4,247	.328	Arabic English	UAE Saudi Arabia Bahrain
Saudi Arabia	829,996	15.5	Arabic	Bahrain Iraq Jordan Yemen Oman UAE Qatar Kuwait
Syria	71,498	11.6	Arabic Kurdish	Lebanon Israel Turkey Iraq Jordan
Tunisia	63,170	7.7	Arabic French	Algeria Libya
United Arab Emirates	32,278	2.0	Arabic English Farsi Hindi Urdu	Qatar Saudi Arabia Oman

AUSTRALIA

DESERTS: GIBSON, GREAT SANDY, GREAT VICTORIA, SIMPSON

AUSTRALIA - GENERAL INFORMATION

Area	2,967,895 sq mi
Population Growth	1.17 %
Population Density	5/sq mi
Capital City	CANBERRA

AUSTRALIA - MAJOR CITIES

	Population	Latitude, Longitude
CANBERRA	274,000	35.18S, 149.08E
Sydney	3,441,000	33.53S, 151.10E
Melbourne	2,949,000	37.45S, 144.58E
Brisbane	1,169,000	27.30S, 153.00E
Perth	1,001,000	31.57S, 115.52E
Adelaide	987,000	34.56S, 138.36E
Newcastle	423,000	17.20S, 133.21E
Wollongong	238,000	17.13s, 137.57E
Gold Coast	208,000	27.59S, 153.22E
Hobart	180,000	42.54S, 147.18E
Geelong	150,000	38.10S, 144.26E

AUSTRALIA - LANGUAGES, ETHNIC GROUPS AND RELIGIONS

Languages	English	
	Aborigine	
Ethnic Groups	White	95%
	Asian	4%
	Aborigine and Other	1%
Religions	Anglican	26%
	Catholic	26%
	Other Christian	24%
	Other	24%

SOUTH AMERICA

CHILE

DESERTS: ATACAMA

CHILE - GENERAL INFORMATION

Area	756,945 sq km
Population Growth	1.49 %
Population Density	17 /sq km

CHILE - MAJOR CITIES

	Population	Latitude, Longitude
SANTIAGO	4,320,000	33.30s, 70.40W
Vina del Mar	320,000	33.02S, 71.35W
Valparaiso	273,000	33.05s, 71.40W
Talcahuano	225,000	36.40S, 73.10W
Conception	281,000	36.50S, 73.03W
Antofagasta	203,000	23.40S, 70.23W
Temuco	175,000	38.45S, 72.40W
Rancagua	157,000	34.10s, 70.45W
Talca	145,000	35.28S, 71.40W
Chillan	130,000	36.37S, 72.10W
Arica	158,000	18.30S, 70.20W
CHILE - LANGUAGES, ETHNIC GROUPS AND RELIGIONS		
Languages	Spanish	
Ethnic Groups	European and Mestizo	95 %
	Native American	3%
	Other	2%
Religions	catholic	89%
	Protestant and Other	11 %

ASIA

CHINA

DESERTS: GOBI

CHINA - GENERAL INFORMATION

Area	9,600,000 Sq km
Population Growth	1.24 %
Population Density	113 /sq km
Capital City	BEIJING

CHINA - MAJOR CITIES

	Population	Latitude, Longitude
BEIJING	6,000,000	39.55N, 116 26E
Shanghai	7,000,000	31.06N, 121 .22E
Tianjin	5,400,000	39.08N, 117 .12E
Shenyang	4,200,000	41.50N, 123 .26E
Wuhan	3,400,000	36.45N, 114 .15E
Guangzhou	3,300,000	23.08N, 113 .20E
Chongqing	2,800,000	29.30N, 106 .35E
Harbin	2,630,000	45.45N, 126 .41E
Chengdu	2,580,000	30.37N, 104 .06E
Nanjing	2,250,000	32.03N, 118.47E

CHINA - LANGUAGES, ETHNIC GROUPS AND RELIGIONS

Languages	Mandarin	
	Cantonesc	
	Shanghainese	
	Fuzhou	
	Minnan and Other	
Ethnic Groups	Han Chinese	**94%**
	Other	**6%**
Religions	Atheist and Eclectic	97 %
	Other	3%

MONGOLIA

DESERTS: GOBI

MONGOLIA - GENERAL INFORMATION

Area	1,564,619 sq km
Population Growth	2.76 %
Population Density	1 /sq km
Capital City	ULAN-BATOR

MONGOLIA - MAJOR CITIES

	Population	Latitude, Longitude
ULAN-BATOR	511,000	47.54N, 106.52E
Darhan	75,000	46.29N, 109.24E
Erdene	46,000	46.18N, 100.35E
Choybalsan	23,000	48.02N, 114.32E
Nalayh	15,000	47.40N, 107.12E
Ulangom	15,000	49.59N, 92.00E
Uliastay	14,000	47.42N, 96.52E

MONGOLIA - LANGUAGES, ETHNIC GROUPS AND RELIGIONS

Languages	Khalkha Mongol	
	Turkish	
	Russian	
	Chinese	
Ethnic Groups	Mongol	90%
	Kazakh	4%
	Chinese	2%
	Russian	2%
	Other	
Religions	Tibetan Buddhist	94%
	Muslim	4%
	Other	1%

PAKISTAN

DESERTS: THAR

PAKISTAN - GENERAL INFORMATION

Area	803,943 sq km
Population Growth	2.70 %
Population Density	134 /sq km
Capital City	ISLAMABAD

PAKISTAN - MAJOR CITIES

	Population	Latitude, Longitude
ISLAMABAD	270,000	33.40N, 73.08E
Karachi	5,400,000	24.51N, 67.02E
Lahore	3,132,000	31.34N, 74.22E
Faisalabad	888,000	31.25N, 73.09E
Rawalpindi	810,000	33.40N, 73.08E
Hyderabad	770,000	25.23N, 68.24E
Multan	750,000	30.10N, 71.36E
Gujranwala	700,000	32.06N, 74.11E
Peshawar	600,000	34.01N, 71.40E
Sialkot	310,000	32.29N, 74.35E
Sargodha	300,000	32.01N, 72.40E

PAKISTAN - LANGUAGES, ETHNIC GROUPS AND RELIGIONS

Languages	Urdu	
	English	
	Punjabi	
	Sindhi	
	Pashtu and Other	
Ethnic Groups	Punjabi	66%
	Sindhi	13 %
	Pashtun	9%
	Baluchi and Other	12%
Religions	Sunni Muslim	77 %
	Shi'a Muslim	20 %
	Other	3%

AFGHANISTAN

DESERTS: KAR KUM

AFGHANISTAN - GENERAL INFORMATION

Area	249,999 sq mi
Population Growth	2.35 %
Population Density	58 /sq mi
Capital City	KABUL

AFGHANISTAN - MAJOR CITIES

	Population	Latitude, Longitude
KABUL	1,127,000	34.31N, 69.12E
Qandahar	195,000	31.35N, 65.45E
Herat	156,000	34.20N, 62.12E
Mazar-i-Sharif	115,000	36.42N, 67.06E
Jalalabad	111,000	34.26N, 70.28E
Qonduz	110,000	36,45N, 68.51E
Baghlan	109,000	36.11N, 68.44E
Maymana	40,000	35.53N, 64.38E
Pul-i-Khomri	35,000	35.55N, 68.45E
Ghazni	32,000	33.33N, 68.26E

AFGHANISTAN - LANGUAGES, ETHNIC GROUPS AND RELIGIONS

Languages	Pashtu	
	Dari	
	Turkish Languages	
	Baluchi	
	Pashai	
Ethnic Groups	Pashtun	50 %
	Tajik	25 %
	Uzbek	9%
	Hazara	9%
	Other	7 %
Religions	Sunni Muslim	74%
	Shi'a Muslim	15 %
	Other	11 %

SOUTHWEST ASIA

EGYPT

DESERTS: SAHARA

EGYPT - GENERAL INFORMATION

Area	1,001,449 sq km
Population Growth	2.65 %
Population Density	53 /sq km
Capital City	CAIRO

EGYPT - MAJOR CITIES

	Population	Latitude, Longitude
CAIRO	11,000,000	30.03N, 31.1

Alexandria	2,905,000	31.13N, 29.55E
Giza	1,640,000	30.01N, 31.10E
Shoubra El-Kheima	497,000	30.06N, 31.15E
El Mahalla El Koubra	355,000	30.59N, 31.10E
Tanta	344,000	30.48N, 31.00E
Port Said	364,000	31.17N, 32.18E
El-Mansoura	260,000	31,03N, 31.23E
Asyut	257,000	27.14N, 31.07E
Zagazig	256,000	30.35N, 31.30E
Suez	274,000	29.59N, 32.33E

EGYPT - LANGUAGES, ETHNIC GROUPS AND RELIGIONS

Languages	Arabic	
	English	
	French	
Ethnic Groups	Eastern Hamitic	90%
	Other	10%
Religions	Sunni Muslim	94%
	Other	6%

IRAN

DESERTS: KAVIR, LUT

IRAN - GENERAL INFORMATION

Area	1,648,000 sq km
Population Growth	3.23 %
Population Density	32 /sq km
Capital City	TEHRAN

IRAN - MAJOR CITIES

	Population	Latitude, Longitude
TEHRAN	6,028,000	35.40N, 51.2 6E
Mashad	1,120,000	36.41N, 52.39E
Isfahan	930,000	32.41N, 51.41E
Tabriz	855,000	38.05N, 46.18E
Shiraz	801,000	29.38N, 52.34E
Bakhtaran	535,000	31.28N, 54.54E
Karaj	527,000	35.48N, 50.58E
Ahwaz	471,000	31.17N, 48.43E
Qom	425,000	34.39N, 50.57E
Abadan	329,000	30.20N, 48.15E

IRAN - LANGUAGES, ETHNIC GROUPS AND RELIGIONS

Languages	Farsi
	Turkish

	Kurdish	
	Arabic	
	English	
Ethnic Groups	Persian	**63 %**
	Turkish	**18%**
	Kurdish	**3%**
	Arab and Other	16%
Religions	Shi'a Muslim	**93%**
	Sunni Muslim	**5%**
	Bahai and Other	**2%**

IRAQ

DESERTS: ARABIAN

IRAQ - GENERAL INFORMATION

Area	434,924 sq km
Population Growth	3.53 %
Population Density	40 /sq km
Capital City	BAGHDAD

IRAQ - MAJOR CITIES

	Population	**Latitude, Longitude**
BAGHDAD	3,250,000	**33,20N, 44.26E**
Basra	1,600,000	3030N,47.50E
Mosul	1,250,000	**36,21N, 43.08E**
Kirkuk	535,000	**35.28N, 44.26E**
An Najaf	141,000	**31.59N, 44.19E**
Erbil	99,000	**36.12N, 44.01E**

IRAQ - LANGUAGES, ETHNIC GROUPS AND RELIGIONS

Languages	Arabic	
	Kurdish	
	Assyrian	
	Armenian	
Ethnic Groups	Arab	**75 %**
	Kurdish	**17 %**
	Other	**8%**
Religions	Shi'a Muslim	**62%**
	Sunni Muslim	**35 %**
	Christian and Other	*3%*

LIBYA

DESERTS: SAHARA

LIBYA - GENERAL INFORMATION		
Area		1,759,540 sq km
Population Growth		3.08 %
Population Density		2 /sq km
Capital City		TRIPOLI

LIBYA - MAJOR CITIES		
	Population	Latitude, Longitude
TRIPOLI	610,000	32.49N, 13.07E
Misurata	320,000	32.07N, 20.04E
Benghazi	103,000	32.23N, 15.06E
Azzawiya	94,000	32.45N, 12.44E
Al-Beida	77,000	32.49N, 12.45E
Agedabia	51,000	30.48N, 20.15E
Dama	48,000	32.40N, 22.35E
Sebha	41,000	27.09N, 14.29E
Tubruq	35,000	32.05N, 20.30E
A1-Marj	30,000	32.25N, 20.30E
Zeleiten	26,000	32.28N, 14.34E

LIBYA - LANGUAGES, ETHNIC GROUPS AND RELIGIONS		
Languages	Arabic	
	Italian	
	English	
Ethnic Groups	Arab Berber	97%
	Other	3%
Religions	Sunni Muslim	97%
	Other	3%

SYRIA

DESERTS: SYRIAN

SYRIA - GENERAL INFORMATION		
Area		185,180 sq km
Population Growth		3.74 %
Population Density		62 /sq km
Capital City		DAMASCUS

SYRIA - MAJOR CITIES		
	Population	Latitude, Longitude
DAMASCUS	1,250,000	**33.30N, 36.19E**
Aleppo	1,200,000	**36.14N, 37.10E**

Helms	430,000	34.44N, 37.17E
Latakaia	240,000	35.31N, 35.47E
Hama	200,000	35.05N, 36.40E
Deir ez-Zor	106,000	35.20N, 40.05E
Rakka	90,000	35.57N, 39.03E
Hasakeh	74,000	36.32N, 40.44E
Tartous	53,000	34.55N, 35.52E
Edleb	52,000	35.56N, 37.21E
Dera'a	50,000	32.37N, 36.06E

SYRIA - LANGUAGES, ETHNIC GROUPS AND RELIGIONS

Languages	Arabic	
	Kurdish	
	Armenian	
	French	
	English	
Ethnic Groups	Arab	90%
	Kurdish and Other	10%
Religions	Sunni Muslim	74 %
	Other Muslim	16 %
	Christian	10%

JORDAN

DESERTS: SYRIAN

JORDAN - GENERAL INFORMATION

Area	91,880 sq km
Population Growth	3.62 %
Population Density	31/sq km
Capital City	AMMAN

JORDAN - MAJOR CITIES

	Population	**Latitude, Longitude**
AMMAN	813,000	**31.57N, 35.56E**
Zarqa	277,000	**32.04N, 36.06E**
Irbid	141,000	**32.33N, 35.51E**
Ajhm	47,000	**32.20N, 35.45E**
Jarash	32,000	32.17N,35.45E
Madaba	26,000	31.44N, 35.48E

JORDAN - LANGUAGES, ETHNIC GROUPS AND RELIGIONS

Languages	Arabic	
	English	
Ethnic Groups	Arab	**98%**
	Circassian	**1%**

	Armenian	1%
Religions	Sunni Muslim	95%
	Christian	5%

DESERTS: ARABIAN

SAUDI ARABIA - GENERAL INFORMATION

Area	2,149,690 sq km
Population Growth	4.16 %
Population Density	7 /sq km
Capital City	RIYADH

SAUDI ARABIA - MAJOR CITIES

	Population	Latitude, Longitude
RIYADH	1,976,000	24.39N, 46.46E
Jeddah	1,084,000	21.30N,39.10E
Mecca	810,000	21.26N, 39.49E
Taif	300,000	21.15N, 40.21E
Medina	250,000	24.30N, 39.35E
Dammam	175,000	26.25N, 50.06E
Hufuf	120,000	25.20N, 49.34E
Tabouk	80,000	28,22N, 36.32E
Buraidah	75,000	26.20N, 43.59E
Al-Mobarraz	60,000	25.26N, 49.37E
Kharmis-Mushait	54,000	18.19N, 42.45E

SAUDI ARABIA - LANGUAGES, ETHNIC GROUPS AND RELIGIONS

Languages	Arabic	
Ethnic Groups	Arab	90%
	Afro-Asian	10%
Religions	Muslim	99%
	Other	1%

SOUTH AFRICA

BOTSWANA

DESERTS: KALAHARI

BOTSWANA - GENERAL INFORMATION

Area	**231,804 sq mi**
Population 1988	1,190,000
Population Growth	3.49 %
Population Density	5 /sq mi
Capital City	GABORONE

BOTSWANA - MAJOR CITIES

	Population	Latitude, Longitude
GABORONE	95,000	24.45S, 25.55E
Francistown	34,000	21.07S, 27.32E
Selebi-Phikwe	33,000	21.58S, 27.48E
Serowe	25,000	22.25S, 26.44E
Mahalapye	22,000	23.05S, 26.51E
Molepolole	21,000	24.25S, 25.30E
Kanye	20,000	24.59S, 25.19E
Lobatse	20,000	25.11S, 25.40E
Mochudi	19,000	24.28S, 26.05E
Mauri	15,000	20.00S, 23.25E
Ramotswa	14,000	24.56S, 25.50E

BOTSWANA - LANGUAGES, ETHNIC GROUPS AND RELIGIONS

Languages	English	
	Setswana	
Ethnic Groups	Batswana	95 %
	Bushmen	4%
	European	1%
Religions	Indigenous Beliefs	50 %
	Christian	50 %

NAMIBIA

DESERTS: KALAHARA

NAMIBIA - GENERAL INFORMATION

Area	318,261 sq mi
Population Growth	5.30 %
Population Density	4 /sq mi
Capital City	WINDHOEK

NAMIBIA - MAJOR CITIES

	Population	Latitude, Longitude
WINDHOEK	100,000	22.34S, 17.06E
Tsumeb	13,000	19.13S, 17.42E
Keetmanshoop	11,000	26.36S, 18.08E
Otjiwarongo	9,000	20.29S, 16.36E
Luderitz	8,000	26.38S, 15.10E
Swakopmund	7,000	22.40S, 14.34E

NAMIBIA - LANGUAGES, ETHNIC GROUPS AND RELIGIONS

Languages	Afrikaans
	German
	English

	African Languages	
Ethnic Groups	Black	**86%**
	White	**7%**
	Mixed	**7%**
Religions	Christian	**60%**
	Indigenous Beliefs	**40%**

SOUTH AFRICA

DESERTS: KALAHARI

SOUTH AFRICA - GENERAL INFORMATION

Area	471,443 sq mi
Population Growth	2.20 %
Population Density	74 /sq mi
Capital City	PRETORIA

SOUTH AFRICA - MAJOR CITIES

	Population	**Latitude, Longitude**
PRETORIA	800,000	**25.45S, 28 12E**
Cape Town	860,000	**33 56S, 18 2 8E**
Durban	810,000	**29 53S, 31 00E**
Johannesburg	770,000	**26 10S, 28 02E**
Port Elizabeth	300,000	**33 58S, 25 3 6E**
Roodepoort	210,000	**26 10S, 27 5 3E**
Umhlazi	180,000	**29 53S, 31 00E**
Germiston	140,000	**26 15S, 28 10E**
Pietermaritzburg	130,000	**29 36S, 30 24E**
Boksburg	120,000	**26 13S, 28 15E**
Bloemfontein	110,000	29.07S,26.14E

SOUTH AFRICA LANGUAGES, ETHNIC GROUPS AND RELIGIONS

Language	Afrikaans	
	English	
	Zulu	
	Xhosa	
	Sotho and Other	
Ethnic Groups	African	**70%**
	White	18%
	Mixed	9%
	East Indian	3%
Religions	Christian	**60%**
	Hindu	2%
	Muslim and Other	**38%**

APPENDIX B

EMPLOYMENT OF AIRCRAFT IN DESERT OPERATIONS

The desert is probably the most severe of all environments in which aviation units must operate. Standard operating procedures for desert operations are different from areas having an abundance of contrasting terrain and substantial vegetation. This appendix describes some special considerations when employing aircraft in desert operations.

AIR OPERATIONS

Air combat operations that rely on heavy vegetation and varying contour terrain need to be flexible enough to incorporate different methods of camouflage and terrain flight techniques. The varying types of sand have a tremendous effect on operations-vast flat areas afford unlimited visibility, dunes are hard to distinguish at night, blowing sand impairs visibility and presents flight and maintenance problems, and surface composition affects the choice of landing zones, maintenance sites, FARPs, and operating bases. Additionally, low-hovering and taxiing aircraft generate blowing sand and dust that can cause aircrews to lose outside visual reference, and, if performed near other equipment, present additional maintenance problems for that equipment.

Air operations are not the only area affected by the desert environment. Aviation ground operations may require flexibility and modification to work around the heat and the effects of the terrain. Aircrews must resort to instrument flight during duststorms. Sand also causes excessive wearing, pitting, and eroding of aircraft components.

In certain areas, the desert, with its relatively level terrain and shallow compartments, contains few highly distinguishable terrain features to mask aviation forces. Formations of two or more aircraft can be seen 10 kilometers away because the dark airframes contrast against the desert sand. Aviation units normally deploy their aircraft along routes and may need to consider widely dispersed formations. Aviation forces can make maximum use of deception techniques during periods of limited visibility.

Air cavalry assets can conduct reconnaissance and security operations over great distances in the desert because of the lack of vegetation and relief. Even when they are sand painted, armor vehicles stand out starkly against the sand. When combined with traditional target acquisition principles, such as dust signature and movement, these factors make it easier to acquire and engage armored and mechanized forces well out of range of their main guns.

Aeroscouts flying nap of the earth (NOE) cannot necessarily find the enemy more easily than ground observers. Stationary targets are the most difficult to see as there is little to draw the observer's attention:

Therefore, aeroscouts must use caution to avoid blundering into enemy air defense weapons. The aircraft should land at a distance of 5-10 kilometers from the area of interest, and the observer should dismount and scan the area for suspected enemy. The observer must remain in contact with the pilot by using a portable radio. The process should be repeated at varying intervals until contact is made.

Attack helicopter battalions (ATKHBs) are a potent force in desert warfare. If they are employed quickly and violently, maximum results can be obtained both in offensive and defensive operations. They are best used where a quick concentration of combat power is needed. A desert environment presents excellent target acquisition and engagement possibilities. Attack assets must remain dispersed to provide security. Mission planning that incorporates flexibility is a key ingredient in the successful employment of ATKHBs.

Terrain flying and desert navigation require continuous concentration. Due to lack of terrain and poor reference points, the aviator may rely on dead reckoning, self-contained navigation equipment, and radio navigational aids. As light decreases, the ability to judge distances accurately is degraded and visual illusions become more common. Because of glare, haze, and frequently blowing sand, it may be difficult to detect changes in terrain and the horizon.

Attack helicopters should move from assembly areas to battle positions (or holding areas if necessary) over attack routes that will provide whatever cover and concealment and prominent terrain features necessary to assist in navigation and to decrease the possibility of detection. Attack helicopters may have multiple routes for ingress and egress. Route reconnaissance, premission planning, and prebriefs will maximize the benefits to route planning in desert operations.

The weather in desert regions can be extremely unpredictable. Sandstorms, accompanied by constantly fluctuating wind speeds, may reduce visibility from in excess of 50 kilometers to zero in less than five minutes. Pilots must be carefully briefed on prevailing weather conditions before takeoff. Warning of any expected variations in conditions must be transmitted immediately to all airborne aircraft.

THE PERFORMANCE OF HELICOPTERS IN HEAT

Aviation personnel must refer to the appropriate aircraft technical manuals to determine aircraft limitations and capabilities in the desert environment. Significant effects on the payload capabilities should be anticipated.

Commanders must develop realistic aircraft utilization procedures based on the environmental effects data provided by aviation staff personnel to obtain the fullest benefit from aviation assets.

Helicopters hovering close to the ground can cause the engine to ingest sand; can cause observation by the enemy due to the formation of dust clouds; or cause disorientation of the pilot due to blowing sand, particularly at night. Helicopters should not be moved under their own power while on the ground, but should be pushed or towed by men or vehicles. Maintenance should be restricted to the minimum time, and should take place on rock or on oiled or wet sand, if available. All apertures (Pitot tubes, for example) of aircraft should be covered when not in use (including helicopter windscreens).

Temperature and humidity have a direct impact on personnel and vehicle performance. Temperature and humidity affect air density. Air density decreases as temperature increases. High temperature and humidity reduce the efficiency of aircraft propulsion and aircraft lift capabilities. Although temperature and humidity may not have a direct effect on a particular operation, extremely high temperatures and humidity will reduce aircraft payloads.

APPENDIX C

TECHNIQUES FOR OPERATING EQUIPMENT IN THE DESERT

The effects of the desert environment on equipment were described in Chapter 1. This appendix describes techniques which, if used while operating equipment in the desert, can save both equipment and lives.

DRIVING

Drivers and track commanders should observe the guidelines in the following paragraphs while operating vehicles in desert areas.

Dusty Conditions

Wear goggles while driving open-hatched regardless of visibility. Clear-lens goggles should be worn at night unless night-vision goggles are used. Bandanas or surgical masks should be worn to avoid breathing heavy dust.

Vehicles in an extended convoy should maintain a dust distance of twice the normal interval, or as specified in the unit SOP to allow time for the dust to dissipate. When driving on extremely dusty roads or trails and if traffic conditions permit, a staggered column formation can be used with vehicles alternately driving on the left and right side of the road.

If the vehicles should become engulfed in dust, the convoy commander should consider adjusting the rate of march accordingly. Any commander of a vehicle engulfed in dust should alert the convoy commander by radio, move to the right side of the road, and stop or slow to allow the dust to dissipate. Extreme caution must be observed to ensure oncoming and following vehicles are not jeopardized. The lead vehicle must warn vehicles to return to column formation if encountering traffic.

Sandy Deserts

Sandy deserts may be relatively flat or interspersed with windblown dunes. When driving in sand, the following techniques should be applied:

- The best time to drive on sand is at night or early morning when the sand is damp and traction is better. However, this is not always the case especially with the newer type military tires with closer tread design. Damp sand packs between the tread in the grooves of these tires resulting in virtually no surface traction.

- Drivers of track vehicles must be wary of a lack of steering response, which indicates sand is building up between the rear sprockets and treads. If the buildup of sand is allowed to continue, it will force the tread off. "Shaking" the vehicle with the steering or backing up will remove the sand.
- Wheel vehicles may gain some traction by reducing the air pressure in the tires. However, prolonged driving on partially deflated bias ply tires will overheat the tires and break down the sidewalls. Vehicles equipped with radial tires or central tire inflation system (CTIS) are not affected by the lower tire pressure if the maximum speed listed in the operator's manual is not exceeded.
- Vehicle loads must be evenly distributed. Rear-wheel drive should be used where possible to prevent the front wheels from digging into the sand and becoming mired.
- Drivers must switch to all-wheel drive or change gears before a vehicle bogs down in the sand.
- Before entering the sand drivers should select a gear that will allow the vehicle to maintain as much torque as possible without causing the wheels to spin and to minimize changing gears.
- Large-wheeled vehicles, such as 5,000-gallon tankers, should have a designated "puller". The designated vehicle should be preconfigured to assist these vehicles when they become bogged down in loose sand.

Some sand areas will be covered by a surface crust. This is caused by chemicals in the ground cementing sand particles together, In many cases it will be possible to drive on top of this crust and minimize dust signature and the chance of bogging down. Consider the following techniques when driving on a crust:

- Use staggered columns to facilitate movement. As a general rule vehicles should not follow one behind the other.
- Ensure vehicles maintain a minimum speed (determined from experience) below which they will break through the crust.
- Avoid sharp turns and abrupt starts or stops that could cause a vehicle to break through the crust.
- Reconnoiter patches of the crust that are a different shade to ensure they are not softer than the surrounding crust.

Crossing Dunes

Crossing dunes requires careful reconnaissance. Normally, the upwind side of the dune will be covered with a crust and have a fairly gradual slope. The downwind side will be steeper and have no crust. Prior to crossing a dune, the driver should climb it on foot checking crust thickness, the angle at the crest to ensure the vehicle will not become bellied at the top, and the degree of slope and softness of the downwind side. If he is satisfied his vehicle can climb the dune, he should drive the vehicle straight up it at the best speed, crest it, and maintain a controlled descent on the other side.

Little hills may be formed by the wind blowing sand around small shrubs. Wheel vehicles should not attempt to move through areas where this has occurred without engineer assistance.

Cacti or thorn bushes will cause frequent tire punctures. Increase the number of tires carried in the unit's PLL when operating in areas covered with this type of vegetation.

Rocky Areas

Rock and boulder-strewn areas, including lava beds, may extend for many miles. Desert rocks, eroded and sharp-edged, vary in size and are so numerous that it is almost impossible to avoid any but the largest. The subsequent harsh jolting fatigues individuals and causes extreme wear on tracks, wheels, springs, and shock absorbers. Rocks and stones can become lodged between the tires on vehicles equipped with dusk that can cause severe damage to tires and brake components. Vehicles can follow one another in this type of terrain and it may be feasible to reconnoiter and mark a route. Drivers should achieve a "rolling" effect as they cross large rocks by braking as the vehicle's wheels ride over a rock so the axle settles relatively gently on the far side.

Salt Marshes

Salt marshes (called sebkha) are normally impassable, the worst type being those with a dry crust of silt on top. Marsh mud used on desert sand will, however, produce an excellent temporary road. Many desert areas have salt marshes either in the center of a drainage basin or near the seacoast. Old trails or paths may cross the marsh, which are visible during the dry season but not in the wet season. In the wet season standing water indicates trails due to the crust being too hard or too thick for the water to penetrate. However, such routes should not be tried by load-carrying vehicles without prior reconnaissance and marking.

RECOVERY

Track vehicle recovery methods are the same in the desert as in temperate climates. The techniques described in the following paragraphs will assist wheel vehicle recovery operations in sand crusts or salt marshes. To assist in recovery, wheel vehicles should carry the following items:

- Steel or aluminum channels, at least for the driving wheels. These are pierced to reduce weight and ribbed for strength. Pierced steel planking (PSP) or galvanized iron maybe used as a substitute.
- Sand mats made of canvas, preferably with lateral strips of metal to give strength and increase the traction of the wheels.
- Jacks and jack blocks.
- Tow rope(s).
- Shovels.

Once a vehicle becomes mired, excavate the ground under the vehicle in a gradual slope towards the direction of recovery to a point where no part of the underside is touching the ground. Channels or spurs and mats are laid under or against the wheels facing the direction of recovery. Tire pressure may be reduced to increase traction, but this also lowers the vehicle. It maybe necessary to lift the wheels with a jack if the vehicle is resting on its frame or axles.

When the vehicle begins to move, any faltering will cause it to sink again. Once out, the driver must maintain speed until the vehicle has reached the nearest hard area. At this point the tires are reinflated, the vehicle inspected for damage, and recovery equipment collected.

Vehicles equipped with winches can winch themselves out using ground anchors. The ground anchor may consist of a tarpaulin full of sand placed in a hole and the winch cable attached to it, or it may be one, or preferably two spare wheels well dug in.

A rubberized fabric balloon may be used on light vehicles to lift them free of broken crust. The balloon is placed under the vehicle and blown up with the vehicle exhaust.

If a lone vehicle breaks or bogs down in the desert, the crew must stay with it. A vehicle is much easier to find than a lone man.

MAINTENANCE

Equipment directly affected by heat, such as aircraft and radios, are equally affected by all deserts. However, power trains and suspension systems are affected in proportion to trafficability and soil texture. Most damage to equipment can be avoided by careful driving and by careful observation by vehicle commanders.

Track tension must be correct as constant driving on rocky plateau deserts will reduce the life of the track. Suspension units will require frequent replacement of torsion bars and suspension arms. To prevent damage to internal parts of the idler and suspension arms caused by the terrain, direct-support maintenance units must be provided with equipment capable of tapping and removing the sheared bolts.

To prevent problems that can result when desert vegetation clogs engine oil coolers and cylinder cooling pins, place a small-mesh wire screen over the top grille doors. It may still be necessary to remove packs about every 10 days to clean the engine cooling fins. The wire screening should be periodically checked, removed, and cleaned.

Maintenance personnel must inspect and adjust transmission bands frequently, especially on vehicles operating in hot, barren mountains. This will help reduce transmission oveheating.

Extra stocks of air-cooled generators are necessary because high-ambient temperatures limit their ability to maintain the proper operating temperature and contribute to premature failure.

NOTE: The M113A1 is especially susceptible to overheating problems in desert conditions. This includes the transmission and the solid-state voltage regulator, which is more prone to overheating

WHEELED VEHICLES

Wheeled vehicles are subject to brake system component failures and power steering leaks on rocky deserts. Vehicles equipped with manual transmissions are prone to clutch failure caused by drivers slipping the clutch. Vehicles with automatic transmissions tend to overheat; therefore, stop frequently to allow the transmission time to cool. The M54 5-ton truck is prone to air hydraulic cylinder failure and power steering seal leaks on rocky deserts. All vehicles of the 1/4- to 5-ton range are prone to clutch failure caused by drivers "riding" the clutch pedal. Tire consumption is very high, so all vehicles must carry one, or preferably two spare tires, and the unit's PLL of tires should be considerably increased. Approximately one vehicle in every three to four should carry slave cables to provide for battery failure.

Vehicles should be equipped with the following:

- Extra fan belts.
- Two spare tires.

Extra oil.

Extra radiator hoses.

Heavy duty tape.

Extra air and fuel filters.

- Jack stand support plate.
- Sand ladders (fabricated) and matting.
- Tow rope/cable.
- Extra water cans.
- Siphoning hose and funnel.
- Slave cables.

and early failure than the older mechanical type.

RADIOS

Radios, regardless of type, must be kept cool and clean. They must be in the shade whenever possible and should be located in a ventilated area (or even in an air-conditioned can). If water is available, wrap the radio in a damp towel, ensuring that the air vents are not blocked. Additional radios should be available in vital communications centers, such as tactical operations centers, to allow immediate replacement if the set in use shows signs of overheating.

It is essential that antennas be cut or adjusted to the length of the operating frequency. Directional antennas must be faced exactly in the required direction; approximate azimuth produced by guesswork will not do. A basic whip antenna relies on the capacitor effect between itself and the ground for efficient propagation. The electrical ground may be very poor, and the antenna performance alone may be degraded by as much as one-third if the surface soil lacks moisture, which is normally the case in the desert. If a ground-mounted antenna is not fitted with a counterpoise, the ground around it should be dampened using any fluid available. Vehicle-mounted antennas are more efficient if the mass of the vehicle is forward of the antennas and is oriented towards the distant station.

Desert operations require dispersion, yet the environment is likely to degrade the transmission range of radios, particularly VHF (FM) fitted with secure equipment. This degradation is most likely to recur in the hottest part of the day, approximately 1200 to 1700 hours.

If stations start to lose contact, especially if the hotter part of the day is approaching, alternative communication plans must be ready. Alternatives include the following:

- Using relay stations, including an airborne relay station (the aircraft must remain at least 4,000 meters behind the line of contact). Ground relay stations or RETRANS are also useful and should be planned in conjunction with the scheme of maneuver.
- Deploying any unemployed vehicle with a radio as a relay between stations.
- Using alternative radio links such as VHF multichannel telephones at higher levels, or HF (SSB) voice.
- Using wire. Normally wire will not be used as operations will be fluid, but it maybe of some value in some static defensive situations.
- Using a unit such as all or part of the task force scout platoon for messenger service, Although it is

undesirable to use such a unit in this manner, it may be necessary to maintain communications.

GENERAL TIPS

General tips for operating equipment in the desert include the following:

- Check track tension daily.
- Check drive belt adjustment frequently.
- Lubricate suspension items daily, and clean grease fittings.
- Reduce sand ingestion by stretching nylon stockings over air cleaners.
- Emphasize proper engine cooldown and shutdown procedures, especially diesels.
- Adjust battery specific gravity to the environment (refer to TMs).
- Set voltage regulators at lower end of specifications.
- Start up vehicles regularly to prevent discharge of batteries.
- Increase stocks of oils and lubricants.
- Use high-grade 20W-50 oil; it serves well under desert conditions.
- Compensate for increased pressure due to severe heat in closed pressurized systems.
- Check lubrication orders and TMs for the correct viscosity of lubricants for higher temperatures.
- Keep lubrication to the absolute minimum on exposed or semiexposed moving parts; this includes working parts of weapons.
- Erect screens against blowing sand in maintenance areas.
- Cover the gap between the fuel nozzle and the fuel tank filler neck opening during refueling operations.
- Protect exposed electrical cables and wires with electrical tape.
- Keep optics covered; clean them with a soft paintbrush or a low-pressure air system (this works well for weapons also).
- Clean sand and dirt from hulls of armored vehicles.
- Check tire pressures and fuel levels at the midpoint of the temperature range during the day.
- Ground all refueling equipment-STATIC ELECTRICITY KILLS.
- Replenish radiators with potable water whenever possible.
- Determine battery shortages early and requisition early.
- Drain fuel lines at night and in the morning due to condensation.
- Increase PLLs for the following parts due to high failure rates:

    - Tires.

    - All track components.

    -All suspension components for both wheel and track vehicles.

    - Brake shoes.

    - Bearings and bushings.

-Plastic and rubber parts, including seals.

- All filters.

- Generator components.

- Deploy with plastic bags to cover weapons and protect other equipment during maintenance or when not in use.
- Bring muzzle plugs.
- Prepare all vehicles for desert operations in accordance with the appropriate TMs.
- Issue small paintbrushes to all soldiers/marines for weapons cleaning and other equipment maintenance. The paintbrush is one of the more valuable tools available to the soldier/marine for maintenance.

APPENDIX D

EFFECTS OF THE ENVIRONMENT ON NBC WEAPONS

The primary purpose of using NBC weapons and smoke is to produce casualties, destroy or disable equipment, and generally disrupt operations. NBC weapons and smoke are employed in coordination with conventional weapons. FM 3-3, FM 3-4, FM 3-5, and FM 3-50 describe NBC defense and smoke operations. This appendix addresses the effects of a desert environment on NBC weapons.

WEATHER AND TERRAIN

Both desert weather and terrain affect the behavior of NBC and smoke weapons, and to some extent influence their tactical employment. The following paragraphs describe the effects of the weather and the terrain on the behavior of NBC weapons.

Weather

The effectiveness of NBC and smoke weapons is directly proportional to air stability. Air stability is a result of temperature variations at different levels of the air. The span of desert day and night temperatures causes extremes of air stability. At night and early morning the desert air is very stable. This may be the best time for NBC and smoke employment because of extensive downwind drift and area coverage. Desert air is very unstable during the late morning through afternoon. This may be the worst time for NBC and smoke employment because of quick and irregular dissipation. This may lead to ineffective target area coverage and possible danger to friendly troops. Temperature differences are determined by comparing the air temperature differences every 3.5 meters above the ground. Three types of temperature differences influence NBC and smoke operations:

- Unstable (lapse). Unstable conditions exist when air temperature decreases with altitude. In the desert, this mostly occurs between late morning and early evening.
- Neutral. Neutral conditions exist when air temperature does not change with altitude, In the desert, this mostly occurs during early morning and early evening.
- Stable (inversion). Stable (inversion) weather conditions exist when the air temperature increases with altitude. In the desert, this mostly occurs between late evening and early morning.

High desert temperatures in the middle of the day result in a decrease in air density, so nuclear blast waves move faster. Some chemical agents that come in contact with the skin are more effective during high temperatures due to perspiration. Desert sunlight is lethal to the effectiveness of most biological agents, but most are unaffected at night.

High winds are common in certain desert seasons and affect the dissemination of biological, chemical, smoke, and radioactive clouds. High winds may break up clouds in an odd fashion and force them in the wrong direction and possibly endanger friendly troops.

Terrain

Flat desert terrain lacking vegetation and under stable air conditions moves NBC and smoke effects evenly and steadily in all directions. Nuclear induced radiation is greater in sandy soil due to the high silica content. Desert soil below the surface crust is a fine powder, so the blast and suction effects of a nuclear burst will cause considerable dust clouds.

NUCLEAR WEAPONS IN THE DESERT

Many potential threat nations maintain nuclear weapons and additional nations continue to obtain nuclear capability. Nuclear weapons can be delivered by missiles or bombs and can be exploded in the air, on the ground, or below the ground. Depending upon the radiation dosage received, lethal effects could be felt immediately or delayed for days. Detonated nuclear weapons release energy that affects military troops and equipment in three forms: blast, nuclear radiation, and thermal radiation (heat and light). These effects are discussed in the following paragraphs.

Blast

Nuclear blasts, even from an air burst, raise considerable amounts of desert sand and dirt, which inhibits observation and maneuver for a long time. Lighter desert air density causes a drop in static overpressure, but a more rapidly expanding blast (shock) front. This increases the danger to aircraft and helicopters flying in the area. The radius of damage is normally smaller in desert climates so dug-in personnel and equipment are safer. Desert trafficability is degraded in the immediate area of the strike, especially for wheeled vehicles, due to the destruction of the sand crust.

Nuclear Radiation

Immediate nuclear radiation is a function of weapon yield and changes little with desert temperature. Residual radiation is high in the case of low air burst or ground burst weapons. Residual radiation in the desert is affected by the time of day and the wind. At night, with no wind, residual radiation may be evenly distributed around the point of burst. At night, with a steady wind, residual radiation may drift downwind for many miles. In the late afternoon, residual radiation may drift in a totally irregular pattern and direction due to desert air instability. Induced radiation is prevalent in desert sand due to the amount of silica in the soil. Constant radiation monitoring and reconnaissance are vital to protect troops, especially from contaminated water sources.

Thermal Radiation

The effective range of thermal radiation (heat and light) increases in the desert where there is little terrain masking. This increases the danger of troops receiving severe burns. Blinding light (dazzle), out to 50 kilometers, must be considered in the desert when determining the radius of warning for friendly troops.

BIOLOGICAL WEAPONS IN THE DESERT

The United States has renounced the use of biological agents, but many potential threat nations maintain biological weapons. Biological agents are living microorganisms that multiply inside the human body and cause disease. Biological agents may be disseminated as a liquid or vapor using rockets, bombs, or aerosol generators.

Depending upon the agent, they can produce lethal effects in 1-24 days from the time of exposure. High desert temperatures (120 degrees Fahrenheit and higher) and sunlight may destroy most biological agents. Cooler nighttime temperatures and the lack of sunlight provide biological agents a chance to enter the body. High desert winds will disperse biological agents more rapidly than low winds. Normally, the most effective wind speeds for effective target coverage are from 8-18 knots (14-32 kmph). Stable desert air conditions (night or early morning) provide the greatest agent concentration and area coverage. Unstable desert air conditions (late morning and afternoon) promote atmospheric mixing and lower agent concentration, reducing effective target coverage.

Biological weapons are best suited for strategic rather than tactical use in the desert. Test water and food sources frequently to ensure they are not contaminated. Sanitation, personal hygiene, and immunizations must be instituted to ensure individual protection.

The United States also renounces the use of toxins but some potential threat countries continue to develop toxins for possible military use. Toxins are extracted from natural biological sources and are not chemically produced. They can be disseminated as a liquid, vapor, or powder, and delivered by aerosol generators, artillery, rockets, or bombs. Depending upon the toxin, they can produce lethal effects in one minute to 12 hours from the time of exposure. In the desert, toxins can effectively be used as strategic or tactical weapons.

CHEMICAL WEAPONS IN THE DESERT

The United States has reserved the right to retaliate in kind with chemical agents against enemies who use them frost. There are six major types of chemical agents: nerve, blood, blister, choking, incapacitating, and tear. They are classified as persistent and nonpersistent. Chemical agents may be delivered by mines, rockets, artillery, or bombs. Depending upon the agent type, lethalities can occur between minutes to an hour from the time of attack.

High daytime temperatures of the desert increase incapacitating effects of liquid agents, which rely on skin penetration. Desert air instability (late morning and early afternoon) will usually cause quick, vertical and irregular dissipation of chemical agents. This reduces the lethal concentration and target area coverage. Desert air stability (early morning and evening) will most likely cause adequate agent dissipation and target area coverage. High desert temperatures also cause liquid agents to evaporate more quickly and decrease the time of hazard. If a liquid chemical agent soaks into the desert sand, it will increase the time of hazard. Figures D 1 and D-2 show the evaporation times of chemical agents (HD, GA, GF, VX) on sand at 104 and 68 degrees Fahrenheit for negligible risk levels. Negligible risk levels cause mild incapacitation among no more than 5 percent of unprotected soldiers/marines crossing a contaminated area.

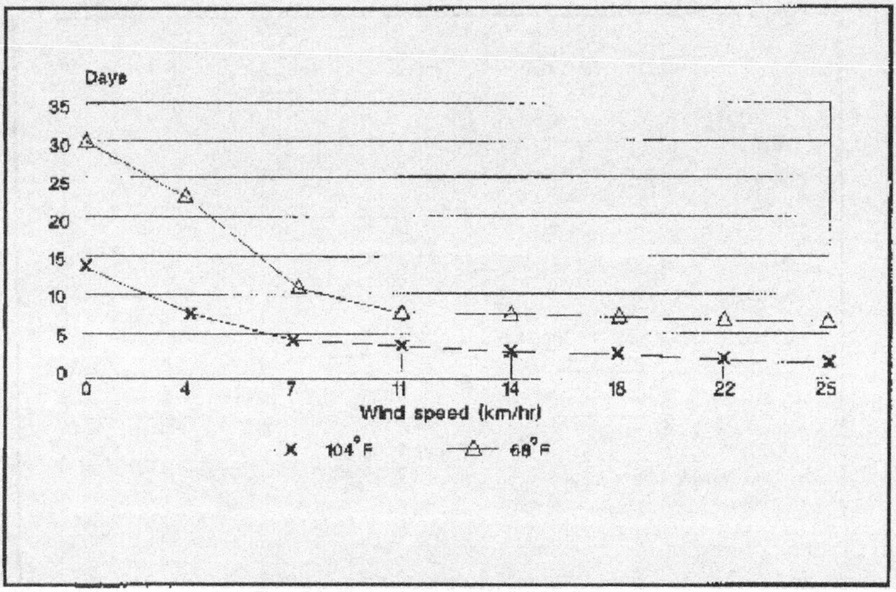

Figure D-1. Evaporation times to negligible risk for VX agents on sand.

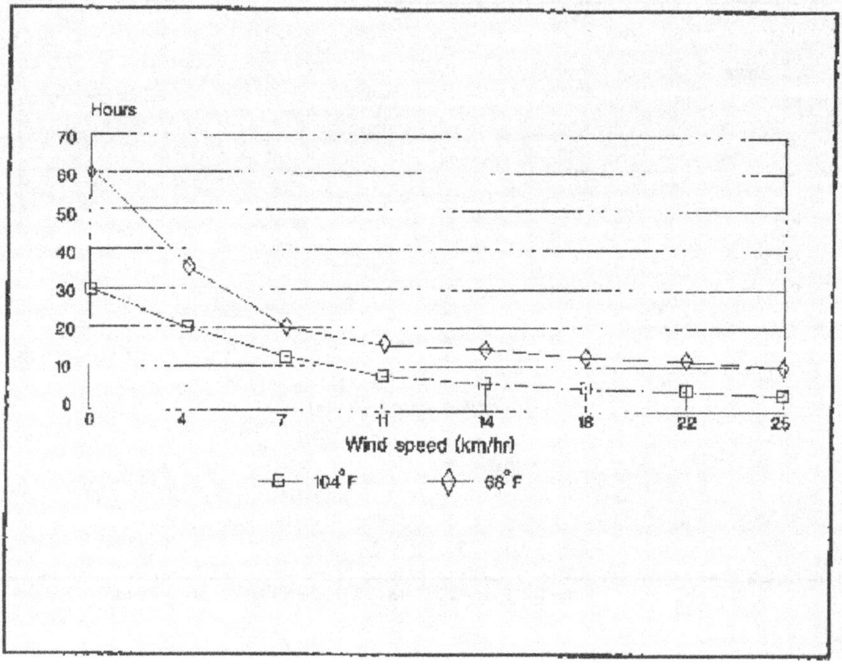

lot GF, HD, and GA agents.

Strong desert winds also increase the evaporation rate of liquid agents. They cause chemical clouds to drift downwind in an irregular direction and concentration. This tends to disperse target area coverage and lethal concentrations, and may endanger friendly troops.

SMOKE SYSTEMS IN THE DESERT

The lack of cover and concealment in flat desert terrain with little vegetation makes the use of smoke more vital to survival.

Desert winds, temperature, humidity, and terrain all affect smoke cloud behavior. The weather condition with the greatest impact on smoke operations is wind. Low desert winds allow smoke to remain on target areas for a longer period of time than high winds, In general, if the wind speed is less than 5 knots (9 kmph) or greater than 10 knots (18 kmph), smoke may not provide good target area coverage.

Temperature

Desert temperature differences have a direct relationship with making effective smoke. Three types of temperature differences influence smoke:

- Unstable (lapse). Lapse desert temperatures tend to break up and dissipate smoke, but are best for producing smoke curtains.
- Neutral. Neutral desert temperatures have limited vertical air currents and are good, but not the best, for producing smoke hazes and blankets.
- Stable. Stable (inversion) desert temperatures have no vertical air currents, and are the best for producing smoke hazes and blankets.

Humidity

Practically all smoke particles absorb moisture from the air. The lack of moisture in the desert air decreases the particle size and density of smoke making it less effective. Desert smoke streamers are shorter and less dense than smoke streamers in humid weather conditions. This increases the time and distance required to build adequate target area coverage.

TACTICS

The effects of NBC weapons and smoke on tactics in desert operations are discussed in the following paragraphs.

Nuclear

Nuclear weapons have a major bearing on tactics because of their ability to contaminate and shape desert terrain by making it impassable. Nuclear weapons can destroy troop and equipment concentrations and command and control centers and are usually considered weapons of mass destruction and combat multipliers; therefore, a smaller force with nuclear weapons may well defeat a much larger force without nuclear weapons.

The high maneuverability of tracked vehicles in the desert produces an endless number of avenues of approach. This creates difficulties for defense operations in the desert. Nuclear explosions could be used to shape desert terrain and canalize an enemy. This would provide profitable targets for other weapon systems. Equally, an attacking force could employ nuclear weapons in the desert to cut off enemy reinforcements, isolating them and making them vulnerable.

Biological Agents/Toxins

High desert temperatures and sunlight prevent the effective use of most biological agents during the daytime. An exception is spore-forming (anthrax) biological agents. During the day, troops crossing or occupying desert terrain face little danger from long-term biological contamination (except from spore-forming agents). But, because of favorable night desert conditions, an aerosol-delivered attack or a liquid biological attack would be effective. During night desert conditions, biological agents could effectively cover hundreds of square kilometers with a small amount of agent. This type of attack could be conducted covertly. These agents could contaminate vast areas of terrain and create mass casualties.

Toxins in the desert have nearly the same military use as chemical agents and are employed in the same manner. However, toxins are more deadly and require lower concentrations for lethalities. Toxins are difficult to detect with today's standard detection equipment.

Chemical Agents

The versatility of chemical agents gives commanders flexibility in desert operations. Commanders must consider how the employment of chemical agents affect offensive and defensive desert operations. Chemical agents can be used to create casualties, degrade performance, slow maneuver, restrict terrain, and disrupt logistical support. An assaulting force could use chemical agents to breach a defense or widen a gap. The best agent to use in this case would be a nonpersistent agent (one that is fast acting and leaves the target area quickly). High explosives could be mixed with the chemical attack to conceal the use of chemicals and complement their effects.

The threat of a chemical attack forces the use of protective masks and clothing. Heat, fatigue, and stress seriously affect the performance of troops. This is especially true with high desert temperatures. Well-trained soldiers/marines tolerate wearing protective gear better than those who are not as well trained. Troops in protective gear fire weapons less accurately, move more slowly, and must rest more often.

The actual or anticipated use of chemical weapons slows down a force and forces troops to take precautions. Desert heat, fatigue, and stress caused by wearing protective equipment slows down unit movement. Chemical agents can be used to create contaminated obstacles to desert maneuver. The best agent to use in this case would be a persistent chemical agent (one that remains on the target area for a time).

Chemical agents could be used to support offensive desert operations. Chemical agents could be used to protect flanks along an axis of advance to slow enemy counterattacks and to slow enemy fire and movement. In defensive desert operations, chemical agents could be used against second echelon forces to separate, slow down, and isolate them.

Chemical agents could be used to supplement conventional obstacles, or they could be used alone to restrict desert terrain. They may slow maneuver and channel attackers into engagement areas. A commander could contaminate a narrow desert mountain pass or bridge with a chemical agent and force the enemy to use an alternate route.

Logistical centers are lucrative targets for desert chemical attacks. Contaminating logistical supplies and equipment reduces the mobility of reinforcements and slows the delivery of supplies and equipment.

Smoke

Smoke is a combat multiplier that enhances the commander's ability to concentrate combat power at the critical time and place. Smoke is a far more significant battlefield factor than ever on flat desert terrain with little cover and concealment. Smoke can defeat enemy binoculars, weapon sites, and laser range finders in the desert. Smoke placed on the enemy at night interferes with enemy operations and observation by defeating enemy night sights and infrared sights.

In the desert offense, smoke can be used to deny the enemy information about the size, composition, and location of friendly maneuver forces. A smoke screen can be placed either to the front or to the flank. When the enemy cannot be screened effectively, obscuring smoke may be required. We must keep the enemy in doubt about the attacking unit's strength, position, activities, and movement. The longer the enemy is in doubt during an operation, the greater the chances are for mission success. Also, smoke can conceal desert breaching operations and river or gap crossings. It can also be used in deception operations.

In the desert defense, smoke is used to deny the enemy information about the size, location, and composition of friendly defensive positions. We must deny the enemy information by concealing the preparation and location of battle positions, artillery units, and reserves. Smoke can be used to support desert defensive positions by slowing enemy maneuver, disrupting command and control, isolating attacking echelons, silhouetting targets, and concealing obstacles.

PROTECTION

The commander of troops in desert operations must choose between the two following options when there is the threat of NBC warfare:

- Troops remain unprotected, with a high-chemical casualty risk, but a lower chance of heat fatigue.
- Troops are fully or partially protected, with a low-chemical casualty risk, but a higher chance of heat fatigue.

A decision on the level of protection is made according to the circumstances. If partial protection is ordered, the pace of physical work will be slower and proficiency reduced. The bulk of strenuous physical activity must be done at night or during the coolest part of the desert day.

In the desert, heat casualties (5 percent minimal) can be expected to occur in 30 minutes while performing heavy work in 90-degree-Fahrenheit temperatures when dressed in MOPP 4. Work/rest periods must be utilized to reduce the chances of heat fatigue. When protective clothing is worn, at least 10 degrees should be added to the WBGT index. Because of higher body temperatures, soldiers/marines in MOPP equipment perspire more than usual. Water must be consumed (2 quarts per hour) during continuous moderate work periods (and in MOPP equipment) when temperatures reach 80 degrees Fahrenheit and above to replace lost fluids or dehydration will follow.

DECONTAMINATION

The main problem of decontamination in the desert is lack of water, Although decontamination takes place as far forward as possible, the lack of water may burden the logistical system. Weathering may be a viable option for chemical contamination. The persistency of nerve agents (GA, GF, VX, GD) and mustard (HD) on chemical agent resistant coating (CARC) painted vehicles is between 4-24 hours at 104 degrees Fahrenheit. Sea water may be used as a substitute for fresh water during normal decontamination operations, but all equipment must eventually be flushed with fresh water to prevent corrosion.

Desert sand can be used for chemical decontamination, but it increases soldier/marine fatigue during its application. Sand removes most, but not all, liquid contamination, and saves valuable water supplies; however, the absorption capacity of desert sand is exhausted in 30-60 seconds after application. Remove the sand by sweeping or brushing the contaminated surface. Chemical agent detection should be conducted to ensure the agent is adequately removed.

APPENDIX E

DESERT CONCEALMENT AND CAMOUFLAGE

In the desert, camouflage problems are encountered that require imagination, ingenuity, and intelligence. The lack of natural overhead cover, the increased range of vision, and the bright tones of the desert terrain place emphasis upon siting, dispersion discipline, and the skillful employment of decoys to achieve deception and surprise. Total concealment is rarely achieved, yet proper camouflage measures can reduce the effectiveness of enemy observation, and consequently enemy operations.

Cover from enemy direct fire may be afforded by dunes, hills, and other irregularities in the desert terrain. Camouflage is an essential part of all operations in the desert and the importance of the concept must be impressed upon fresh units and individual replacements upon their arrival in theater. Poor camouflage may also compromise a high-level plan and lead to an operational failure. One poorly concealed vehicle can compromise an entire task force. Improvisation of available assets is just as important as being able to properly use camouflage systems. As previously described, deserts generally do not offer much natural concealment or means for camouflage; therefore, make maximum use of any artificial means available.

VEHICLES AND AIRCRAFT

Movement of vehicles produces dust, diesel plumes, and distinctive track marks. The slower the speed, the less dust that is produced; however, the need for speed must be balanced against the amount of dust that may be produced. Drivers must avoid harsh use of accelerators, the main cause of diesel plumes.

Shine from optical instruments (which should be kept shaded), and matte paint that has been polished by continual wear, or from tracks, particularly if rubber blocks have been removed, are difficult to camouflage during the desert day. See Figure E-l for shading optics. Running gear on tracks that has been polished by wear should be covered with burlap when stationary. Windscreens and windows should be removed or lowered to prevent reflection of the sun and heat. Vehicle silhouettes can be reduced in the forward areas by removing cabs and tops.

Figure E-1. shade optics to prevent shine

Disruptive pattern painting for vehicles and aircraft is described in FM 20-3. Local materials can also be used. The color and texture of the local terrain is best represented by placing dirt on vehicles and using a little water to make it stick.

The effects are increased by covering a vehicle with a wide-mesh net and using foliage brackets to attach local vegetation. Twine or wire may be used as an alternative to the mesh net, provided vegetation is available.

Some or all of the equipment listed in the following paragraphs should be available for every vehicle and aircraft, although aircraft will not necessarily be able to carry all of it.

The preferred net is the lightweight camouflage screen system (LCSS), desert version, which provides concealment against visual, near IR, and radar target acquisition/surveillance sensor devices. Additionally, the transparent version of the LCSS allows US units to camouflage radars (less CW type radars) without degrading operations. A desert camouflage net should be a complete cover, as it depends on its limitation of the ground surface, and both color and texture, for its effect. The alternatives to the LCSS in order of priority include the following:

- The specially produced desert-pattern net of the lightweight screen system.
- An open-weave cloth (colored as appropriate to the soil or "patched) stitched to an ordinary wide-mesh net and used with the string uppermost. This provides both color and texture and can be suitably garnished with radar-scattering plastic, such as that used in the lightweight screening system, and with any local vegetation.
- A cover of close-weave cloth, colored as appropriate.
- A standard net garnished solid, threaded in long straight strips that have been colored to harmonize with the terrain. The garnishing must be maintained.

The number of nets issued depends on the size of the equipment to be covered, but should be sufficient to allow a gradual slope of not more than 15 degrees from the top of the equipment to the earth. Each company-size unit should be equipped with a spray gun and various tints of paint to provide for temporary variations in net color to match the terrain.

When using nets for stationary equipment-

- Do not allow nets to touch sensitive items such as helicopter rotor heads and radio antennas which may cause a net to catch fire.
- Do not pull nets so tight that each supporting pole stands out.
- Ensure the net does not prevent the equipment from fulfilling its primary task. In some equipment, such as helicopters, a net must be easily removable to reduce reaction time.
- Avoid straight-edged patterns on the ground, which indicate something is there.
- Use burlap spray-painted in a nondescript desert color to cover all reflecting surfaces (excluding fire control optics) and shadow-producing areas under vehicle bodies, including tank suspensions. Aircraft equipped with windscreen covers will not require it.
- Cut desert scrub in the immediate area.
- Use poles, natural or man-made, to raise the nets from the equipment, thereby hiding its shape. They must be brought into the area of operations by the force and are extremely difficult to replace in the desert if lost or damaged.
- Make a "mushroom" out of thin iron tubing locally, It resembles an open umbrella without its cloth cover and with the end of the spokes joined together. Slotted into a socket that has been welded onto the top of a tank, self-propelled gun, or personnel carrier, it lifts the net above the vehicle, concealing its shape, increasing air circulation, and permitting the crew or team to use the top hatches.
- Hook and hold a camouflage net to the ground away from the vehicle by using wooden pegs or long steel pins, depending on soil consistency.
- Use mallets to drive pegs and pins into the ground.

After dismounting local security, camouflage is the first priority when a vehicle halts. Actions to be taken are-

- Site in vegetation or shadow, if available.
- Cover shiny surfaces and shadow areas with burlap screens.
- Drape the net.
- Add any available vegetation to the net.
- Blot out vehicle tracks for 50 meters behind vehicles.

Stationary aircraft take a relatively long time to conceal as they are fragile in comparison with other equipment, have a considerable heat signature, and must also be readily accessible for maintenance. The more they are concealed, the greater their response time is likely to be. Tactical flying is discussed in Appendix B, but take the following actions in sequence when approaching a landing site where aircraft will stay for some time:

- Ensure aircraft approach the site terrain-masked from enemy surveillance.

- Clos e down aircraft as soon as possible.
- Cove r all reflective surfaces.
- Mov e aircraft into shadow if it can be towed or pushed.
- Shif t the main rotor (depending on the type) until it is at a 45-degree angle with the fuselage and drape a net over the rotor and fuselage. The rotor must be picketed to the ground.
- Concea l the remainder of the aircraft.

POSITION SELECTION

Position selection is critical at all levels. One of the fundamentals of camouflage in any environment, but particularly the desert, is to fit into the existing ground pattern with a minimum of change to the terrain. A wadi bottom with vegetation or a pile of boulders that can be improved with grey burlap and chickenwire are good examples. Sites chosen must not be so obvious that they are virtually automatic targets for enemy suppressive frees, and antennas must be screened against the enemy, if possible.

Shadows, particularly in the morning and evening, identify objects; so equipment must be placed in total shadow (rarely found), or with its maximum vertical area facing the sun so that minimum shadow falls on the ground ("maximum vertical area" is the rear of a 5-ton truck with canopy, but the front of an M88, for example). See Figure E-2 for the effects of shadows. The shadow can be broken up, which is normally achieved by siting equipment next to scrub or broken surfaces, such as rocks. Equipment should not be sited broadside to the sun, and it is usually necessary to move as the sun moves. Digging in reduces the length of any shadow that is cast (on the principle that the lower the object, the shorter the shadow).

Figure E-2.Camoutlags: the effects of shadows.

Vehicles passing over pebbles or heavy ground surfaces press the pebbles or gravel into the soil, causing track marks to be prominent when viewed from the air. Avoid such areas if possible. Use existing trails and blend new trails into old ones whenever possible.

Soil texture suitable for digging must be a consideration when reconnoitering for battle positions. Holes must be covered to avoid shadows being cast. If vehicles will be in position for more than a day, trenches should be dug for them.

In forward areas, tactical operations centers are probably the most difficult positions to hide although their need for concealment is great. They require strict camouflage discipline. Vehicles and aircraft should not be allowed to approach closer than 300-400 meters. They must be dispersed and concealed so nets may have to be readily available for aircraft. Pay special attention to lights and noise at night.

Generators will have to be dug-in and allowed adequate air space for cooling. Radios and antenna systems must be remoted as far out as possible, and in different directions. Whenever possible, dig in the entire command post. Use engineer assets to build a berm around the perimeter and to help break up the silhouette and to enhance security. Other equipment should not be placed too close to minimize the possibility of the enemy's attention being attracted to the site.

Engineer activity often precedes operations, which makes it important that such work be concealed from enemy surveillance. The following guidelines should be used to conceal engineer activity:

- Employ the minimum number of equipment and personnel.
- Keep equipment well away from the site, and dispersed and concealed if not in use.
- Complete all possible preparations well away from the site.
- Follow the ground pattern, if possible.

Combat service and support assets must rely on concealment for most of their protection. The following guidelines will assist unit commanders in concealing trains while stationary or on the move:

- All vehicles of a given type should look alike. This will make it difficult for an enemy to pick out critical vehicles, such as water and fuel trucks, in a column. Canopies over fuel trucks disguise them and prevent radiant heat from striking the fuel containers.
- Vehicles should follow the tracks of the preceding vehicle if it is possible to do so without breaking through the crust, as this reduces the possibility of an enemy intelligence analyst to calculate how many vehicles have passed.
- Screen exhaust systems to reduce heat signature.
- Vehicles must never form a pattern, either when stationary or on the move.

SUPPLY POINTS

A supply point is likely to be in a location where its main threat of detection will be either by the eye or by photograph. Normally, greater emphasis can be placed on selecting supply positions from the point of view of concealment rather than for tactical efficiency, particularly in situations where air defense cover may be limited. The following guidelines should be used when setting up supply points:

- The location should be selected where trails already exist. Vehicles must use existing trails to the extent possible.

- Stocks should be irregularly spaced, both in length and depth, to the maximum extent possible so that there is no definite pattern.
- Stocks should be piled as low as possible and preferably dug-in. For example, a stack of gasoline cans should be only one can high.
- The shape of the area should not be square or rectangular, but should follow the local ground pattern.
- Stocks should be covered with sand, gravel, burlap, netting, or anything else that harmonizes with the local terrain, and the sides should be gradually sloped with soil filled to the top of the dump.
- The contents of each supply point should be mixed so that the destruction of one supply point will not cause an immediate shortage of a particular commodity.

APPENDIX F

OPERATIONS IN MOUNTAINS

Mountain operations are described in detail in FM 90-6. This appendix describes special conditions associated with operating in mountains such as those in the southern Sinai and on the shores of the Red Sea. It does not address tactics and techniques for mountain operations that are equally applicable to all mountains, except for the purpose of clarity.

TERRAIN

Mountains are high and rugged, with very steep slopes. Valleys running into a range become more and more narrow with the sides becoming gradually steeper. Valleys are usually the only routes that allow ground movement of men and equipment at any speed or in any quantity. Water is nonexistent on hilltops and unusual in valleys except during flash floods after rains. Lateral ground communications are limited unless the force is moving across the spines of mountain ranges. Navigation may be difficult, as maps are likely to be inaccurate.

PERSONNEL

Troops operating in mountainous country must be in peak physical condition. Regardless of their normal physical condition, personnel operating in mountainous areas require additional stamina and energy. They must also possess the ability to conduct sustained physical exertion and recover from it quickly.

Acclimatization is described in Chapter 1. Acclimatization to height, which varies much more among individuals than that for heat, must also be considered for operations in mountains. Lack of oxygen at high altitudes can cause unacclimatized troops to lose up to 50 percent of their normal physical efficiency when operating in altitudes over 6,000 feet. Mountain sickness may occur at altitudes over 7,800 feet and is usually characterized by severe headache, loss of appetite, nausea and dizziness, and may last from 5 to 7 days. Troops can acclimatize by appropriate staging techniques. It may take several weeks to become completely acclimatized, depending on altitude and the individual's personal physical reactions.

The risk of sunburn, particularly to the uncovered face, is greater in mountains than on the desert floor due to thinner atmosphere. Use antisunburn ointment and keep the face in shade around midday, using face nets or sweat rags. An individual camouflage net or scarf is particularly useful for this purpose. Recognition of heat illnesses in higher altitudes may not be as apparent as at lower altitudes because sweat evaporates very quickly. Measures to avoid dehydration and salt loss are extremely important. Daily temperature variations may be considerable making it necessary to ensure troops do not become chilled at night. Layering of clothing is essential. Troops who have been sweating heavily before the temperature starts to drop should take their wet shirts off and place them over relatively dry shirts and sweaters. Soldiers/marines should add layers of clothing as it gets colder and remove them as needed. This may have to be leader supervised and disciplined in the same manner as water consumption.

Requirements for hygiene areas important in mountainous areas as in the desert itself. Normal rocky ground will make it extremely difficult to dig any form of latrine so cover excrement with rocks in a specially marked area.

GENERAL CONSIDERATIONS

Infantry is the basic maneuver force in mountains. Mechanized infantry is confined to valleys and foothills (if these exist), but their ability to dismount and move on foot enables them to reach almost anywhere in the area. Airmobile infantry can also be extensively used. Consideration should be given to modifying the TOE of infantry units operating in barren mountains. A strong antitank platoon may not be necessary. However, the infantry requires extra radars and radios for the number of observation posts and separate positions that they may expect to occupy.

Mountains are not a good environment for tank and armored cavalry operations, because tanks and armored cavalry are unable to maximize their mobility, flexibility, and firepower.

Avenues of approach at ground level are few. Roads or trails are limited and require extensive engineer effort to maintain. Off-road trafficability varies from relatively easy to very difficult. Most movement and maneuver in this type of terrain is either by air or on foot. Unnecessary vertical foot or vehicle movement should be avoided. Rock slides and avalanches, although not as common as in high cold mountains, do exist and can restrict movement.

Air cavalry is the major reconnaissance means but they must guard against being ambushed by ground troops located at their own altitude or even higher. Security of units must include observation, especially at night, of all avenues of approach including those within the capabilities of skilled mountaineers.

It is relatively easier to conceal troops in barren mountains than on the desert floor due to rugged ground, deep shadows (especially at dawn and dusk), and the difficulties an observer encounters when establishing perspective. Carefully placed rocks can be used to hide equipment however, rocks can chip and splinter under small arms fire. The normal type camouflage net, which breaks up outline by shadow, maybe used rather than the overall cover normally used in the desert.

Helicopter units of all types can be used, although they maybe slightly inhibited by altitude and rugged terrain. Payloads and endurance are degraded due to density and attitude. Winds are turbulent with considerable fluctuations in air flow strength and direction, particularly on the lee side of mountains. These winds, combined with the terrain, produce extra strain on crews as they have little margin for error. Flight crews should receive training in these conditions before flying in operations under these conditions.

When using men on foot for navigation, use all available maps, the lensatic compass, and a pocket altimeter. The pocket altimeter is essentially a barometer, measuring height by means of varying air pressure. If a navigator can only establish his location in the horizontal plane by resection on one point, the altimeter tells him his height, and thus his exact position. The instrument must be reset at every known altitude as it is affected by fluctuations of air pressure. Air photographs can also be helpful if they are scaled and contoured.

Supply of water and ammunition and the evacuation of wounded, especially if helicopters cannot land, can complicate operations. Water and ammunition may have to be transported by unit or civilian porters using A-frames or other suitable devices, or even by animal transport such as camel or mule.

OFFENSIVE OPERATIONS

The objective in mountainous areas of operations is normally to dominate terrain from which the enemy can be pinned down and destroyed. Avenues of approach are normally few, with very limited lateral movement except by helicopter. Reconnaissance must be continuous using all available means, as enemy defensive positions will be difficult to find. Observation posts are emplaced on high ground, normally by helicopter.

When contact is made, airmobile infantry can be used to outflank and envelop the enemy while suppressive fires and close air support are placed on all suspected positions, especially on dominating ground. Engineers should be well forward to assist in clearing obstacles. If airmobile infantry is unable to outflank the enemy, it will be necessary to launch a deliberate attack.

Frontal attacks in daylight, even with considerable supporting frees, have a limited chance of success against a well-emplaced enemy. Flank attacks on foot take a lot of time. The best opportunity is at night or in very poor visibility, but progress of men on foot will be slow and objectives should be limited.

The force should make every effort to secure ground higher than enemy positions to allow the attack to be downhill. Mobile forces should select objectives to the enemy's rear to kill the enemy as they reposition or counterattack. Foot mobile forces must seek adequate terrain (restrictive) to equalize the enemy's mounted mobility advantage.

Air superiority is required to allow a continuous flow of supplies and combat support by helicopter. Friendly mobile units must concentrate to destroy enemy command and control, artillery, service support, and air defense assets. It may be possible to infiltrate to a position behind the enemy, preferably using the most difficult, and hence unlikely route. Although this is very slow, it normally has the advantage of surprise.

The importance of dominating terrain, together with the enemy's knowledge that troops on the objective will be physically tired and dehydrated, makes an immediate counterattack likely. Supporting weapons must be brought forward at once, preferably by helicopter, and casualties removed by the same method.

Airmobile and attack helicopter units are well suited for pursuit operations. They can be used to outflank retreating enemy, and set up positions overlooking likely withdrawal routes. Small engineer parties can be emplaced to block defiles and interdict trails. Close air support and field artillery are used to reinforce airmobile and attack helicopter units and to counter efforts by enemy engineers to create obstacles.

DEFENSIVE OPERATIONS

A defense from a series of strongpoints is normal in hot mountains due to the need to hold dominating terrain and restrictions on ground mobility. Due to the amount of rock in the soil, it takes more time to prepare positions and normally requires engineer support.

It is necessary to hold terrain dominating avenues of approach. Any terrain that dominates a friendly position must either be held, or denied to the enemy by fire. It may be necessary to stock several days' supplies, especially water, ammunition, and medical equipment in a position in case helicopters or supply vehicles are unable to reach it.

When a covering force is used, it is organized around cavalry reinforced with attack helicopters, supported by field and air defense artillery. Airmobile infantry operates on ridge lines. If the enemy closes on a battle position it is difficult to extract airmobile infantry, so sheltered landing sites nearby should be available. In any event, extractions must be covered by air or ground suppressive fires. Stay-behind observers should be used to call down field artillery fires on targets of opportunity or to report enemy activity. When tanks are a threat and terrain is suitable, the covering force is reinforced with tank-heavy units and antitank weapon systems.

Combat in the main battle area is usually a series of isolated actions fought from strongpoints on ridge lines and in valleys. Patrols are used extensively to harass the enemy and prevent infiltration; all possible routes must be covered. If the enemy attempts to outflank the friendly force, he must be blocked by attack helicopters, if available, or airmobile infantry.

Reserves should be kept centrally located and deployed by air to block or counterattack. If this is not possible, reserves may have to be split up and placed behind key terrain where they are available for immediate counterattack.

If retrograde operations are necessary, mountainous terrain is as good a place to conduct them as anywhere. More time is required to reconnoiter and prepare rearward positions, and they should be prestocked as much as possible. Unlike the desert floor where movement between positions is likely to cover relatively great distances, movement in these conditions is usually from ridge to ridge. Routes must be covered by flank guards, especially at defiles or other critical points, as the enemy will attempt to block them or cut off rear guards.

COMBAT SUPPORT

It may be difficult to find good gun positions at lower altitudes due to crest clearance problems-so high-angle fire is often used. The best weapons are light field artillery and mortars that are airmobile and can be manhandled so they can be positioned as high as possible.

Field artillery observation posts are emplaced on the highest ground available, although in low-cloud conditions it will be necessary to ensure that they are staggered in height. Predicted fire may be inaccurate due to rapidly changing weather conditions making observed fire a more sure method for achieving the desired results.

Like field artillery, there is limited use for self-propelled weapons in this environment, although some may be used in valleys. Airmobile towed weapons allow employment throughout the mountainous area of operations.

Major tasks for engineer, even in an airmobile force, are: construction, improvement, and route repair, and their denial to the enemy. Mining is important due to the limited number of routes. Lines of communication require constant drainage and possibly bridging to overcome the problem of flash flooding.

Because of the frequent interdictions of mountainous roadways, military police will experience multiple defile operations. Use temporary traffic signs to expedite traffic movement to the front. The number of stragglers may be expected to increase in this environment. Because of difficulty in resupply, the supply points for water, POL, food, and ammunition will become especially lucrative targets for enemy attack. Military police rear area security elements must develop plans for relief and for augmenting base defense forces.

COMBAT SERVICE SUPPORT

Air transportation is the best CSS means in mountain operations due to its mobility. It may be limited by the weather, enemy activity, or the scarcity of landing sites, so there should be alternative means available. Terrain permitting, wheel-vehicle transportation should be employed as far forward as possible, using high-mobility vehicles off main mutes. Beyond the limits of wheel transport the only alternatives to CSS transport are animals (which may need to be acclimatized) or porters.

The composition and employment of trains in mountain operations are described in FM 90-6. Brigade trains should locate near an airstrip that can handle USAF tactical airlift. They are an obvious target for enemy air attack or artillery, or raids by enemy deep patrols, so adequate air defense and a coordinated area defense plan are necessary. Guards must be placed on all dominating terrain around the area, equipped with ground surveillance radars and STANO devices, and patrols should be employed outside the perimeter.

Supply points may be set up in the brigade trains area to operate distribution points for Class I, III, and V supplies. However, where routes are limited it may be necessary to resupply totally by air from the DISCOM area.

The variations of supplies in demand in the desert are very much the same as for those in temperate climates and are described in Chapter 4. The differences are described below:

- Class I. Mess trucks are not practical in this terrain. Food is either eaten cold, or heated on can heaters. Each soldier/marine should carry a one-day supply of emergency rations to be used if the daily resupply does not arrive.
- Class II. There is a high demand for footwear. Combat boots may be expected to last approximately two weeks in the harsh rocky terrain.
- Class III. Individual vehicle consumption will be greater than normal. Aircraft fuel requirements are greater, but it should be possible for much of their refueling and servicing to take place well to the rear where resupply is relatively easy.
- Water continues to carry a high priority. Demand for water is approximately 9 quarts per day, per man, as a minimum, and sometimes considerably more. Troops should carry four canteens of water, and every effort should be made to prestock water in positions or along routes.

First aid at squad arid platoon level is very important as medics will not necessarily be able to reach individual isolated positions. It is easy to lose casualties in this terrain so a buddy system to keep watch on each individual should be a matter of SOP. Medical evacuation is most often by air. It is a comparatively long distance to the nearest helicopter landing site, so teams of stretcher bearers will be required.

RDL Table of Document Download
Homepage Contents Informing Instructions

APPENDIX G

WATER USAGE IN DESERT OPERATIONS

Combat operations in the desert pose a number of unique problems. Because there is so little water and because our troops and much of our equipment cannot survive without it, water is a critical item of supply in the desert. Forces trying to survive in the desert without adequate water supplies have always met with disaster. Finding and keeping water sources may be the most crucial issue in desert conflicts. At the very least, water sources will be critical.

PROPER USES OF WATER

Water must be used to support immediate and future missions. There will be times when there is enough water to fill all requirements, but there will also be times when command direction and considerable thought are necessary to decide how to make the best use of available water. The first priority must go to the survival of the force and accomplishment of the immediate mission, and second priority to the maintenance and sustainment of the force. A general priority for uses of water is--

- Personnel (drinking only).
- Medical treatment.
- Vehicle and equipment cooling systems.
- Personnel (uses other than drinking).
- Decontamination.
- Food preparation.
- Laundry.
- Construction.

How far the available water will stretch depends upon your evaluation of the local situation and mission; how you set priorities for water; and how careful you are in using your limited supply.

INTEGRATED BATTLEFIELD EFFECTS ON WATER USE

Water needs increase dramatically on a nuclear or chemical battlefield. Decontamination of men and equipment requires large quantities of water. Being "buttoned up" in vehicles or in NBC protective gear locks in heat and perspiration and makes men sweat more. This increases the need for drinking water to avoid dehydration and heat casualties. Replacement of fluids lost through sweat is a critical use of water and generally comes ahead of all other uses.

LEADER RESPONSIBILITIES

Your primary responsibility is to accomplish your mission. Water is essential to do that. You must estimate how much you need and when you need it, just as you do for any type of essential supply. When the supply is limited, you must adjust your plans. During the accomplishment of your mission you must monitor the supply and ensure that it is used according to your plan.

One of the biggest water-related problems you may face is that as troops recognize how valuable water is to their survival, they may hoard it and not drink enough to sustain the efforts you expect of them. Before periods of activity, have your troops drink as much as they can. During the activity, take positive steps to make soldiers/marines replace the water lost by sweating. Thirst is a poor indicator of the body's need for water and may be ignored during hard work or in the heat of battle. Squad and section leaders must make their men drink regularly.

EFFECTS OF HEAT AND LACK OF WATER

Objects absorb heat from the sun and the air. In the desert, heat from these two sources is extreme. The clear, low-humidity air lets most of the heat from the sun through. As the sunlight strikes an object, such as a soldier/marine or the ground, much of the heat is absorbed. The ground, heated by the sun, in turn heats the air, often to temperatures well over 100 degrees Fahrenheit. A man, or any other object in the sun, absorbs heat from both sources. A man in the shade only has to contend with the heat from the air. The third major source of heat is the body itself. Like an engine, the body generates excess heat as it functions. The more work performed, the more heat that is generated. There are four ways that heat leaves an object:

- Radiation heat radiates from an object to a cooler object through a medium.
- Conduction heat flows from a hot object to a cooler object through direct contact.
- Convection heat flows from a hot object into a cooler surrounding medium like air.
- Evaporation heat is absorbed in changing a liquid (like water) into a vapor.

The heat losses from radiation and conduction are relatively small for people and equipment. If the air is much cooler than the surface of an object, such as a vehicle radiator or a person's skin, convection can remove significant amounts of heat. As the two temperatures get closer, this loss becomes smaller. If the air is hotter than the object, as is often the case in the desert, heat is gained from the air. In hot, dry desert climates, the primary method of losing excess heat from the body is the evaporation of sweat. The rate of sweating depends on the amount of excess heat the body needs to lose. Hard work in hot climates can result in 1-1/2 to 2-1/2 quarts of sweat lost per hour.

Supporting medical units measure the combined effects of the sun, air temperature, and humidity on dismounted troops in open terrain. This combined effect, measured as the Wet Bulb Globe Temperature (WBGT) or the Wet Globe Temperature (WGT), is translated by the supporting medical unit to a heat condition.

Men cannot be expected to routinely perform more than about 5 hours of heavy work per day in heat condition BLACK. However, using a 20-minute work/40-minute rest cycle, it will take 15 hours to do 5 hours of work. Similarly, in heat condition RED, a maximum of 6 hours of heavy work can be expected; in heat condition YELLOW, 7 hours; and in heat condition GREEN, 8 hours.

The heat condition experienced by individual soldiers/marines may differ from the general heat conditions in the area. The activity level in closed vehicles (such as tanks and armored personnel carriers) will probably be less than that of troops outside, but the air temperature may be 20 to 30 degrees Fahrenheit higher, thus increasing the soldier's heat gain from the air and, as a result, his sweat rate and need to drink water. As a general rule, increase the water intake to that for two or three heat conditions higher for troops in enclosed vehicles. On the other hand, troops performing light duty in the shade are not as severely stressed and will need less water and less rest time.

In practice, keeping track of the current heat condition and applying it to troops working in a wide variety of conditions will be difficult, if not impossible. This data gives you an understanding of how much water is required by troops when they work hard in the desert heat and to point out the limits of an individual's efforts in extreme heat. Work done in the heat of the day takes much longer and is more fatiguing than work done under relatively cooler conditions. Work schedules planned to take advantage of the cooler times of day-early morning, late evening, and night-not only increase productivity and reduce water use, but they are also easier on the men and better for their morale.

EFFECTS OF WATER LOSS

The body has a small reserve of water and can lose some without any effects. After a loss of about 2 quarts (which represents about 2.5 to 3.0 percent of body weight), effectiveness is impaired. Soldiers/marines may begin to stumble, become fatigued and unable to concentrate clearly, and develop headaches. Thirst will be present but not overpowering. So unless well trained, or reminded or goaded to drink, troops may not replace the water loss.

As dehydration continues, the effects will become more pronounced. The soldier/marine will become less and less effective and more likely to become a heat casualty. Some soldiers/marines will experience heat cramps, others will develop heat exhaustion or heatstroke. Heat cramps and heat exhaustion can be treated with good success and the soldier/marine returned to duty in a few days; however, without prompt medical attention, heatstroke can be fatal. Even if the man survives, he will probably not be returned to duty. In any case, a heat casualty is lost for some time. Preventing casualties is much easier than treating and replacing the casualties.

WATER SUPPLY PLANNING

Water planning is complicated because water is heavy (about 8.3 pounds per gallon) and may be considered perishable. Water stored in small containers gets hotter than water stored in large containers. As water gets hotter, it loses its disinfectant and becomes less desirable to drink. These facts make it difficult to carry an adequate supply of water, and frequent resupply is often required. The following questions must be answered when planning the unit's water supply:

- How much water is needed?
- Where is it needed?
- When is it needed?
- How will water get to the unit?
- How does water supply affect the mission?
- How does the mission affect water requirements?
- What measures need to be taken to ensure water is properly used?

There are several requirements for water. Some requirements, such as water for radiators, are reasonably constant. Some, such as water for food preparation or showers, are prescribed by the situation. Others, such as water for drinking and personal hygiene, depend on how the mission is accomplished. Planning water requirements for centralized service support functions (shower, laundry, medical treatment, maintenance, and construction) is the responsibility of the supporting organization. The largest and most critical planning factor is drinking water. The quantity required depends on the environment and the difficulty and intensity of individual activity.

When calculating water requirements for a whole day, you need to consider other requirements, such as shaving, brushing teeth, and helmet baths. On the average, these functions require almost 2 gallons of water per man per day. When B rations are issued, plan for 1 gallon of water per meal for the mess kit laundry, and 0.5 gallon per meal for food preparation and kitchen cleanup if the unit prepares its own B rations. The water used to heat individual combat rations can be reused for washing and shaving.

When calculating water requirements for individual details, plan to use 2 quarts of water per hour of hard work per man (including rest periods) during the heat of the day, or 1 quart of water per hour of hard work per man in the cooler parts of the day. These quantities are intended to satisfy requirements for drinking water as well as for the water that men will pour over their heads when they are hot. Experience with local conditions and the work performed may change these estimates. It is important to remember that water lost by sweating will be replaced sometime during the day, but men work best if the water is replaced as it is lost.

WATER REQUIREMENTS

Some examples of water planning calculations are provided in the following paragraphs.

Example 1. A 10-man squad performing heavy work that will take about 2 hours at a Continental United States (CONUS) base.

If the squad works in the late morning, with a 30-minute work, 30-minute rest cycle (heat condition RED), the work will take almost 4 hours. At 2 quarts of water per man per hour, the squad will require 20 gallons of water or 8 quarts (2 gallons) per man.

If the squad works at night or in the very early morning (heat condition GREEN or cooler), with a 50-minute work, 10-minute rest cycle, the work will take about 2 hours. At 1 quart of water per man per hour, the squad will require 5 gallons of water or 2 quarts (0.5 gallon) per man.

Example 2. A 40-man platoon doing a variety of work over an entire day.

The platoon requires 4 gallons of water per man per day for drinking, and 2.5 gallons per man per day for personal hygiene, or 6.5 gallons per man per day. For the whole platoon, 260 gallons (6.5 x 40) per day are required. (Meals not considered.)

Example 3. A 160-man company doing a variety of work and eating two B. ration meals and one MRE or MCI ration per day delivered from the battalion consolidated kitchen.

The company requires 4 gallons of water per man per day for drinking, 2.5 gallons per man per day for personal hygiene, and 1 gallon per meal for the mess kit laundry.

6.5 gallons (4 + 2.5) per man per day x 160 men	= 1,040 gallons
+ 1 gallon per B-ration meal x 320 rations*	= 320 gallons
	Total 1,360 gallons per day

*Not required if an expedient means can be used and utensils to avoid having to wash plates

NOTE: If company personnel operate their own field kitchen, 0.5 gallon per B-ration meal is required. In this example, an additional 160 gallons (0.5 x 320) would be needed.

Example 4. A 750-man battalion performing a variety of tasks, operating a battalion field kitchen, and feeding two B-ration meals and one MRE or MCI per day.

The battalion requires 6.5 gallons of water per man per day for drinking and personal hygiene, and 1.5 gallons per B-ration meal served.

6.5 gallons per man per day x 750	= 4,875 gallons
+ 1.5 gallons per B-ration meal x 1,500	= 2,250 gallons
	Total 7,125 gallons per day

The situation will usually dictate whether an element will pick up its own water or have it delivered to it. Normally, water is either produced at or delivered to a water supply point in the brigade rear. Forward units pick up their water at the water supply point and move it forward using organic transportation. Water distribution is normally planned and coordinated at company or battalion level.

Long distances and increased consumption compound water transportation problems in the desert. Several specialized pieces of equipment are available to overcome these problems. They have been designed and allocated to permit units to move and store the larger than normal volumes of water required in the desert. They are also capable of delivery by ground and air transportation, airdrop, or low altitude parachute extraction system (LAPES). Air transportation, while possible, is usually limited by aircraft availability and weight of water required. It is normally used only when ground transportation is not feasible. Water shortages can severely limit your unit's mission capabilities. If you cut water use to absolutely essential requirements, you can temporarily overcome limited water shortages, but severe shortages will limit your unit's mobility and capability. Water shortages may make some daylight operations or hard work infeasible or unsupportable. Each course of action must be analyzed with respect to water support requirements and the capacity of troops to sustain their efforts under severe heat conditions. Water supplies are equally important to the enemy. Taking and keeping water sources, or denying or destroying enemy water supplies can critically alter the options available to the enemy.

You and your unit can do several things to use water to your advantage. First, make sure that enough water is available for the most critical use-drinking. Furthermore, water must be available right where it is needed. Troops working in the desert should not have to walk more than a short distance to a water source. During rest periods some soldiers/marines would rather sit than walk to get water, especially if they are not thirsty. It must be easy for them to get water. Similarly, beverages must be readily available in dining facilities. There should be enough beverage dispensers that troops do not have to wait long in lines.

In addition to having water readily available, you must also make sure that soldiers/marines actually drink all the water they need. Since the signs of dehydration are not obvious until a person is close to heat injury, leaders at the squad and platoon level must keep track of water consumed and take the following measures:

- Have troops drink an extra quart of water before hard work. Storing water in your stomach gives you an extra quart of reserve capacity.
- Keep track of how much water each man drinks at the squad level; at the platoon level, monitor the use rate at each squad.
- Have troops take breaks as often as the heat condition requires, and during breaks remind or require the troops to drink.
- Make sure water is kept as cool as possible so it will be as palatable as possible to drink.
- Watch the troops for the first signs of heat stress and reduced effectiveness, such as stumbling and slurred speech.
- Have your soldiers/marines check their urine. A lack of the need to urinate and dark-colored urine are signs of dehydration.
- Use the buddy system within the squad to help ensure soldiers/marines are drinking enough.
- Make sure troops wear their uniform correctly. Shirts should be on with sleeves rolled down, scarf around neck, and hat on. The uniform should be worn as loosely as possible.

In order for water to be useful for its most critical purpose in the desert, it must be protected not only from enemy action but also from heat and contamination. The larger quantities of water required for drinking in the desert increase the importance of the quality of the water. Water can carry minerals, microbiological organisms, and toxic materials. The body can handle only so much of these contaminants before its natural defense mechanisms become swamped and health and effectiveness deteriorate.

When water has been treated and distributed to water points, it has already been checked for contamination by water purification unit operators and medics; however, these checks do not ensure the water will not become contaminated somewhere in the unit distribution system before a soldier/marine drinks it.

When water is purified and distributed, it is usually disinfected by adding chlorine to a level prescribed by the Command Surgeon. The chlorine not only kills the microbiological organisms presently in the water, but some also remains in the water to kill any bacteria that might get into the water later. Such contamination can be minimized by using common sense at the unit level. Store water only in clean containers intended for water, and do not let anything get in the water that you would not want to drink.

However, even if your unit is very careful, water may eventually become contaminated. As the water is handled or gets hot, the chlorine's disinfecting power disappears. Your company's field sanitation team measures chlorine levels in unit water containers. The team also has ampules of a chlorine compound that are used to replace chlorine in small containers. Each soldier/marine is issued a small bottle of iodine tablets to disinfect water in his canteen if he must take water from an expedient, untreated source.

Heat is another contaminant of water in the desert. When water is warmer than 75 or 80 degrees

Fahrenheit, it becomes difficult to drink. Bad tastes in water become more pronounced as water becomes warmer and people will not want to drink it. Small-unit leaders will have a very difficult task trying to get troops to drink all the water they need to replace losses if the water tastes bad. Water tastes best, and it is easier to drink large quantities of it, if it is between 50 and 70 degrees Fahrenheit. There are three ways to avoid the problems of drinking hot water:

- Drink it before it gets hot.
- Keep it in a cool place or in the shade.
- Cool it to a palatable temperature.

Drinking water before a period of heavy work or before leaving the unit area on a mission gets valuable water into the body before the water has a chance to heat up. It also provides an additional reserve of water that is easy to carry.

Water in containers absorbs heat from three sources-the air, the ground, and sunlight. There is no easy way to cut absorption from the air, but you can reduce heat absorbed from the ground and sun by keeping containers in as much shade as possible. There are certainly many ways of getting shade for water containers; with a little American ingenuity and available materials, it can be done.

Small, uninsulated water containers heat up more quickly than larger or insulated containers. Fortunately, the water in small containers can be used more quickly before it heats up. Water in uninsulated 5-gallon cans starting at a cool 60 degrees Fahrenheit can heat up to unpalatable temperatures in three to four hours on a hot (greater than 90 degrees Fahrenheit) day in the full sun, but it will take seven to eight hours if kept in the shade. Unshaded 55-gallon drums will heat to unpalatable temperatures in about one day of full sun, but will stay drinkable for two days if shaded. Insulated 400-gallon water trailers in or out of the sun will keep water cool for several days.

Water must be cool to start with if it is to be kept cool. One of the supplemental items available to company-size units is a small mobile water chiller. The chiller is designed to provide cool (about 60 degrees Fahrenheit), palatable water for company-size units. It can be used to cool and dispense water from any container into canteens, water cans, or any other container at a rate of at least 0.5 gallon per minute. It should be used in all cases where individuals must be encouraged to drink large quantities of water, and to cool water being taken with or to elements working away from the company area.

SURVIVAL TECHNIQUES

The water support logistics structure is designed to provide enough water to retain maximum force effectiveness, especially on a battlefield. It would be foolish to expect that the large quantities of water required for a combat force will always be available. Water is a critical supply in desert combat. Having less than full supply for all needs does not necessarily spell disaster for operations, but it may force changes in plans. There are several options available to you if you are forced to operate on reduced water supplies. The first part of this chapter outlines those options. The second part (described in the following paragraphs) describes actions and techniques for small groups or individuals to take when totally cut off from normal water supply.

The first and most obvious option is to eliminate, reduce, or postpone water uses not immediately required for survival or mission accomplishment. Showers, laundry, personal hygiene, and B-rations can usually be eliminated for several weeks without severe impact on troop health or combat effectiveness, although this tactic will eventually reduce troop health and morale. Also, construction requiring water can often be avoided.

The one water use that cannot be denied without significant risk is individual drinking. However, significant reductions in individual drinking requirements are possible by-

- Limiting all but absolutely essential work to the cooler parts of the day (early morning, late evening, and night).
- Keeping individuals in the shade as much as possible.
- Severely limiting all activity.

All of these actions reduce the effectiveness or capability of a force, but they help it to survive.

What can individuals or small groups do when they are totally cut off from normal water supply? If you are totally cut off from the normal water supply, the first question you must consider is whether you should try to walk to safety or stay put and hope for rescue. Walking requires 1 gallon of water for every 20 miles covered at night, and 2 gallons for every 20 miles covered during the day. Without any water and walking only at night, you may be able to cover 20 to 25 miles before you collapse. If your chance of being rescued is not increased by walking 20 miles, you may be better off staying put and surviving one to three days longer. If you do not know where you are going, do not try to walk with a limited supply of water.

If you decide to walk to safety, follow the following guidelines in addition to the general conservation practices listed in the next section:

- Take as much water as you have and can carry, and carry little or no food.
- Drink as much as you can comfortably hold before you set out.
- Walk only at night.

Whether you decide to walk or not, you should follow the principles listed below to conserve water in emergency situations:

- Avoid the sun. Stay in shade as much as possible. If you are walking, rest in shade during the day. This may require some ingenuity. You may want to use standard or improvised tents, lie under vehicles, or dig holes in the ground.
- Cease activity. Do not perform any work that you do not have to for survival.
- Remain clothed. It will reduce the water lost to evaporation.
- Shield yourself from excessive winds. Winds, though they feel good also increase the evaporation rate.
- Drink any potable water you have as you feel the urge. Saving it will not reduce your body's need for it or the rate at which you use it.
- Do not drink contaminated water from such sources as car radiators or urine. It will actually require more water to remove the waste material. Instead, in emergencies, use such water to soak your clothing as this reduces sweating.
- Do not eat unless you have plenty of water.

Do not count on finding water if you are stranded in the desert. Still, in certain cases, some water can be found. It does rain sometimes in the desert (although it may be 20 years between showers) and some water will remain under the surface. Signs of possible water are green plants or dry lake beds. Sometimes water can be obtained in these places by digging down until the soil becomes moist and then waiting for water to seep into the hole. Desert trails lead from one water point to another, but they may be further apart than you can travel without water. Almost all soils contain some moisture.

APPENDIX H

FRATRICIDE REDUCTION

The problem of fratricide is as old as warfare itself. It is a complex problem that defies simple solutions. Fratricide is defined as "the employment of friendly weapons and munitions, with the intent to kill the enemy or destroy his equipment or facilities that results in unforeseen and unintentional death or injury to friendly personnel." This is obviously a broad definition. This appendix focuses on actions leaders can take with current resources to reduce the risk of fratricide.

MAGNITUDE OF THE PROBLEM

The modern battlefield is more lethal than any in history. The pace of operations is rapid, and the non-linear nature of the battlefield creates command and control challenges for all unit leaders.

Our ability to acquire targets using thermal imagery exceeds our ability to accurately identify targets as friend or foe. The accuracy and lethality of modem weapons make it possible to engage and destroy targets at these extended acquisition ranges.

Added to this is the problem of battlefield obscuration. Rain, dust, fog, smoke, and snow degrade the ability to identify targets by reducing the intensity and clarity of thermal images. The effects of battlefield obscuration must be considered when thermal identification is relied upon.

On the battlefield, positive visual identification cannot be the sole engagement criteria at ranges beyond 1,000 meters. Situational awareness is key and must be maintained throughout an operation.

The following are recommended actions to take at crew and leader level in the event the crew are victims of friendly fires:

- React to contact until you recognize friendly fire.
- Cease fire.
- Report on the next higher unit net¬

  - That you are receiving friendly fire.
  - The location and direction of the firing vehicle.
- Provide a visual recognition signal to cease fire.
- Protect troops, request medical assistance as needed.
- Do not return fire when you positively identify the firing unit as friendly.

The following are recommended actions to take at crew and leader level when the crew are engaging friendly forces :

- Cease fire.
- Report on next higher net¬
  - The engaged friendly force (if unknown, report number and type of vehicles).
  - The location.
  - The direction and distance to victim.
  - The type of fire.
  - The target effects.

The following are recommended actions to take at crew and leader level in the event the crew observes a friendly fire incident:

- Seek cover and protect self.
- Report on next higher net¬
  - The friendly force engaged.
  - The location of the incident.
  - The direction and distance to victim/firer.
  - The type of fire.
  - The target effects.
- Provide a visual friendly recognition signal.
- Provide assistance (when safe to do so) as needed.

Leader actions should focus on identifying and stopping the friendly fire incident and establishing controls to prevent its recurrence. Some recommended actions for identifying and stopping friendly fire incidents are-

- Find and stop firing.
- Conduct in-stride risk assessment.
- Implement controls to preclude recurrence.

PREVENTIVE MEASURES

Reduction of fratricide risk begins with the planning phase of an operation and continues through the execution of the operation. The following are considerations for identifying fratricide risks in the planning, preparation, and execution phases of a given operation:

- Planning phase. A good plan that is well understood helps to minimize fratricide risk. The following considerations help indicate the potential for fratricide in a given operation:
  - The clarity of the enemy situation.
  - The clarity of the friendly situation.

- The clarity of the commander's intent.
- The complexity of the operation.
- The planning time available to all levels.

• Preparation phase. The following additional fratricide risks may become evident during rehearsals:

- Number and type of rehearsals.
- Training and proficiency levels of unit/individuals.
- The habitual relationships between units conducting the operation.
- The endurance of the troops conducting the operation.

• Execution phase. During execution, in-stride risk assessment and reaction are necessary to overcome unforeseen Fratricide risk situations. The following are factors to consider when assessing fratricide risks:

- Intervisibility between adjacent units.
- Amount of battlefield obscuration.
- Ability or inability to positively identify targets.
- Equipment similarities and dissimilarities between enemy and friendly vehicles.
- Vehicle density on the battlefield.
- The tempo of the battle.

Graphics are a basic tool that commanders at all levels use to clarify their intent, add precision to their concept, and communicate their plan to subordinates. As such, graphics can be a very useful tool in reducing the risk of fratricide. Commanders at all levels must understand the definitions and purpose of operational graphics and the techniques of their employment. See FM 100-5-1 for the definitions of each type of graphic control measure.

Briefbacks and rehearsals are primary tools in identifying and reducing fratricide risk. The following are some considerations on briefbacks and rehearsals to aid in reducing fratricide:

• Briefbacks ensure subordinates understand their commander's intent. They often highlight areas of confusion, complexity, or planning errors.
• The type of rehearsal conducted impacts on the risks identified.
• Rehearsals should extend to all levels of command and involve all key players.
• Use briefbacks or rehearsals to ensure subordinates know where fratricide risks exist, and what to do to reduce or eliminate the risk.

Maintaining situational awareness at all levels is key to fratricide reduction. Units must develop techniques to gain and maintain situational awareness in SOPs. Techniques could include-

• Eavesdropping on next higher net.
• Cross talk on radio between units.
• Accurate position reporting and navigation.

- Training and use/exchange of LOS.

Risk assessment must be conducted at all levels during the planning, preparation, and execution phases of all operations. Identification of fratricide risk factors is conducted at every level and the results should be clearly communicated up and down the chain of command.

Figure H-1 provides a worksheet for considering fratricide risk in the context of mission requirements. The worksheet lists 26 mission-accomplishment factors that affect the risk of fratricide. Assess the potential risk in each area as low, medium, or high, and assign a point value to each (one point for low risk, two for medium risk, three for high risk). Add the point values for the overall fratricide assessment score. Use the resulting score only as a guide, however. Your final assessment must be based both on observable risk factors like those on the worksheet and on your "feel" for the intangible factors affecting the operation. Note that descriptive terms are listed only in the low- and high-risk columns of the worksheet. Your assessment of each factor will determine whether the risk matches one of these extremes or lies somewhere between them as a medium risk.

The following fratricide reduction measures are provided as reminders for prudent/appropriate actions to reduce fratricide risk. They are not directive in nature, nor intended to restrict initiative. Apply the following measures to METT-T situations as appropriate:

- Identify and assess potential fratricide risk in estimate of the situation. Express this risk in the OPORD or FRAGO.
- Maintain situational awareness-current intelligence; unit locations/dispositions; denial areas (minefields/FASCAM); contaminated areas (e.g., ICM and NBC); SITREPs; and METT-T.
- Ensure positive target identification. Review vehicle/weapons ID cards. Know at what ranges and under what conditions positive ID of friendly vehicles/weapons is possible.
- Establish a command climate that stresses fratricide prevention. Enforce fratricide prevention measures; use doctrinally sound tactics, techniques, and procedures to ensure constant supervision of execution of orders and performance to standards.
- Recognize the signs of battlefield stress. Take quick effective action to deal with it to maintain unit cohesion.
- Conduct individual and collective (unit) fratricide awareness training; target identification/recognition training; fire discipline; and leader training.
- Develop a simple decisive plan.
- Give complete and concise mission orders.
- Use SOPS that are consistent with doctrine to simplify mission orders. Periodically review and change SOPS as needed.
- Strive for maximum planning time for you and your subordinates.
- Use common language/vocabulary and doctrinally correct standard terminology and control measures, such as, fire support coordination line (FSCL), zone of engagement, restrictive fire line (RFL), and others.
- Ensure thorough coordination is performed.
- Plan for and establish good communications.
- Plan for collocation of command posts, as appropriate to the mission (e.g., passage of lines, and so

forth).

- Establis h and designate liaison officers (LO) as appropriate.
- Mak e sure rules of engagement (ROE) are clear.
- Conside r the effect of fratricide on key elements of terrain analysis (observation and fields of fires, cover and concealment, obstacles and movement, key terrain, and avenues of approach).
- Conduc t rehearsals whenever the situation allows time to do so.
- B e in the right place at the right time. Us e position location/navigation (GPS) devices; know your location and the locations of adjacent units (left, right, leading and follow on); and synchronize tactical movement.
- Includ e fratricide incidents in after-action reviews (AAR).

FACTORS	LOW (1)	MEDIUM (2)	HIGH (3)
1 UNDERSTAND PLAN			
Commander's Intent	Clear		Foggy
– Complexity	Simple		Complex
– Enemy Situation	Known		Unknown
Friendly Situation	Clear		Unclear
– ROE	Clear		Unclear
2 ENVIRONMENT			
– Intervisibility	Favorable		Unfavorable
– Obscuration	Clear		Obscured
– Battle tempo	Slow		Fast
– Positive target ID	100%		0%
3 CONTROL MEASURES			
– Command relationships	Organic		Joint/Combined
– Audio	Loud/Clear		Jammed
– Visual	Well Seen		Obscured
– Graphic	Standard		Not understood
– SOPs	Standard		Not used
– LOs	Proficient		Untrained
Location/Navigation	Sure		Unsure
4 EQUIPMENT (Compared to US)			
– Friendly	Similar		Different
– Enemy	Different		Similar
5. TRAINING			
Individual proficiency	MOS Qual		Untrained
Unit proficiency	Trained		Untrained
– Rehearsal	Multiple		None

[illegible]	[illegible]		[illegible]
– Habitual relationship	Yes		No
– Endurance	Alert		Fatigued
6 PLANNING TIME (1/3 -2/3 Rule)			
– Higher HQ	Adequate		Inadequate
– Own HQ	Adequate		Inadequate
– Lower HQ	Adequate		Inadequate
OVERALL FRATRICIDE ASSESSMENT	LOW 26–46%*	MEDIUM 42–62%*	HIGH 58–78%*

* Commander may use numbers as the situation dictates. Numbers alone may not give accurate fratricide risk.

Figure H-1 Fratricide risk assessment worksheet.

FRATRICIDE RISK CONSIDERATIONS

This format, which parallels the five-paragraph OPORD, contains key factors and considerations in fratricide reduction, This is not a change to the OPORD format; rather, it should be used during OPORD development to ensure fratricide reduction measures are included in the order. It is not a strict guide. The factors and considerations are listed where they would likely appear in the OPORD, but they may warrant evaluation during preparation of other paragraphs.

1. Situation.

a. Enemy forces.

(1) Are there similarities between enemy and friendly equipment and uniforms that could lead to fratricide?

(2) What languages do enemy forces speak? Could these contribute to fratricide risk?

(3) What are the enemy's deception capabilities and its past record of deception activites?

(4) Do you know the locations of enemy forces?

b. Friendly forces.

(1) Among the allied forces, are there differences (or similarities with enemy forces) in language, uniform, and equipment that could increase fratricide risk during combined operations?

(2) Could differences in equipment and uniforms among US armed forces increase fratricide risk during joint operations?

(3) What differences in equipment and uniforms can be stressed to help prevent fratricide?

(4) What is the friendly deception plan?

(5) What are the locations of your unit and adjacent units (left, right, leading, follow-on)?

(6) What are the locations of neutrals and noncombatants?

c. Own forces.

(1) What is the status of training activities? What are the levels of individual, crew, and unit proficiency?

(2) Will fatigue be a factor for friendly forces during the operation? Has an effective sleep plan been developed?

(3) Are friendly forces acclimatized to the area of operations?

(4) What is the age (new, old, or mix) and condition of equipment in friendly units? What is the status of new equipment training (NET)?

(5) What are the expected MOPP requirements for the operation?

d. Attachments and detachments.

(1) Do attached elements know the above information regarding enemy and friendly forces?

(2) Are detached elements supplied the above information by their gaining units?

e. Weather.

(1) What are the expected visibility conditions (light data and precipitation) for the operation?

(2) What effect will heat and cold have on troops, weapons, and equipment?

f. Terrain.

(1) Do you know the topography and vegetation (such as urban, mountains, hilly, rolling, flat, desert, swamp/marsh, prairie/steppe, jungle, dense forest, open woods) of the expected area of operations?

(2) Have you evaluated the terrain using the factors of OCOKA?

2. Mission. Is the mission, as well as all associated tasks and purposes, clearly understood?

3. Execution.

a. Task organization.

(1) Has the unit worked under this task organization before?

(2) Are SOPS compatible with the task organization (especially with attached units)?

(3) Are special markings or signals (for example, cats' eyes, chemlites, or panels) needed for positive identification of uniforms and equipment?

(4) What special weapons and/or equipment are to be used? Do they look or sound like enemy weapons and/or equipment?

b. Concept of the operation.

(1) Maneuver. Are main and supporting efforts identified to ensure awareness of fratricide risks and prevention?

(2) Fires (direct and indirect).

(a) Are priorities of fires identified?

(b) Have target lists been developed?

(c) Has the fire execution matrix/overlay been developed?

(d) Have locations of denial areas (minefields/FASCAM) and contaminated areas (ICM, NBC) been identified?

(e) Are the locations of all supporting fires targets identified in the OPORD/OPLAN overlays?

(f) Are aviation and CAS targets clearly identified?

(g) Has the direct-fire plan been developed?

(h) Have final protective fires (FPF) been designated?

(i) Have you identified and verified sector limits?

(3) Engineer tasks.

(a) Are friendly minefield, including FASCAM and ICM dud-contaminated areas, known?

(b) Are obstacles identified, along with the approximate time needed for reduction/breaching of each?

(4) Tasks to each subordinate unit. Are friendly forces identified, as appropriate, for each subordinate maneuver element?

(5) Tasks to CS/CSS units. Have locations of friendly forces been reported to CS/CSS units?

(6) Coordinating instructions.

(a) Will a rehearsal be conducted? Is it necessary? Are direct and indirect fires included?

(b) Is a briefback necessary?

(c) Are appropriate control measures clearly explained and illustrated in the OPORD and overlays? Have they been disseminated to everyone who has a need to know? What is the plan for using these control measures to synchronize the battle and prevent fratricide?

(d) Have target/vehicle identification drills been practiced?

(e) Do subordinate units know the immediate action, drill, or signal for "cease fire" or "I am friendly" if they come under unknown or friendly fire? Is there a backup action?

(f) Is guidance in handling dud munitions (e.g., ICMs and CBUs) included?

4. Service Support.

a. Are trains locations and identification markings known by everyone?

b. Do medical and maintenance personnel know the routes between train units?

5. Command and Signal.

a. Command.

(1) What is the location of the commander and key staff?

(2) What is the chain of command?

b. Signal.

(1) Do instructions include signals for special and emergency events?

(2) Do instructions include how to identify friendly forces to aircraft?

(3) Do instructions include backup code words and visual signals for all special and emergency events?

(4) Are signal operation instructions (SOI) distributed to all units with a need to know (e.g., higher, lower, adjacent, leading, follow-on).

A2C2	Army airspace command and control
AA	assembly area
AAA	antiaircraft artillery
AAFAD	all army for air defense
AFCE	Allied Air Farces Central Europe
A AG	army artillery group
AAR	after-action review
AC	analysis console
ACA	airspace coordination areas
ACC	air component commander
ACP	Allied communication Publication
ACR	armored cavalry regiment
ACT	air cavalry troop
AD	armored division
ADA	air defense artillery
ADCOORD	air defense coordinator
ADW	air defense warming
AH	Air Force
AFAC	airborne forward air controller
AFCENT	Allied Forres Central Europe
ARO	aerial fire support officer
AFV	armored fighting vehicle
AG	adjutant general
AGOS	air-ground operations system
AGOSOP	Air-Ground Operations Standard Operating Procedures
AHB	attack helicopter battalion
AI	air interdiction
A/L	administrative/logistics
ALO	air liason officer
AM	amplitude modification
ANGLICO	air and naval gunfire liaison company
AO	area of operation
AOE	authorized organization equipment
AP	antipersonnel (mine)
APC	armored personnel carrier
APERS	antipersonnel (animation)

APT	armor-piercing incendiary
ARNG	Army National Guard
ARSOA	Army special operations aviation
ARSOF	Army special operations forces
ARTEP	Army Training and Evaluation Program
ASOC	air support operations center
ASP	ammunition Supply point
AT	antitank
ATGM	antitank guided missile
ATHS	automatic target hand-off system
ATKHB	attack helicopter battalion

ATP ammunition transfer point

AVLJJ	armored vehicle launched bridge
AVUM	aviation unit maintenance
AXP	ambulance exchange point
BAT	battlefield air interdiction
RAS	battalion aid station
B-BAG	battle bags
BDA	battle damage assessment
BDAR	battle damage assessment and repair
BRI	brigade
BDO	battle dress overgarment
BFV	bradley fighting vehicle
BHL	battle handover line
BII	basic issue items
BMNT	beginning morning nautical twilight
BMD	battalion maintenance officer
bn	battalion
Bn/TF	battalion/task force
BP	battle position
BSA	brigade support area

C2	command and control
C2 node	command and control junction

C3 command, control and communication

CA counterair

CAA	combined arms Army
CAB	combat aviation brigade
CAM	chemical agent monitor
CARC	chemical agent resistant coating

CAS	close air support
CBR.	chemical, biological, radiological
CBU	cluster bomb unit
CM	commander
CE	chemical energy; communication-electronics
CEB	clothing exchange and bath
CESO	communications electronics signal officer
CEV	combat engineer vehicle
CFA	covering force area
CFL	coordinated fire line
CFV	cavalry fighting vehicle
CH	cargo helicopter
CLAMMS	cleared lane mechanical making system
cm	centimeter(s)
CMT	company maintenance team
COA	change of assignment course of action
COLT	combat observation lasing team
COMMZ	communications zone
COMSEC	communications security
CONUS	continental United States
COP	command and observation post
COSCOM	corps support command
CP	checkpoint: command post
CPX	command post exercise
CRP	combat reconnaissance patrol
CRT	combat ray tube
	mbat support
	s support battalion
CSM	command sergeant major
CSR	controlled supply rate
CSS	combat service support
CSSE	combat service support element
CTA	common table of allowances
CW	continuous wave
D-day	deployment day
DA	Department of the Army
DAG	divisional artillery group
DAO	division ammunition officer
DBDU	desert battle dress uniform
DD	Department of Defense
DEW	director! energy wanton

DISCOM	**division support command**
DKIE	**decontamination kit individual equipment**
DLIC	**detachment left in conflict**
DMD	**digital message device**
DMMC	**division material management center**
DNVT	**digital nonsecure voice terminal**
DP	**decision point**
DPICM	**dual-purpose improved conventional munition**
DS	**direct support**
DSA	**division support area**
DST	**decision support template**
DSVT	**digital subscriber voice terminal**
DTG	**date-time group**
DTOC	**division tactical operations center**
DZ	**drop zone**

EA	**engagement area**
EAC	**echelons above corps**
EC	**electronic combat**
ECM	**electronic counter measures**
EEFI	**essential elements of friendly information**
EEI	**essential elements of information**
EENT	**erd evening nautical twilight**
EMP	**electromagnetic pulse**
EN	**enemy (in illustration)**
EPW	**enemy prisoner of war**
ETAC	**enlisted terminal attack controller**
ETL	**effective translational lift**

EW	**electronic warfare**
EXTAL	**extra time allowance**

1SO	**first sergeant**
FA	**field artillery**
FAAO	**field artillery air observer**
FAC	**forward air controller**
FAAR.	**forward area alerting radar**
FARP	**forward arming refuel point**
FASCAM	**family of scatterable mines**
FAX	**facsimile**
FDC	**fire direction center**
FEB A	**forward edge of battle area**

forward entry devices
R folding fin aircraft rocket
T fire support team

FIST-V fire support team vehicle
FKSM fort knox supplemental Material
FLIR forward looking infrared
FLOT forward line of own troops
FM field manual; frequency modulation
FO forward observer
FOOGAS petroleum-based expedient Inflammable material
FPF final protective fire
FRAG-HE fragmentary high-explosive (ammunition)
FRAGO fragmentary order
FRG Federal Republic of Germany
FS fire support
FSB forward support battalion
FSCL Fire support coordination line
FSCOORD Fire support coordinator
FSE fire support element
FSO fire support officer
ft feet

G1 assistant chief of staff (personnel)
ssistant chief of staff (operations and plans)
ir air operations and planning officer
G4 assistant chief of staff (logistics)
GA gas agent
Gator (a mine delivered by fixed-wing aircraft)
GBU guided bomb unit
GEMSS ground em placed mine scattering system
GLD ground laser designator
GMRD ground motorized rifle division
GPS global positioning systems
GS general support
GSR ground surveillance radar

HE high explosive
HE-APERS high explosive antipersonnel (ammunition)
HEI high explosive incendiary
HEMTT heavy expanded mobile tactical truck

HET	**heavy equipment transport**
HF	**high frequency**
HHC	**headquarters and headquarters company**
HHT	**headquarters and headquarters troop**
HIMAD	**hi-to-medium-altitude air defense**
HMMWV	**high mobility multipurpose wheeled vehicle**
hq	**headquarters**
HTO	**high technology observer**
ICM	**improved conventional munitions**
ID	**identification**
IDS	**infrared discrimination system**
IEW	**intelligence and electronic warfare**
IEWSE	**intelligence and electronic warfare support element**
IFF	**identification, friend or foe**
IFFN	**identification, friend, foe, or neutral**
INTSUM	**intelligence summary**
IP	**initial point**
IPB	**intelligence preparation of the battlefield**
IR	**infrared**
ITV	**improved TOW vehicle**
I2	**I squared (image intensification)**
JAAT	**joint air attack team**
JTIDS	**joint tactical information distribution system**
KE	**kinetic energy**
KIA	**killed in action**
km	**kilometer(s)**
kph	**kilometer(s) per hour [as a unit of measure only)**
kmih	**kilometers in the hour**
kmph	**kilometer(s) per hour (as a unit of measure that indicates motion)**
KY	**Kentucky**
LANTIRN	**low-altitude navigation and targeting infrared for night**
LAPES	**low-altitude parachute extraction system**
LBE	**load-bearing equipment**
LC	**line of contact**
LCC	**land component commander**
LCSS	**lightweight camouflage screen system**
LD	**line of departure; low drag**
LD/LC	**line of departure is line of contact**

LGB	**laser guided bomb**
LO	**liaison officer**
LQA	**light observation aircraft limit of advance**
LOC	**line(s) of communication**
LOG	**logistics**
LOGPAC	**logistics package**
LRF	**laser range finder**
LRP	**logistics release point**
MAC	**military air command**
MAGTF	**Marine air around task force**
medic(s)	**medical person: medical personnel**
MEF	**Marine expeditionary force**
MEMO	**mission-essential maintenance only**
METT-T	**mission, enemy, terrain, troops, and time available**
MEU	**Marine expeditionary unit**
MHE	**material-handling equipment**
Ml	**military intelligence**
MIC	**mid-intensity conflict**
MICLIC	**mine clearing line charge**
mib	**miles in the boar**
MLC	**main line of communication**
MLRS	**multiple-launch rocket system**
mm	**millimeter(s)**
MMS	**multi mission ship**
MOPP	**mission oriented protection posture**
MORT	**mortar (in illustration)**
MOS	**military occupational specialty**
MOUT	**military operations on urbanized Terrain**
MP	**military police**
MPF	**maritime pre- positioning force**
mph	**miles per hour**
MPP	**mobile pre-positioning**
MPSRON-1	**maritime pre-positioning squadron one**
MPSRON-2	**maritime pre- positioning squadron two**
MPSRON-3	**maritime pre- positioning squadron three**
MPT	**mobile pay team**
MR	**motorized rifle**
MRB	**motorized rifle battalion**
MRC	**motorized rifle company**

MRD	motorized rifle division
MRE	meals, ready to eat
MRL	multiple rocket launcher
MRP	motorized rifle platoon
MRR.	motorized rifle regiment
MRS	motorized rifle squad
MRS	muzzle reference system
MSB	main support battalion
MSE	mobile subscriber equipment
MSR	main supply route.
MSRT	mobile subscriber radiotelephone terminal
MST	maintenance support team
MTF'	medical treatment facilities
MTLR	moving target locating radar
MTOE	modification table of organization and equipment
MTP	mission training plan
MULE	modular universal laser equipment
NA1	named areas of interest
NATO	North Atlantic Treaty Organization
NBC	nuclear, biological, chemical
NBCWRS	NBC warning and reporting system
NCO	noncommissioned office
NCOIC	non commissioned officer in charge
NCS	net control station
NET	new equipment training
NGLO	naval gunfire liaison officer
NLT	no later than
NOE	night observation device
NOR	nap of the earth
NVD	night-vision device
NVG	night vision goggles
NZO	standing barrier fire
OCOKA	observation and fields of fire, cover and concealment, obstacles, key terrain, avenues of approach
Ol	operations and intelligence
OIC	officer in charge

OIR	**other intelligence requirements**
OMG	**operational maneuver group**
OP	**observation post**
OPCON	**operational control**
OPLAN	**operational plan**
OPORD	**operation order**
OPSEC	**operations security**
ORP	**objective rally point**
P&A	**personnel and administration**
PAC	**Personnel and Administrative Center**
PADS	**position and azimuth determining system**
PDS	**personnel daily summary reports**
PERINTREP	**periodic intelligence report**
PEWS	**platoon early warning system**
PIR	**priority intelligence requirement**
PL	**phase line(s)**
PHI (	**a hybrid PLRS)**
PLL	**prescribed land list**
PLRS	**position location reporting system**
plt	**platoon**
POL	**petroleum, oils, and lubricants**
PMCS	**preventive maintenance checks and services**
POMCUS	**prepositioning of materiel configured to unit sets**
PP	**passage point(s)**
PRF	**pulse repetition frequency**
PSG	**platoon sergeant**
PSNC0	**personnel services noncommissioned officer**
PSO	**personnel services officer**
PSP	**pierced-steel planking**
PSS	**personnel service support**
PST	**pass time**
PWRS	**prepositioned war reserve stock**
PX	**post exchange**
PZ	**pick up zone**
PZO	**rolling harrier fine**
QSTAG	**Quadripartite Standardization Agreement**

RAG	regimental artillery group
RAOC	rear area operations center
RAP	rocket-assisted projectile
RAU	radio access units
REG	repair and evacuation group
REMS	remotely employed sensor
RETRANS	relay station; retransmit
RFL	restrictive fire line
ROE	rules of engagement
ROWPU	reverse osmosis water purification unit
RP	release point
RPG	rocket-propelled grenade
R&S	reconnaissance and surveillance
RSR	required supply rate
RT	right; route
RX	repairable, exchange
2IC	second in command
S&S	supply and service
S1	adjutant
S2	intelligence officer
S3	operation and training officer
S3-Air	assistant battalion S3 (air operations)
S4	supply officer
S AFAD	small arms far air defense
SAM	surface-to-air missile
SCT	scout (in illustration)
SGM	sergeant major
SITREP	situation report
SLGR	small lightweight GPS receiver
SME	subject matter expert
SOI	signal operation instructions
SOP	standing operating procedure
SP	shut point
SPG	self-propelled grenade
sq km	square kilometer(s)
S&S	supply and service
SSB	single side hand
STANAG	Standardisation Agreement
STB	supertropical bleach

TAACOM Theater Army Air Defense Command
TAC-A tactical air coordinator-airborne
TACCP tactical command post
TACCS Tactical Army Combat Service Support (CSS) Computer System
TACFIRB tactical line direction system
TACP tactical air control party
TAI target areas of interest
TAMMS The Army Maintenance Management System
TAR tactical air reconnaissance
TSC training support center
TC tank commander
TCF tactical combat force
TCP tactical computer processor
TCT tactical computer terminal
TD tank division
TDIS time-distance
TEWT tactical exercise without troops
TF task force
TIRS terrain index reference system
TTS thermal imaging system
TOC tactical operations center
TOE table(s) of organization and equipment
TOT Time on target
TOW tube-launched, optically tracked, wire-guided missile
TPU tank pump unit
TR tank regiment
TRP target, reference points
TSOP tactical SOP

UAV unmanned aerial vehicle.
UH Utility helicopter

UHF	ultra high frequency
UMCP	unit maintenance collection point
UMT	unit maintenance team; unit ministry team
US	United States (of America)
USAF	United States Air Force
USMC	United States Marine Corps
USN	United States Navy
VFMED	variable format message entry device
VHF	very high frequency
VINSON	encryption device
vpkm	vehicle per kilometer
VTR	vehicle/tank retriever
WCS	weapons control status
WESS	weapons effect signature simulator
WIA	wounded in action
WO	warning order
WP	white phosphorus
WSM	weapon system manager
WSRO	weapons systems replacement operations
WTR	wide temperature range
XO	executive officer

REFERENCES

DEPARTMENT OF DEFENSE FORMS (DD Form)

DD Form 565 Statement of Recognition of Deceased. August 1984.

DEPARTMENT OF THE ARMY FORMS (DA Form)

DA Form 581 Request for Issue and Turn-in of Ammunition. August 1989.

DA Form 1155 Witness Statement of Individual. June 1966.

DA Form 1156 Casualty Feeder Report. June 1966.

DA Form 2404 Equipment Inspection and Maintenance Worksheet. April 1979.

DA Form 2765 Request for Issue or Turn-in. April 1966.

DA Form 2765-1 Request for Issue or Turn-in. April 1966.

DA Form 5368-R Quick Fire Plan. December 1984.

FIELD MANUALS (FM)

FM 3-3 NBC Contamination Avoidance. May 1987.

FM 3-4 NBC Protection. October 1985.

FM 3-5 NBC Decontamination. June 1985.

FM 3-50 Smoke Operations. December 1990.

FM 3-100 NBC Defense, Chemical Warfare, Smoke, and Flame Operations. May 1991.

FM 5-34 Engineer Field Data. September 1987.

FM 5-100 Engineer Combat Operations. November 1988.

FM 5-101 Mobility. January 1985.

FM 5-102 Countermobility. March 1985.

FM 5-103 Survivability. June 1985.

FM 6-20 Fire Support in the AirLand Battle. May 1988.

FM 6-20-1 Tactics, Techniques, and Procedures for the Field Artillery Cannon Battalion.

November 1990 .

FM 6-20-40 Tactics, Techniques, and Procedures for Fire Support for Brigade Operations (Heavy). January 1990 .

FM 7-7J The Mechanized Infantry Platoon and Squad (Bradley). February 1986 .

FM 7-10 The Infantry Rifle Company. December 1990 .

FM 7-20 The Infantry Battalion. April 1992 .

FM 7-30 Infantry, Airborne, and Air Assault Brigade Operations. April 1981 .

FM 7-70 The Light Infantry Platoon/Squad. September 1986 .

FM 7-90 Tactical Employment of Mortars. June 1985 .

FM 17-15 Tank Platoon. October 1987 .

FM 17-95 Cavalry Operations. September 1991 .

FM 17-98 The Scout Platoon. October 1987 .

FM 17-98-1 Scout Leader's Handbook. September 1990 .

FM 20-3 Camouflage. November 1990 .

FM 20-32 Mine/Countermine Operations. December 1985 .

FM 21-18 Foot Marches. June 1990 .

FM 21-75 Combat Skills of the Soldier. August 1984 .

FM 21-76 Survival. March 1986 .

FM 22-100 Military Leadership. July 1990 .

FM 31-12 Army Forces in Amphibious Operations (The Army Landing Forces). March 1961 .

FM 34-1 Intelligence and Electronic Warfare Operations. July 1987 .

FM 34-2 Collection Management. October 1990 .

FM 34-3 Intelligence Analysis. March 1990 .

FM 34-10 Division Intelligence and Electronic Warfare Operations. November 1986 .

FM 34-80 Brigade and Battalion Intelligence and Electronic Warfare Operations. April 1986 .

FM 34-130 Intelligence Preparation of the Battlefield. May 1989 .

FM 43-5 Unit Maintenance Operations. September 1988 .

FM 43-11 Direct Support Maintenance Operations (Nondivisional). September 1991.

FM 43-12 Division Maintenance Operations. November 1989.

FM 44-1 U S Army Air Defense Artillery Employment. May 1983.

FM 44-3 Air Defense Artillery Employment Chaparral/Vulcan/Stinger. June 1984.

FM 44-16 Chaparral/Vulcan/Stinger Platoon Combat Operations. May 1987.

FM 44-100 United States Army Air Defense Operations. November 1988.

FM 63-3J Combat Service Support Operations - Corps. August 1985.

FM 63-20 Forward Support Battalion. February 1990.

FM 71-1 The Tank and Mechanized Infantry Company Team. November 1988.

FM 71-2 The Tank and Mechanized Infantry Battalion Task Force. September 1988.

FM 71-3 The Armored and Mechanized Infantry Brigade. May 1988.

FM 71-100 Division Operations. June 1990.

FM 71-123 Tactics and Techniques for Combined Arms Heavy Forces: Armored Brigade, Battalion/Task Force, and Company/Team. August 1992.

FM 90-2 Battlefield Deception. October 1988.

FM 90-4 Air Assault Operations. March 1987.

FM 90-6 Mountain Operations. June 1980.

FM 90-10 (HTF) Military Operations in Urban Terrain (MOUT) (How to Fight). August 1979.

FM 90-10-1 (HTF) An Infantryman's Guide to Urban Combat (How to Fight). September 1982.

FM 90-13 (HTF) River Crossing Operations (How to Fight). November 1978.

FM 90-13-1 Combined Arms Breaching Operations. February 1991.

FM 90-14 Rear Battle. June 1985.

FM 90-26 Airborne Operations. December 1990.

FM 100-2-1 The Soviet Army: Operations and Tactics. July 1984.

FM 100-2-2 The Soviet Army: Specialized Warfare and Rear Area Support. July 1984.

FM 100-2-3 The Soviet Army: Troops, Organization, and Equipment. June 1991.

FM 100-5 Operations. May 1986.

FM 100-25 Doctrine for Army Special Operations Forces. December 1991.

FM 100-26 The Air-Ground Operations System. March 1973.

FM 100-103 Army Airspace Command and Control in a Combat Zone. October 1987

FM 101-5 Staff Organization and Operations. May 1984.

FM 101-5-1 Operational Terms and Symbols. October 1985.

FM 101-10-1/2 Staff Officers Field Manual-Organizational, Technical, and Logistical Data, Planning Factors (Volume 2). July 1990.

FLEET MARINE FORCE MANUALS (FMFM)

FMFM 1-5 Maritime Pre-positioning Force (MPF) Operations. Coordinating Draft.

FMFM 4-1 Combat Service Support. September 1981.

FMFM 5-1 Organization and Function of Marine Aviation. October 1991.

FMFM 5-35 Assault Support Helicopter Tactical Manual. March 1992.

FMFM 6-1 Marine Division. March 1978.

FLEET MARINE FORCE REFERENCE PUBLICATIONS (FMFRP)

FMFRP 0-52 Remote Sensing Field Guide, Desert. October 1990.

FMFRP 0-53 A foot in the Desert. October 1990.

FMFRP 0-54 The Persian Gulf Region, A Climatological Study. October 1990,

FMFRP 0-55 Desert Water Supply. November 1990.

FMFRP 0-56 Southwest Asia: Environment and its Relationship to Military Activities. November 1990.

FMFRP 0-57 A Study of Windborne Sand and Dust in Desert Areas. October 1990.

FMFRP 0-58 Problems in Desert Warfare. October 1990.

FMFRP 0-59 The Environment and its Effects on Materiel, Personnel, and Operations with Special Emphasis on the Middle East. October 1990.

FMFRP 0-60 General Design and Construction Criteria for Kuwait.

FMFRP 0-63 Troop Construction in the Middle East. December 1990.

FMFRP 2-12 MAGTF: A Global Capability. April 1991.

GRAPHIC TRAINING AIDS

GTA 3-6-5 NBC Warning and Reporting System.

STANDARDIZATION AGREEMENTS (STANAG)

STANAG 2025 Basic Military Road Traffic Regulations.

STANAG 2050 Statistical Classification of Diseases, Injuries, and Causes of Death.

STANAG 2128 Medical and Dental Supply Procedures.

TRAINING CIRCULARS (TC)

TC 90-1 Military Operations on Urbanized Terrain Training. June 1986.

TECHNICAL MANUALS

TM 3-6665-312-12&P Operator's and Organizational Maintenance Manual Including Repair Parts and Special Tools List for M1A1 Automatic Chemical Agent Alarm. March 1985.

COMMAND PUBLICATIONS

Command publications cannot be obtained through Armywide resupply channels. Determine availability by contacting the installation shown.

FKSM 71-2-8 Battalion Task Force Combat Semite Support Doctrinal Sketch. May 1986.

FKSM 71-8 Close Combat Heavy (CCH) Maneuver Organizations. February 1989.

FKSM 71-10 Movement Coordination Exercise (MCX). August 1989.

ST 17-17 Combined Arms Heavy Battalion/Risk Force SOP. May 1990.

ST 100-3 Battle Book.

For information on the preceding references, contact:

Commander, USAARMC
ATTN: ATZK-IMO-RA
Fort Knox, KY 40121-5000

SH 7-176 Infantry Reference Data Book

For information on the preceding reference, contact:

Commandant, USAIS
ATTN: ATSH-AT
Fort Benning, GA 31905-5000

ST 100-9 The Command Estimate

For information on the preceding reference, contact:

Commander, USACGSC
ATTN: ATZL-SWN-L
Fort Leavenworth, KS 66027-6900

Made in United States
Orlando, FL
04 November 2026

90207236R00315